Introduction to
CLASSICAL MECHANICS

Chapman and Hall: books of related interest

VIBRATIONS AND WAVES
A.P. French

Vibrations and Waves covers that portion of the introductory university level physics course for science and engineering students which deals with mechanical vibrations and waves. The text assumes some knowledge of mechanics and a reasonable familiarity with the dynamics of a single particle is required for later chapters. A number of exercises and problems are provided for students to reinforce and test their learning. The MKS system of units is used throughout.

326 pages   1971   234 × 156 mm   Paperback   0 442 30784 5

AN INTRODUCTION TO QUANTUM PHYSICS
A.P. French and E.F. Taylor

An Introduction to Quantum Physics covers the fundamentals of quantum physics through the full Coulomb theory of the hydrogen atom, a semi-quantitative analysis of the helium atom and a qualitative description of the build up of the periodic table, concluding with a brief discussion of radiation by atoms. Throughout, emphasis has been placed on the experimental and theoretical underpinnings of quantum mechanics. Examples have been selected to show how the theory works in practice. The text is enriched with careful explanations and a full use of precise experimental data. The data are not presented as vague graphs showing general results but as summaries of the original research presented in a clear, uniform style.

688 pages   1979   234 × 156 mm   Paperback   0 412 37580 X

SPECIAL RELATIVITY
A.P. French

A self-contained and clear introduction to special relativity for students who have completed an introduction to classical mechanics. The book covers the transition from Newtonian to Einsteinian behaviour for electrons, the relativistic expressions for mass, momentum and energy of a particle. Later chapters cover the Lorentz transformations, the laws of kinematics and dynamics according to special relativity. The approach taken is traditional in that it does not rest heavily on electromagnetic theory. However, the final chapter deals with some of the insights that relativity can provide with regard to the relationship between electricity and magnetism. Throughout, discussion is closely tied to real examples and there is an abundance of interesting problems for students to tackle. Answers are given.

278 pages   1968   226 × 153 mm   Paperback   0 412 34320 7

Introduction to CLASSICAL MECHANICS

A.P. FRENCH
Massachusetts Institute of Technology

M.G. EBISON
The Institute of Physics, London

CHAPMAN AND HALL
LONDON · NEW YORK · TOKYO · MELBOURNE · MADRAS

UK	Chapman and Hall, 11 New Fetter Lane, London EC4P 4EE
USA	Van Nostrand Reinhold, 115 5th Avenue, New York NY10003
JAPAN	Chapman and Hall Japan, Thomson Publishing Japan, Hirakawacho Nemoto Building, 7F, 1−7−11 Hirakawa-cho, Chiyoda-ku, Tokyo 102
AUSTRALIA	Chapman and Hall Australia, Thomas Nelson Australia, 480 La Trobe Street, PO Box 4725, Melbourne 3000
INDIA	Chapman and Hall India, R. Sheshadri, 32 Second Main Road, CIT East, Madras 600 035

First edition 1986
Reprinted 1987, 1989, 1990

© 1986   A.P. French and M.G. Ebison

Typeset in 10/12 pt Times by Colset Private Ltd, Singapore
Printed and bound in Hong Kong

ISBN 0 412 38140 0 (PB)

All rights reserved. No part of this publication may be reproduced or transmitted, in any form or by any means, electronic, mechanical, photocopying, recording or otherwise, or stored in any retrieval system of any nature, without the written permission of the copyright holder and the publisher, application for which shall be made to the publisher.

Contents

Preface	ix
Chapter 1 Space, time and motion	
What is motion?	1
Frames of reference	2
Coordinate systems	3
Combination of vector displacements	5
Scalar product of vectors	7
Units and standards of length and time	9
Velocity	10
Relative velocity and relative motion	11
Acceleration	13
Straight-line motion	13
Uniform circular motion	15
Velocity and acceleration in polar coordinates	16
Problems	18
Chapter 2 Forces	
Forces in equilibrium	25
Action and reaction in the contact of objects	26
Rotational equilbrium: torque	27
Inertia	31
Force and inertial mass: Newton's second law	32
Some comments on Newton's second law	32
The invariance of Newton's second law; relativity	34
Concluding remarks	36
Problems	38
Chapter 3 Using Newton's laws	
Some examples of $F = ma$	44
Circular paths of charged particles in uniform magnetic fields	49
The fracture of rapidly rotating objects	52
Motion against resistive forces	54
Detailed analysis of resisted motion	56
Motion governed by viscosity	58
Growth and decay of resisted motion	60
Simple harmonic motion	62
Problems	66
Chapter 4 Universal gravitation	
The discovery of universal gravitation	70
Kepler's third law	70
The moon and the apple	73
The gravitational attraction of a large sphere	75

Other satellites of the earth	78
The value of G, and the mass of the earth	79
Local variations of g	80
Inertial and gravitational mass	81
Weight	81
Weightlessness	83
The discovery of Neptune	84
Gravitation outside the solar system	87
Einstein's theory of gravitation	90
Problems	92

Chapter 5 Collisions and conservation laws

The conservation of linear momentum	95
Action, reaction, and impulse	96
Extending the principle of momentum conservation	99
Jet propulsion	101
Rockets	102
The zero-momentum frame	105
Kinetic energy of a two-body system	108
Kinetic energy changes in collisions	108
Interacting particles subject to external forces	116
The neutrino	118
Problems	119

Chapter 6 Energy conservation in dynamics; vibrational motions

Introduction	124
Work, energy, and power	125
Energy conservation in one dimension	126
The energy method for one-dimensional motions	129
Some examples of the energy method	131
The harmonic oscillator by the energy method	137
Small oscillations in general	140
The linear oscillator as a two-body problem	141
Problems	145

Chapter 7 Conservative forces and motion in space

Extending the concept of conservative forces	152
Object moving in a vertical circle	154
The simple pendulum	157
The pendulum as a harmonic oscillator	159
The simple pendulum with larger amplitude of swing	160
Universal gravitation: a conservative central force	161
A gravitating spherical shell	165
A gravitating sphere	168
Escape speeds	171
More about the criteria for conservative forces	173
Fields	176
Motion in conservative fields	179
The effect of dissipative forces	183
Problems	186

Chapter 8 Inertial forces and non-inertial frames

Motion observed from unaccelerated frames	193
Motion observed from an accelerated frame	194

Accelerated frames and inertial forces	196
Accelerating frames and gravity	197
Centrifugal force	200
General equation of motion in a rotating frame	203
The earth as a rotating reference frame	207
The tides	210
Tidal heights; effect of the sun	214
The search for a fundamental inertial frame	217
Problems	219

Chapter 9 Motion under central forces

Basic features of the problem	223
The conservation of angular momentum	225
Energy conservation in central force motions	228
Use of the effective potential-energy curves	229
Bounded orbits	232
Unbounded orbits	233
Circular orbits in an inverse-square force field	236
Elliptic orbits: analytical treatment	239
Energy in an elliptic orbit	242
Possible orbits under a $1/r^2$ force	243
Rutherford scattering	247
Problems	250

Chapter 10 Extended systems and rotational dynamics

Momentum and kinetic energy of a many-particle system	255
Angular momentum	258
Angular momentum as a fundamental quantity	262
Conservation of angular momentum	263
Moments of inertia of extended objects	267
Two theorems concerning moments of inertia	268
Kinetic energy of rotating objects	273
Angular momentum conservation and kinetic energy	274
Torsional oscillations and rigid pendulums	278
Linear and rotational motions combined	282
Background to gyroscopic motion	284
Gyroscope in steady precession	289
Atoms and nuclei as gyroscopes	291
The precession of the equinoxes	293
Problems	295

Solutions to problems	300
Index	305

Preface

This book is, in essence, an updated and revised version of an earlier textbook, *Newtonian Mechanics*, written about fifteen years ago by one of us (APF) and published in 1971. The book has been significantly changed in emphasis as well as length. Our aim has been to produce a mechanics text, suitable for use at beginning university level, for students who have a background typified by the British sixth-form level in physics and mathematics. We hope, however, that the book will also be found useful in the teaching of mechanics at the upper levels of the secondary schools themselves. Calculus is freely used from the outset.

In making the present revision we have drastically cut down on the amount of historical and more discursive material. Nevertheless, our goal has been to present classical mechanics as physics, not as applied mathematics. Although we begin at the beginning, we have aimed at developing the basic principles and their applications as rapidly as seemed reasonable, so that by the end of the book students will be able to feel that they have achieved a good working knowledge of the subject and can tackle fairly sophisticated problems. To help with this process, each chapter is followed by a good number of exercises, some of them fairly challenging.

We shall be very grateful to receive comments and corrections from those who use this book.

A.P. French
Maurice Ebison

1 Space, time and motion

What is motion?

You are undoubtedly familiar with motion in all kinds of manifestations, but what would you say if you were asked to *define* it? The chances are that you would find yourself formulating a statement in which the phrase 'a change of position with time', or something equivalent to that, expressed the central thought. For it seems that our ability to give any precise account of motion depends in an essential way on the use of the separate concepts of space and time. We say that an object is moving if it occupies different positions at different instants, and any stroboscopic photograph gives vivid expression to this mental picture.

All of us grow up to be good Newtonians in the sense that our intuitive ideas about space and time are closely in harmony with those of Newton himself. The following paragraphs are a deliberate attempt to express these ideas in simple terms. The description may appear natural and plausible, but it embodies many notions which, on closer scrutiny, will turn out to be naïve, and difficult or impossible to defend. So the account below (set apart with square brackets to emphasize its provisional status) should not be accepted at its face value but should be read with a healthy touch of scepticism.

[*Space*, in Newton's view, is absolute, in the sense that it exists permanently and independently of whether there is any matter in the space or moving through it.

Space is thus a sort of stationary three-dimensional matrix into which one can place objects or through which objects can move without producing any interaction between the object and the space. Each object in the universe exists at a particular point in space and time. An object in motion undergoes a continuous change of its position with time and we simply have the task of finding a practical way of marking these positions. Our physical measurements appear to agree with the theorems of Euclidean geometry, and space is thus assumed to be Euclidean.

Time, in Newton's view, is also absolute and flows without regard to any physical object or event. One can neither speed up time nor slow down its rate, and this flow of time exists uniformly throughout the universe. If we imagine the instant 'now' as it occurs simultaneously on every planet and star in the universe, and an hour later mark the end of this 60-minute interval, we assume that such a time interval has been identical for every object in the universe, as could (in principle) be verified by observations of physical, chemical or biological processes at various locations. As an aid to measuring time intervals, it would be possible, in principle, to place identical clocks at each intersection of a three-dimensional framework and to synchronize these clocks so that they indicate the same time at a common, simultaneous instant. Being identical clocks, they would thereafter correctly mark off the flow of absolute time and remain synchronized with each other.

Space and time, although completely independent of each other, are in a sense interrelated insofar as we find it impossible to conceive of objects existing in space for no time at all, or existing for a finite time interval but 'nowhere' in space. Both space and time are assumed to be infinitely divisible—to have no ultimate structure.]

The preceding four paragraphs describe in everyday language some commonsense notions about the nature of space and time. Embedded in these notions are many assumptions that we adopt, either knowingly or unconsciously, in developing our picture of the universe. It is fascinating, therefore, that, however intuitively correct they seem, many of these ideas have consequences that are inconsistent with experience. This first became apparent in connection with motions at very high speeds, approaching or equalling the speed of light, and with the phenomena of electromagnetism; and it was Einstein, in his development of special relativity theory, who exposed some of the most important limitations of classical ideas, including Newton's own ideas about relativity, and then showed how they needed to be modified, especially with regard to the concept of time.

The crux of the matter is that it is one thing to have abstract concepts of absolute space and time, and it is another thing to have a way of describing the actual motion of an object in terms of measured changes of position during measured intervals of time. If there is any knowledge to be gained about absolute space, it can only be by inference from *relative* measurements. Thus our attention turns to the only basis we have for describing motion—observation of what a given object does in relation to other objects.

Frames of reference

If you should hear somebody say 'That car is moving', you would be quite certain that what is being described is a change of position of the car with respect to the earth's surface and any buildings and the like that may be nearby. Anybody who announced 'There is relative motion between that car and the earth' would be rightly regarded as a tiresome pedant. But this does not alter the fact that it takes the pedantic statement to express the true content of the colloquial one. We accept the local surroundings—a collection of objects attached to the earth and therefore at rest relative to one another —as defining a *frame of reference* with respect to which the changes of position of other objects can be observed and measured.

It is clear that the choice of a particular frame of reference to which to refer the motion of an object is entirely a matter of taste and convenience, but later we shall see that there are powerful *theoretical* reasons for preferring some reference frames to others. The 'best' choice of reference frame becomes ultimately a question of dynamics, i.e. dependent on the actual laws of motion and force. But the choice of a particular reference frame is often made without regard to the dynamics, and for the present we shall just concern ourselves with the purely kinematic problems of analysing positions and motions with respect to any given frame.

Coordinate systems

A frame of reference, as we have said, is defined by some array of physical objects that remain at rest relative to one another. Within any such frame, we make measurements of position and displacement by setting up a *coordinate system* of some kind. In doing this we have a free choice of origin and of the kind of coordinate system that is best suited to the purpose at hand. Since the space of our experience has three dimensions, we must in general specify three separate quantities in order to fix uniquely the position of a point. The most generally useful coordinate systems are the three-dimensional rectangular (Cartesian) coordinates (x, y, z), and spherical coordinates (r, θ, φ). These are both illustrated in Fig. 1.1.

The Cartesian system is almost always chosen to be right-handed, by which we mean that the positive z direction is chosen so that, looking upward along it, the process of rotating from the positive x direction toward the positive y direction corresponds to that of a right-handed screw. It then follows that the cyclic permutations of this operation are also right-handed—from $+y$ to $+z$ looking along $+x$, and from $+z$ to $+x$ looking along $+y$.

We shall on various occasions be making use of *unit vectors* that represent displacements of unit length along the basic coordinate directions. In the rectangular (Cartesian) system we shall denote the unit vectors in the x, y and z directions by i, j and k respectively. The position vector r can then be written as the sum of its three Cartesian vector components:

$$r = xi + yj + zk \qquad (1.1)$$

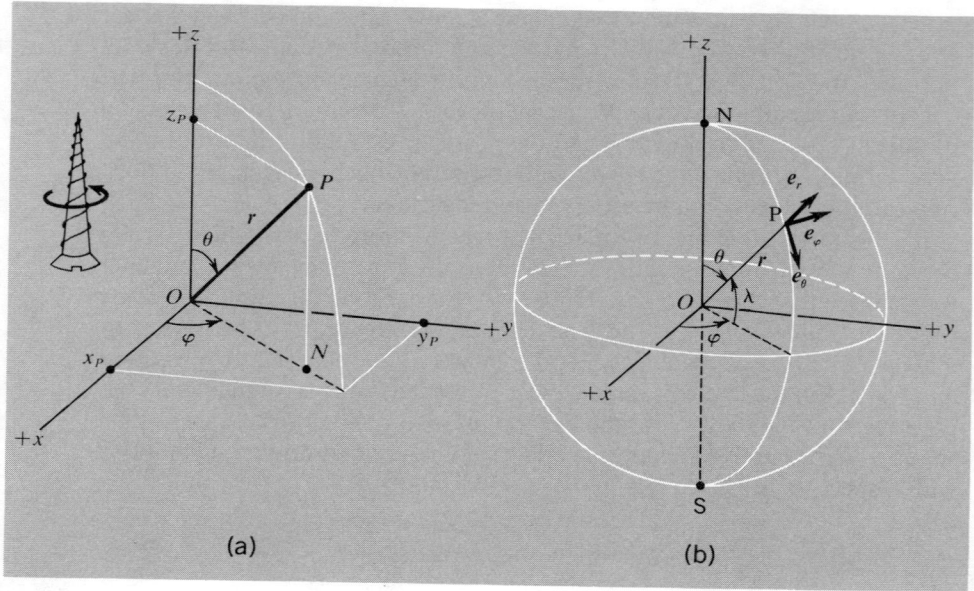

Fig. 1.1 (a) Coordinates of a point in three dimensions, showing both spherical polar and right-handed Cartesian coordinates. (b) Point located by angular coordinates (latitude and longitude) on a sphere, and the unit vectors of a local Cartesian coordinate system at the point in question.

The description of the position or displacement in spherical polar coordinates makes use of one distance and two angles. (Notice that three dimensions requires three independent coordinates, whatever particular form they may take.) The distance is the distance r from the chosen origin. One of the angles (the one shown as θ in Fig. 1.1a) is simply the angle between the vector r and the positive z axis; it is known as the *polar angle*. The other angle represents the angle between the zx plane and the plane defined by the z axis and r. It can be found by drawing a perpendicular PN from the end point P or r onto the xy plane and measuring the angle between the positive x axis and the projection ON. This angle (φ) is called the *azimuth*. The geometry of the figure shows that the rectangular and spherical polar coordinates are related as follows:

$$x = r \sin \theta \cos \varphi$$
$$y = r \sin \theta \sin \varphi \qquad (1.2)$$
$$z = r \cos \theta$$

By setting $\theta = \pi/2$ and $z = 0$ we obtain equations applicable to two dimensions in the xy plane:

$$x = r \cos \varphi$$
$$y = r \sin \varphi \tag{1.3}$$

In the polar coordinate system we use the symbol e_r to denote a unit vector in the direction of increasing r at constant θ and φ, the symbol e_θ to denote a unit vector at right angles to r in the direction of increasing θ at constant φ, and the symbol e_φ to denote a unit vector at right angles to r in the direction of increasing φ and constant θ. These unit vectors become important as soon as we consider motions rather than static displacements, for motions will often have components perpendicular to r.

We have all grown up with one important use of spherical polar coordinates, the mapping of the earth's surface. This is indicated in Fig. 1.1(b). The longitude of a given point is just the angle φ, and the latitude is an angle, λ, equal to $(\pi/2) - \theta$. (This entails calling north latitudes positive and south latitudes negative). At any given point on the earth's surface a set of three mutually orthogonal unit vectors defines for us a local coordinate system; the unit vector e_r points vertically upward, the vector r_θ points due south, parallel to the surface, and the third unit vector, e_φ, points due east, also parallel to the surface. The vector r is given simply by re_r.

Combination of vector displacements

Suppose we were at a point P_1 on a flat horizontal plane (Fig. 1.2a) and wished to go to another point P_2. Imagine that we chose to make the trip by moving only east and north (represented by $+x$ and $+y$ in the figure). We know there are two particularly straightforward ways of doing this: (1) travel a certain distance s_x due east and then a certain distance s_y due north, or (2) travel s_y due north, followed by s_x due east. The order in which we take these two component displacements does not matter; we reach the same point P_2 in either case. This familiar property of linear displacements is an essential feature of all those quantities we call vectors and is not confined to combinations at right angles. Thus, for example, in Fig. 1.2(b) we illustrate how three vector displacements, A, B and C, placed head to tail, can be combined into a single vector displacement S drawn from the original starting point to the final end point. This is what we mean by *adding* the vectors A, B and C. The order in which vectors are added is of no consequence.

We shall often be concerned with forming a numerical multiple of a given vector. A positive multiplier, n, means that we change the length of the vector by the factor n without changing its direction. The negative of a vector (multiplication by -1) is defined to mean a vector of equal magnitude but in the opposite direction, so that added to the original vector it gives zero. A negative multiplier, $-n$, then defines a vector reversed in direction and changed in length by the factor n. These operations are illustrated in Fig. 1.3.

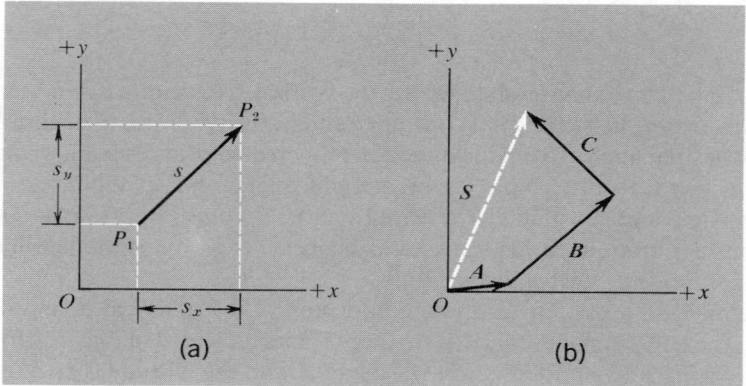

Fig. 1.2 (a) Successive displacements on a plane; the final position is independent of the order in which the displacements are made. (b) Addition of several displacement vectors in a plane.

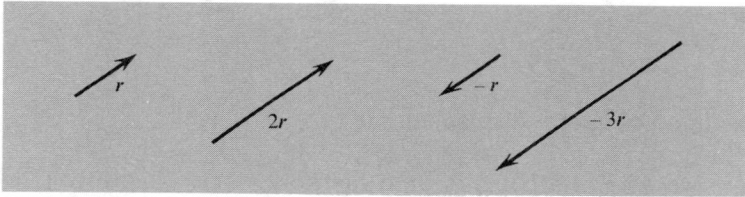

Fig. 1.3 Scalar multiples of a given vector *r*, including negative multiples.

Subtracting one vector from another is accomplished by noting that subtraction basically involves the addition of a negative quantity. Thus if vector ***B*** is to be subtracted from vector ***A***, we form the vector $-\boldsymbol{B}$ and add it to ***A***:

$$\boldsymbol{A} - \boldsymbol{B} = \boldsymbol{A} + (-\boldsymbol{B})$$

In Fig. 1.4 we show both the sum and the difference of two given vectors. We have deliberately chosen the directions of ***A*** and ***B*** to be such that the vector ***A*** − ***B*** is *longer* than the vector ***A*** + ***B***; this will help to emphasize the fact that vector combination is something rather different from simple arithmetical combination.

The evaluation of the vector distance from a point P_1 to a point P_2, when originally the positions of these points are given separately with respect to an origin O (see Fig. 1.5), is a direct application of vector subtraction. The position of P_2 relative to P_1 is given by the vector r_{12} such that

$$r_{12} = r_2 - r_1$$

SCALAR PRODUCT OF VECTORS

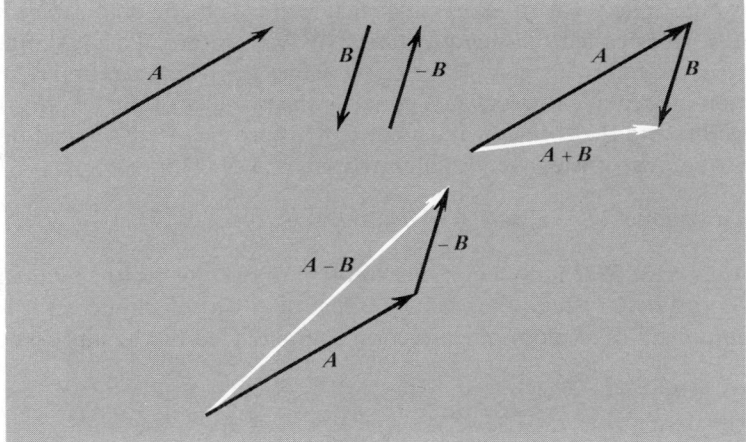

Fig. 1.4 Addition and subtraction of two given vectors. Note that the magnitude of the vector difference may be (as here) larger than the sum.

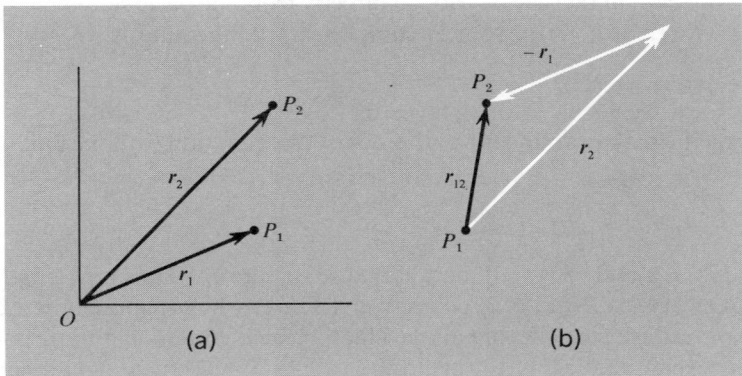

Fig. 1.5 Construction of the relative position vector of one point (P_2) with respect to another point (P_1).

Similarly, the position of P_1 relative to P_2 is given by the vector $r_{21} = r_1 - r_2$. Clearly $r_{21} = -r_{12}$.

Scalar product of vectors

The total vector A in Fig. 1.6 can be written

$$A = A_x \mathbf{i} + A_y \mathbf{j}$$
$$ = (A \cos \alpha)\mathbf{i} + (A \cos \beta)\mathbf{j}$$

A very convenient way of expressing such results is made possible by introducing what is called the *scalar product* of two vectors. This is defined in general in the following way: If the angle between any two vectors, A and B, is θ, then the scalar product, S, is equal to the product of the lengths of the two vectors and the cosine of the angle θ. This product is also called the *dot product* because it is conventionally written as $A \cdot B$. Thus we have

$$\text{scalar product } (S) = A \cdot B = AB \cos \theta$$

If for the vector B we now choose one or other of the unit vectors of an orthogonal coordinate system, the scalar product of A with the unit vector is just the component of A along the direction characterized by the unit vector:

$$A_x = A \cdot i \; ; \quad A_y = A \cdot j$$

Thus the vector A can be written as follows:

$$A = (A \cdot i)i + (A \cdot j)j$$

This result can, in fact, be developed directly from the basic statement that A can be written as a vector sum of components along x and y:

$$A = A_x i + A_y j$$

Forming the scalar product of both sides of this equation with the unit vector i, we have

$$A \cdot i = A_x(i \cdot i) + A_y(j \cdot i)$$

Now $(i \cdot i) = 1$ and $(j \cdot i) = 0$, because these vectors are all of unit length and the values of θ are 0 and $\pi/2$, respectively. Thus we have a more or less automatic procedure for selecting and evaluating each component in turn.

If one were to take this no further, the above development would perhaps seem pointlessly complicated. Its value becomes more apparent if one is interested in relating the components of a given vector in different coordinate systems. Consider, for example, the second set of axes (x', y') shown in Fig. 1.6; they are obtained by a positive (counterclockwise) rotation from the original (x, y) system. The vector A then has two equally valid representations:

$$A = A_x i + A_y j = A_x' i' + A_y' j'$$

If we want to find A_x' in terms of A_x and A_y, we just form the scalar product with i' throughout. This gives us

$$A_x' = A_x(i \cdot i') + A_y(j \cdot i')$$

Looking at Fig. 1.6, we see the following relationships:

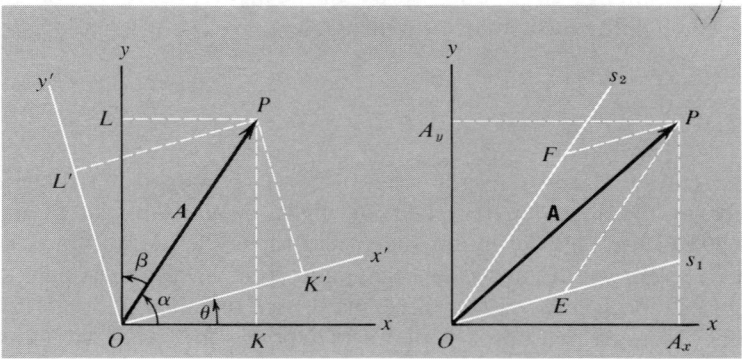

Fig. 1.6 Components of a given vector in two different rectangular coordinate systems related by an angular displacement θ in the xy plane.

$$\boldsymbol{i}\cdot\boldsymbol{i}' = \cos\theta \;;\quad \boldsymbol{j}\cdot\boldsymbol{i}' = \cos\left(\frac{\pi}{2} - \theta\right) = \sin\theta$$

Hence

$$A_x' = A_x \cos\theta + A_y \sin\theta$$

Similarly,

$$\begin{aligned} A_y' &= A_x(\boldsymbol{i}\cdot\boldsymbol{j}') + A_y(\boldsymbol{j}\cdot\boldsymbol{j}') \\ &= A_x \cos\left(\frac{\pi}{2} + \theta\right) + A_y \cos\theta \end{aligned}$$

Therefore,

$$A_y' = -A_x \sin\theta + A_y \cos\theta$$

This procedure avoids the need for tiresome and sometimes awkward considerations of geometrical projections of the vector A onto various axes of coordinates.

Units and standards of length and time

Most of our discussion of motion will be in terms of unspecified positions and times, represented symbolically by r, t, and so on. It should never be forgotten, however, that the description of actual motions involves the numerical measures of such quantities and the use of universally accepted units and standards. Our choice of acceptable standards of both distance and time is the result of a continuing search for the highest degree of consistency and reproducibility in such measurements.

Length

The current standard of length—the metre—is now (since October 1983) defined as the distance travelled by light in vacuum in a time of 1/299 792 458 s. This definition supersedes the earlier definition (1960) in which the metre was defined as a certain number of wavelengths of light from a designated spectral source (^{86}Kr). The reason for this change is that time and frequency measurements exceed all others in precision, and the development of laser technology has brought such measurements into the optical region. The speed of light is now *defined* to be 299 792 458 m s^{-1} exactly. Since, however, the speed of light in vacuum is a natural and universal constant, this definition is in effect a definition of the metre.

Time

The process of defining a standard of time involves a feature that sets it significantly apart from the establishment of a material standard of length. This is that, as Allen Astin has remarked: 'We cannot choose a particular sample of time and keep it on hand for reference.' We depend upon identifying some recurring phenomenon and *assuming* that it always supplies us with time intervals of the same length.

In 1967, the use of atomic vibrations to specify a time standard was adopted by international agreement; it defines the second as corresponding to 9 192 631 770 cycles of vibration in an atomic clock controlled by one of the characteristic frequencies associated with atoms of the isotope caesium 133.

Velocity

The central concept in the quantitative description of motion is that of *velocity*. It is a vector. Our way of designating velocities—miles per hour, metres per second, and so on—is a constant reminder of the fact that velocity is a *derived* quantity, based on separate measures of space and time. In nature itself, things seem to be otherwise, for although we have not as yet identified anything that is directly recognizable as a fundamental natural unit of length or time, we do find a fundamental unit of velocity—the magnitude (c) of the velocity of light in empty space:

$$c = (2.99792458) \times 10^8 \text{ m s}^{-1}$$

It has become customary in high-energy particle physics to express velocities as fractions of c. And in a comparable way, in connection with high-speed flight, the Mach number is used to express the speed of an aircraft as a fraction or multiple of the speed of sound in air. But this does not alter the fact that our basic description of velocities is in terms of the number of units of distance per unit of time.

The measurement of a velocity requires at least two measurements of the position of an object and the two corresponding measurements of time. Let us denote these measurements by (r_1, t_1) and (r_2, t_2). Using these we can deduce the magnitude and direction of what we can loosely call the average velocity between those points:

$$v_{av} = \frac{r_2 - r_1}{t_2 - t_1}$$

However, this average velocity is not, in most cases, a very interesting quantity. Of much more importance in physics is the concept of *instantaneous velocity*, obtained by considering the ratio of displacement to time interval as both approach zero. Using the standard calculus notation, we write

$$\text{instantaneous velocity } v = \lim_{\Delta t \to 0} \frac{\Delta r}{\Delta t} = \frac{dr}{dt} \tag{1.4}$$

Relative velocity and relative motion

Since the vector velocity is the time derivative of the vector displacement, the velocity of one object relative to another is just the vector difference of the individual velocities. Thus if one object is at r_1 and another object is at r_2, the vector distance R from object 1 to object 2 is given by

$$R = r_2 - r_1$$

The rate of change of R is then the velocity, V, of object 2 relative to object 1, and we have

$$V = \frac{dR}{dt} = \frac{dr_2}{dt} - \frac{dr_1}{dt}$$

i.e.

$$V = v_2 - v_1 \tag{1.5}$$

This relative velocity V is the velocity of object 2 in a frame of reference attached to object 1.

In discussing frames of reference earlier in this chapter, we pointed out how the choice of some particular frame of reference may be advantageous because it gives us the clearest picture of what is going on. Nothing could illustrate this better than the practical problems of navigation and the avoidance of collisions at sea or in the air. Imagine, for example, two ships that at some instant are in the situation shown in Fig. 1.7(a). The vectors v_1 and v_2 represent their velocities (which we take to be constant) with respect to the body of water in which they both move. The paths of the ships, extended along the directions of motion from the initial points A and B, intersect at a point P. Will the ships collide, or will they pass one another at a safe distance? The answer to this question is not at all clear if we stick to the ocean frame, but if we describe things from the standpoint of one of the two ships the analysis becomes very straightforward. Let us imagine that we are standing on the deck of the ship marked A. Putting ourselves in that frame of reference means giving ourselves the velocity v_1 with respect to the water. But from *our* standpoint it is as if the water, and everything else, were given a velocity equal and opposite to v_1. Thus to every motion as observed in the ocean frame we add the vector $-v_1$, as implied by eqn (1.5). This automatically, and by definition, brings A to rest, as it were, and shows us that the velocity of the ship B, relative to A, is obtained by combining the vectors v_2 and $-v_1$, as shown in Fig. 1.7(b). The vector *distance* between the ships is unaffected by this change of viewpoint. So now we can see the whole picture. B follows the straight line shown, as indicated by several successive positions in the diagram. It will miss A by the distance AN, the perpendicular distance from A to the line of V. The time at which this closest approach occurs is equal to the distance BN divided by the magnitude of V. Thus B seems to sweep across

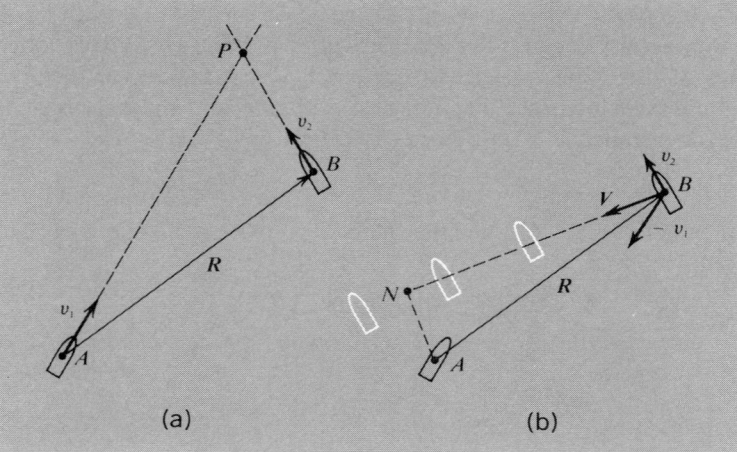

Fig. 1.7 (a) Paths of two ships moving at constant velocity along courses that intersect. (b) Path of ship B relative to ship A, showing that they do not collide even though their paths cross.

STRAIGHT-LINE MOTION

A's bow, more or less sideways. If you have had occasion to observe a close encounter of this sort, especially if it is out on the open water with no landmarks in sight, you will know that it can be a curious experience, quite disturbing to the intuitions, because the observed motion of the other ship seems to be unrelated to the direction in which it is pointing.

Acceleration

The crucial quantity for discussing how motion is governed by forces is *acceleration*, the rate of change of velocity with time:

$$\text{instantaneous acceleration } a = \lim_{\Delta t \to 0} (\Delta v/\Delta t) = dv/dt = d^2r/dt^2 \quad (1.6)$$

Since we must be ready to take into account variation of velocity in direction as well as magnitude, acceleration is a *vector* quantity. Figure 1.8 shows the instantaneous velocity vectors at two neighbouring instants so that the instantaneous vector acceleration a is given by

$$a = \lim_{\Delta t \to 0} (\Delta v/\Delta t) = dv/dt = d^2r/dt^2 \quad (1.7)$$

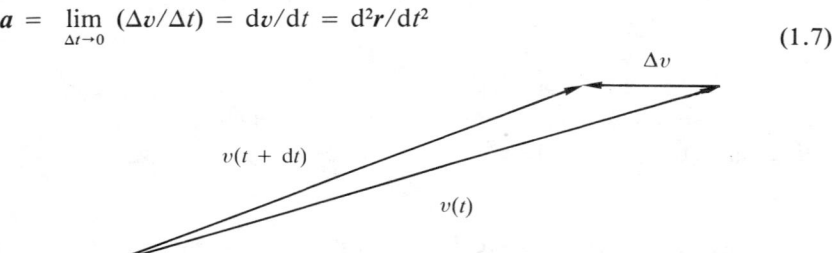

Fig. 1.8 Small change of a velocity that is changing in both magnitude and direction.

Straight-line motion

From eqn (1.6)

$$\Delta v = a\Delta t$$

Hence

$$v_2 - v_1 = \int_{t_1}^{t_2} a(t)\, dt \quad (1.8)$$

where we write $a(t)$ to show that the acceleration is to be considered a specific function of time. Often this integral is evaluated up to some indefinite time t starting from some chosen zero of time at which the velocity is v_0. Thus

$$v - v_0 = \int_0^t a(t)\,dt \tag{1.9}$$

In like manner from the equation

$$\Delta s = v \Delta t$$

where s is distance along the straight line, we obtain

$$s - s_0 = \int_0^t v(t)\,dt \tag{1.10}$$

For a *constant, non-zero* acceleration, a, eqn. 1.9 gives

$$v - v_0 = a \int_0^t dt = at \tag{1.11}$$

Similarly, eqn (1.10) with $v(t)$ as given by eqn (1.10) yields

$$s - s_0 = \int_0^t (v_0 + at)\,dt = v_0 t + \tfrac{1}{2} at^2 \tag{1.12}$$

It is sometimes convenient to remove all explicit reference to time, by combining eqns (1.11) and (1.12):

$$v^2 = v_0^2 + 2a(s - s_0) \tag{1.13}$$

In the case of *constant a* we have the following familiar kinematic equations:

$$v = v_0 + at \tag{1.14a}$$
$$v^2 = v_0^2 + 2a(s - s_0) \tag{1.14b}$$
$$s = s_0 + v_0 t + \tfrac{1}{2} at^2 \tag{1.14c}$$

Although these mathematical expressions for accelerated motion are tidy and extremely useful, it should be remembered that a truly constant acceleration is never maintained indefinitely. For example, the problems that everyone learns to solve on free fall under gravity, using a constant acceleration g, really do not correspond to the facts, because air resistance causes the acceleration to become less as the velocity increases. For low velocities the error may not be big enough to worry about, but it is there. Later we shall be dealing with situations in which the acceleration varies in some mathematically well-defined way with position or time. Thus the emphasis will shift away from eqn (1.14) and toward the more general statements expressed in eqns (1.9) and (1.10).

Uniform circular motion

Probably the most interesting direct application of the vector definitions of velocity and acceleration is to the problem of motion in a circular path at some constant speed. In this case, if the centre of the circle is chosen as an origin, the vector r always has the same length and simply changes its direction at a uniform rate. The instantaneous velocity is always at right angles to r, and its magnitude v is constant. From this we can readily calculate the acceleration. For during a short time, Δt, the distance travelled is $v\,\Delta t$, from P_1 to P_2 along a circular arc (Fig. 1.9a). The angle $\Delta\theta$ between the two corresponding directions of r is therefore given by

$$\Delta\theta = \frac{v\,\Delta t}{r}$$

Imagine that the bisector of this angle is drawn (Fig. 1.9b) and consider the changes in velocity parallel and perpendicular to this bisector. Initially the velocity has a component $v\sin(\Delta\theta/2)$ away from O, and $v\cos(\Delta\theta/2)$ transversely. Subsequently it has a component $v\sin(\Delta\theta/2)$ *towards* O, and again $v\cos(\Delta\theta/2)$ transversely in the same direction as before. Thus the change of velocity is of magnitude $2v\sin(\Delta\theta/2)$ *towards* O. Figure 1.9(c) shows how this same result comes from considering a vector diagram in which Δv is defined as that vector which, added to $v(t)$, gives $v(t + \Delta t)$.

As $\Delta\theta$ is made vanishingly small, we have

$$|\Delta v| = 2v\sin(\Delta\theta/2) \to v\,\Delta\theta$$

But $\Delta\theta = v\,\Delta t/r$, so we have

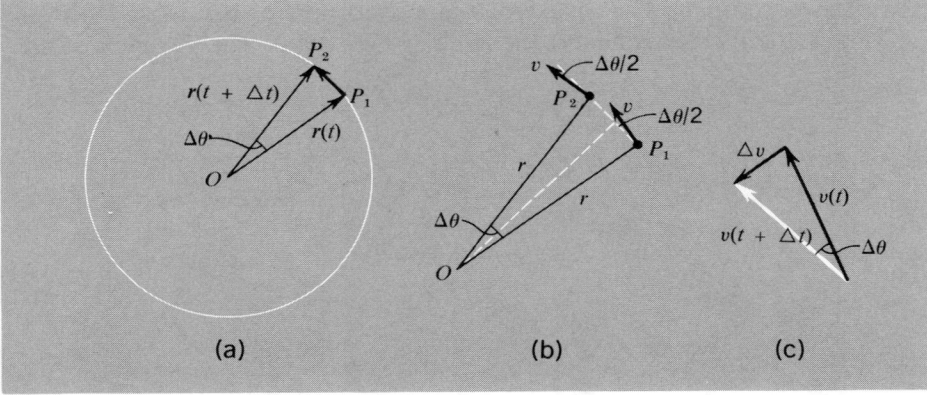

Fig. 1.9 (a) Small displacement (P_1P_2) in a uniform circular motion. (b) Velocity vectors at the beginning and end of the short element of path. (c) Vector diagram for the evaluation of the change of velocity, Δv.

$|\Delta v| = v^2 \Delta t / r$

Hence the magnitude of the acceleration is given by

(Uniform circular motion) $\quad |a| = \dfrac{v^2}{r}$ (1.15)

and its direction is radially inward, regardless of whether the circular path is being traced out clockwise or counterclockwise. This is called the *centripetal acceleration* associated with any circular motion. The need for a dynamical means of supplying this acceleration to an object is an essential feature of any motion that is not strictly straight, because any change in the direction of the path implies a component of Δv perpendicular to v itself.

Velocity and acceleration in polar coordinates

The result of the last section, and other results of more general application, are very nicely developed with the help of polar coordinates in the plane. The use of this type of analysis is particularly appropriate if the origin represents a centre of force of some kind, e.g. the sun, acting on an orbiting planet. The starting point is to write the position vector r as the product of the scalar distance r and the unit vector e_r:

$$r = re_r \qquad (1.16)$$

We now consider the change of r with time. This can arise from a change of its length, or from a change of its direction, or from a combination of both. For the present we shall limit ourselves to circular motion, in which the length of r remains constant. The change of r in a short time Δt is then as shown in Fig. 1.10(a) which is almost the same as Fig. 1.9(a). The *direction* of this

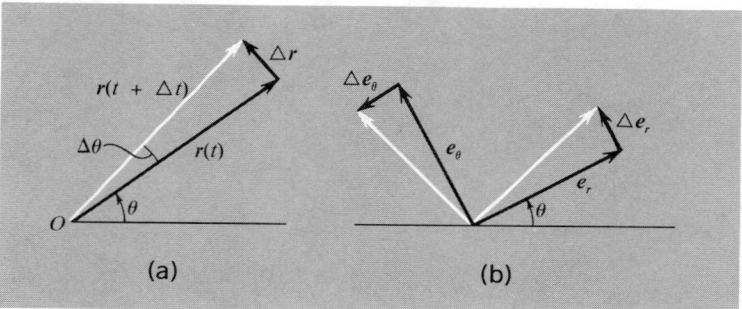

Fig. 1.10 (a) Vector change of displacement, Δr, during a short time Δt in a uniform circular motion. (b) Changes in the unit vectors e_r and e_θ during Δt, showing how Δe_r is parallel to e_θ and Δe_θ is parallel (but opposite) to e_r.

VELOCITY AND ACCELERATION IN POLAR COORDINATES

change ($\Delta \mathbf{r}$) is in the direction of the unit vector $\mathbf{e}_\theta$ drawn at right angles to $\mathbf{e}_r$ as shown in Fig. 1.10(b). Its magnitude, as is clear from Fig. 1.10(a), is equal to $r \, \Delta \theta$. Thus we can put

$$\Delta \mathbf{r} = r \, \Delta \theta \, \mathbf{e}_\theta$$

Dividing by Δt, and letting Δt tend to zero, we then have the result

(Circular motion) $\quad v = \dfrac{dr}{dt} = r \dfrac{d\theta}{dt} \mathbf{e}_\theta \quad$ (1.17a)

If we designate $d\theta/dt$ by the single symbol ω, for angular velocity (measured in rad s^{-1}), we have

(Circular motion) $\quad v = \omega r \mathbf{e}_\theta = v \mathbf{e}_\theta \quad$ (1.17b)

The derivation of the above result embodies the important fact that the unit vector $\mathbf{e}_r$ is changing with time. Although its length is by definition constant, its direction changes in accord with the direction of $\mathbf{r}$ itself. In fact, we can obtain the explicit expression of its rate of change as a special case of eqn (1.17a), with $r = 1$:

$$\frac{d}{dt}(\mathbf{e}_r) = \frac{d\theta}{dt} \mathbf{e}_\theta = \omega \mathbf{e}_\theta \quad (1.18a)$$

In an exactly similar way, as Fig. 1.10(b) shows, a change of θ implies a change of the other unit vector, $\mathbf{e}_\theta$. If the change of θ is positive, as shown, it can be seen that the change of $\mathbf{e}_\theta$ is in the direction of $-\mathbf{e}_r$; it is given by the equation

$$\frac{d}{dt}(\mathbf{e}_\theta) = -\frac{d\theta}{dt} \mathbf{e}_r = -\omega \mathbf{e}_r \quad (1.18b)$$

This possible time dependence of the unit vectors in a polar-coordinate system is a feature that has no counterpart in rectangular coordinates, where the unit vectors $\mathbf{i}, \mathbf{j}$ and $\mathbf{k}$ are defined to have the same directions for all values of the position vector $\mathbf{r}$.

Once we have eqns (1.17a) and (1.17b) we can proceed to calculate the acceleration by taking the next time derivative. If we limit ourselves to the case of *uniform* circular motion, both r and ω are constant, so we have

(Uniform circular motion) $\quad \mathbf{a} = \omega r \dfrac{d}{dt}(\mathbf{e}_\theta) = -\omega^2 r \mathbf{e}_r = -\dfrac{v^2}{r} \mathbf{e}_r \quad$ (1.19)

Thus the result expressed by eqn (1.15) falls out automatically, together with its correct direction. If we label this acceleration specifically as a *radial* acceleration of magnitude a_r, we can put

$$a_r = -\omega^2 r$$

If, still restricting ourselves to motion in a circle, we remove the condition that the motion be uniform, then the acceleration vector $\mathbf{a}$ has a transverse component also. Starting from eqn (1.17b), we have

$$\text{(Arbitrary circular motion)} \quad \mathbf{a} = \frac{dv}{dt}\mathbf{e}_\theta - v\frac{d\theta}{dt}\mathbf{e}_r \quad (1.20)$$

The radial component of $\mathbf{a}$ is the same as we obtained for uniform circular motion (since $d\theta/dt = v/r = \omega$), but it is now joined by a transverse component, a_θ. Thus we have

$$\text{(Arbitrary circular motion)} \quad \begin{cases} a_r = -\dfrac{v^2}{r} = -\omega v = -\omega^2 r \\[2mm] a_\theta = \dfrac{dv}{dt} = r\dfrac{d\omega}{dt} = r\dfrac{d^2\theta}{dt^2} \end{cases} \quad (1.21)$$

(where $\omega = d\theta/dt$)

Problems

1.1 The scalar (dot) product of two vectors, $\mathbf{A}\cdot\mathbf{B}$, is equal to $AB\cos\theta_{AB}$, where θ_{AB} is the angle between the vectors.

(a) By expressing the vectors in terms of their Cartesian components, show that

$$\cos\theta_{AB} = \frac{A_x B_x + A_y B_y + A_z B_z}{AB}$$

(b) By using the relation between rectangular and spherical polar coordinates (eqn 1.2), show that the angle θ_{12} between the radii to two points (R, θ_1, φ_1) and (R, θ_2, φ_2) on a sphere is given by

$$\cos\theta_{12} = \cos\theta_1 \cos\theta_2 + \sin\theta_1 \sin\theta_2 \cos(\varphi_2 - \varphi_1)$$

(Note that the distance between the two points as measured along the great circle that passes through them is equal to $R\theta_{12}$, where θ_{12} is expressed in radians. This can be used, for example, to calculate mileages between points on the earth's surface.)

1.2 (a) Starting from a point on the equator of a sphere of radius R, a particle travels through an angle α eastward and then through an angle β along a great circle toward the north pole. If the initial position of the point is taken to correspond to $x = R$, $y = 0$, $z = 0$, show that its final coordinates are $R\cos\alpha\cos\beta$, $R\sin\alpha\cos\beta$ and $R\sin\beta$. Verify that $x^2 + y^2 + z^2 = R^2$.

(b) Find the coordinates of the final position of the same particle if it first travels through an angle α northward, then changes course by 90° and travels through an angle β along a great circle that starts out eastward.

(c) Show that the straight-line distance Δs between the end points of the displacements in (a) and (b) is given by
$$\Delta s^2 = 2R^2(\sin\beta - \sin\alpha\cos\beta)^2$$

PROBLEMS

1.3 If you found yourself transported to an unfamiliar planet, what methods could you suggest
  (a) To verify that the planet is spherical?
  (b) To find the value of its radius?

1.4 The radius of the earth was found more than 2000 years ago by Eratosthenes through a brilliant piece of analysis. He lived at Alexandria, at the mouth of the Nile, and observed that on midsummer day, at noon, the sun's rays were at 7.2° to the vertical (see the figure). He also knew that the people living at a place 500 miles south

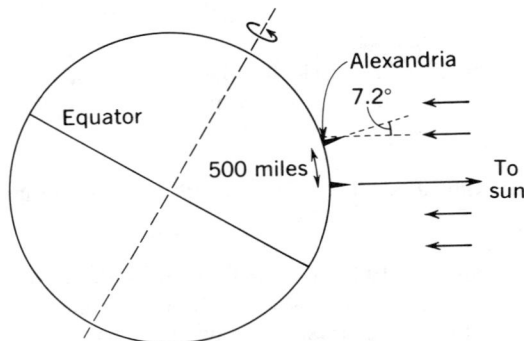

of Alexandria saw the sun as being directly overhead at the same date and time. From this information, Eratosthenes deduced the value of the earth's radius. What was his answer?

1.5 A particle is confined to motion along the x axis between reflecting walls at $x = 0$ and $x = a$. Between these two limits it moves freely at constant velocity. Construct a space-time graph of its motion:
  (a) If the walls are perfectly reflecting, so that upon reaching either wall the particle's velocity changes sign but not magnitude.
  (b) If upon each reflection the magnitude of the velocity is reduced by a factor f (i.e. $v_2 = -fv_1$).

1.6 A particle that starts at $x = 0$ at $t = 0$ with velocity $+v$ (along x) collides with an identical particle that starts at $x = x_0$ at $t = 0$ with velocity $-v/2$. Construct a space-time graph of the motion before and after collision
  (a) For the case that the particles collide elastically, exchanging velocities.
  (b) For the case that the particles stick together upon impact.

1.7 A particle moves along the curve $y = Ax^2$ such that its x position is given by $x = Bt$.
  (a) Express the vector position of the particle in the form $r(t) = xi + yj$.
  (b) Calculate the speed v ($= ds/dt$) of the particle along this path at an arbitrary instant t.

1.8 The refraction of light may be understood by purely kinematic considerations. We need to assume that light takes the shortest (in time) path between two points (Fermat's principle of least time). Referring to the figure, let the speed of light in

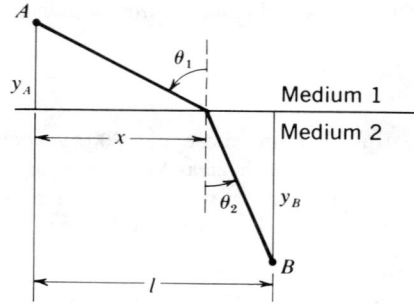

medium 1 be v_1 and in medium 2, v_2. Calculate the time it takes light to go from point A to point B as a function of the variable x. Minimize with respect to x. Given that

$$v_1 = c/n_1 \quad \text{and} \quad v_2 = c/n_2$$

where the n values are known as indices of refraction, prove Snell's law of refraction:

$$n_1 \sin \theta_1 = n_2 \sin \theta_2$$

1.9 At 12:00 hours ship A is 10 km east and 20 km north of a certain port. It is steaming at 40 km/h in a direction 30° east of north. At the same time ship B is 50 km east and 40 km north of the port, and is steaming at 20 km/h in a direction 30° west of north.

(a) Draw a diagram of this situation, and find the velocity of B relative to A.

(b) If the ships continue to move with the above velocities, what is their closest distance to one another and when does it occur?

1.10 The distance from A to B is l. A plane flies a straight course from A to B and back again with a constant speed V relative to the air. Calculate the total time taken for this round trip if a wind of speed v is blowing in the following directions:

(a) Along the line from A to B.
(b) Perpendicular to this line.
(c) At an angle θ to this line.

Show that the time of the round trip is always increased by the existence of the wind.

1.11 A ship is steaming parallel to a straight coastline, distance D offshore, at speed V. A coastguard cutter, whose speed is v ($< V$) sets out from a port to intercept the ship.

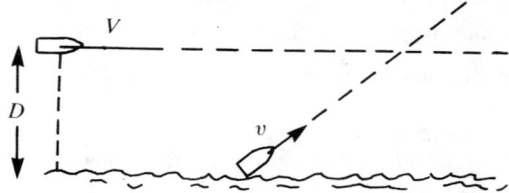

(a) Show that the cutter must start out before the ship passes a point a distance $D(V^2 - v^2)^{1/2}/v$ back along the coast (*Hint:* Draw a vector diagram to show the velocity of the cutter as seen from the ship.)

(b) If the cutter starts out at the latest possible moment, where and when does it reach the ship?

1.12 The astronomer Aristarchus had the idea of comparing the distances of the sun and the moon from the earth by measuring the angular separation θ between them when the moon was exactly half full (see the figure). Using our present knowledge of what these distances are, criticize the feasibility of the method. Aristarchus found $\theta = 87°$. What result would this imply? Calculate what the angle really is and what error would be introduced in the distance if this angle were uncertain by $\pm 0.1°$.

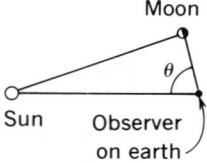

1.13 At $t = 0$ an object is released from rest at the top of a tall building. At the time t_0 a second object is dropped from the same point.

(a) Ignoring air resistance, show that the time at which the objects have a vertical separation l is given by

$$t = \frac{l}{gt_0} + \frac{t_0}{2}$$

How do you interpret this result for $l < \frac{1}{2}gt_0^2$?

(b) The above formula implies that there is an optimum value of t_0 so that the separation l reaches some specified value at the earliest possible value of t. Calculate this optimum value of t_0, and interpret the result.

1.14 Two cars are travelling, one behind the other, on a straight road. Each has a speed of 70 ft/s (about 50 mph) and the distance between them is 90 ft. The driver of the rear car decides to overtake the car ahead and does so by accelerating at 6 ft/s² up to 100 ft/s (about 70 mph) after which he continues at this speed until he is 90 ft ahead of the other car. How far does the overtaking car travel along the road between the beginning and end of this operation? If a third car were in sight, coming in the opposite direction at 88 ft/s (60 mph), what would be the minimum safe distance between the third car and the overtaking car at the beginning of the overtaking operation? (If you are a driver, take note of how large this distance is.)

1.15 In *Paradise Lost*, Book I, John Milton describes the fall of Vulcan from Heaven to earth in the following words:

> ... from Morn
> To Noon he fell; from Noon to dewy Eve,
> A Summer's day; and with the setting Sun
> Dropt from the Zenith like a falling Star. . . .

(It was this nasty fall that gave Vulcan his limp, as a result of his being thrown out of Heaven by Jove.)

(a) Clearly air resistance can be ignored in this trip, which was mostly through outer space. If we assume that the acceleration had the value g (9.8 m/s^2) throughout, how high would Heaven be according to Milton's data? What would have been Vulcan's velocity upon entering the top of the atmosphere?

(b) (Much harder) One really should take account of the fact that the acceleration varies inversely as the square of the distance from the earth's center. Obtain revised values for the altitude of Heaven and the atmospheric entry speed.

1.16 A particle moves in a vertical plane with constant acceleration. Below are values of its x (horizontal) and y (vertical) coordinates at three successive instants of time:

t, s	x, m	y, m
0	4.914	4.054
2×10^{-2}	5.000	4.000
4×10^{-2}	5.098	3.958

Using the basic definitions of velocity and acceleration ($v_x = \Delta x/\Delta t$, etc.), calculate

(a) The x and y components of the average velocity vector during the time intervals 0 to 2×10^{-2} s and 2×10^{-2} to 4×10^{-2} s.

(b) The acceleration vector.

1.17 (a) The figure shows a parabolic atomic-beam trajectory in vacuum, passing through two narrow slits, a distance L apart on the same horizontal level, and travelling an additional horizontal distance L to the detector. Verify that the atoms arrive at the detector at a vertical distance y below the first slit, such that $y \approx gL^2/v^2$, where v is the speed of the atoms. (You can assume $y \ll L$.)

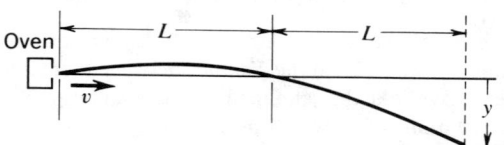

(b) A beam of rubidium atoms (atomic weight 85) passes through two slits at the same level, 1 m apart, and travels an additional distance of 2 m to a detector. The maximum intensity is recorded when the detector is 0.2 mm below the level of the other slits. What is the speed of the atoms detected under these conditions?

1.18 (a) Galileo, in his book *Two New Sciences* (1638), stated that the theoretical maximum range of a projectile of given initial speed over level ground is obtained at a firing angle of 45° to the horizontal, and furthermore that the ranges for angles 45° ± δ (where δ can be any angle < 45°) are equal to one another. Verify these results if you have not been through such calculations previously.

(b) Show that for any angle of projection θ (to the horizontal) the maximum height reached by a projectile is half what it would be at the same instant if gravity were absent.

1.19 A perfectly elastic ball is thrown against a house and bounces back over the head of the thrower, as shown in the figure. When it leaves the thrower's hand, the

PROBLEMS

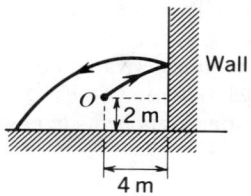

ball is 2 m above the ground and 4 m from the wall, and has $v_{0x} = v_{0y} = 10$ m/s. How far behind the thrower does the ball hit the ground? (Assume that $g = 10$ m/s^2.)

1.20 A man stands on a smooth hillside that makes a constant angle α with the horizontal. He throws a pebble with an initial speed v_0 at an angle θ above the horizontal (see the figure).

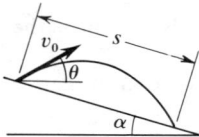

(a) Show that, if air resistance can be ignored, the pebble lands at a distance s down the slope, such that

$$s = \frac{2v_0^2 \sin(\theta + \alpha) \cos \theta}{g \cos^2 \alpha}$$

(b) Hence show that, for given values of v_0 and α, the biggest value of s is obtained with $\theta = 45° - \alpha/2$ and is given by

$$s_{max} = \frac{v_0^2 (1 + \sin \alpha)}{g \cos^2 \alpha}$$

(Use calculus if you like, but it is not necessary.)

1.21 A stopwatch has a hand of length 2.5 cm that makes one complete revolution in 10 s.

(a) What is the vector displacement of the tip of the hand between the points marked 6 s and 8 s? (Take an origin of rectangular coordinates at the centre of the watch-face, with a y axis passing upward through $t = 0$.)

(b) What are the velocity and acceleration of the tip as it passes the point marked 4 s on the dial?

1.22 Calculate the following centripetal accelerations as fractions or multiples of g (≈ 10 m/s^2):

(a) The acceleration towards the earth's axis of a person standing on the earth at 45° latitude.

(b) The acceleration of the moon towards the earth.

(c) The acceleration of an electron moving around a proton at a speed of about 2×10^6 m/s in an orbit of radius 0.5 Å (the first orbit of the Bohr atomic model).

(d) The acceleration of a point on the rim of a bicycle wheel of 26 in. diameter, travelling at 25 mph.

1.23 A particle moves in a plane; its position can be described by rectangular coordinates (x, y) or by polar coordinates (r, θ), where $x = r \cos \theta$ and $y = r \sin \theta$.

(a) Calculate a_x and a_y as the time derivatives of $r \cos \theta$ and $r \sin \theta$, respectively, where both r and θ are assumed to depend on t.

(b) Verify that the acceleration components in polar coordinates are given by

$$a_r = a_x \cos \theta + a_y \sin \theta$$
$$a_\theta = -a_x \sin \theta + a_y \cos \theta$$

Substitute the values of a_x and a_y from (a) and thus obtain the general expressions for a_r and a_θ in polar coordinates.

1.24 A particle oscillates along the x axis according to the following equation: $x = 0.05 \sin(5t - \pi/6)$, where x is in metres and t in s.

(a) What are its velocity and acceleration at $t = 0$?

(b) Make a drawing to show this motion as the projection of a uniform circular motion.

(c) Using (b), find how long it is, after the particle passes through the position $x = 0.04$ m with a negative velocity, before it passes again through the same point, this time with positive velocity.

2 Forces

Newton's great achievement in creating the science of mechanics was to develop quantitative relationships between the forces acting on an object and the changes in the object's motion. More than that, he declared that the main task of mechanics was to learn about forces from observed motions. But this does not alter the fact that the idea of *force* exists independently of the quantitative laws of motion and comes initially from very subjective experiences—the muscular effort involved in applying a push or a pull. We shall begin from this point of view, and rather than plunge at once into dynamics, we shall first take a look at forces in balance.

Forces in equilibrium

The static equilibrium of a given object entails two distinct conditions:

1. The object shall not be subject to any net force tending to move it bodily; it is in what we call *translational equilibrium*.
2. The object shall not be subject to any net influence tending to twist or rotate it; it is in what we call *rotational equilibrium*.

The first condition involves, in general, the combination of forces acting in different directions. It has been known since long before Newton's time that forces are *vectors*. It is a basic property of vectors that the order of addition is immaterial. Thus, if we have a large number of forces applied to the same object, it is possible to represent their addition in many different ways; Fig. 2.1 gives an example. The one essential feature is that, in every case, the force vectors form a closed polygon (i.e. they add up to zero) if equilibrium exists. Since forces may be applied in any direction in three-dimensional space, the force polygon is not necessarily confined to a plane, and the single statement that the force vectors add up to zero will in general be analysable into three separate statements pertaining to three independently chosen directions—usually, although not necessarily, the mutually orthogonal axes of a rectangular coordinate system. Geometrically, one can think of this as the projection of the closed vector polygon onto different planes; regardless of the distortions of shape, the projected polygon remains a closed figure.

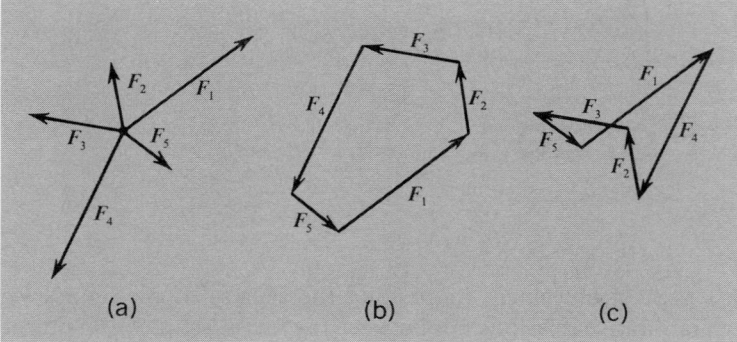

Fig. 2.1 (a) Several forces acting at the same point. (b) The force vectors form a closed polygon, showing equilibrium. (c) Equivalent vector diagram to (b).

When written out in algebraic terms, the projection involves a statement of the analysis of an individual force vector into components, or resolved parts, along the chosen directions. Thus the first condition of equilibrium—equilibrium with respect to bodily translation—can be written as follows:

Vector statement:

$$F = F_1 + F_2 + F_3 + \ldots = 0 \tag{2.1a}$$

Component statement:

$$\begin{aligned} F_x &= F_{1x} + F_{2x} + F_{3x} + \ldots = 0 \\ F_y &= F_{1y} + F_{2y} + F_{3y} + \ldots = 0 \\ F_z &= F_{1z} + F_{2z} + F_{3z} + \ldots = 0 \end{aligned} \tag{2.1b}$$

Action and reaction in the contact of objects

An important phenomenon is expressed by the familiar statement: Action and reaction are equal and opposite. In its general form this says that, regardless of the detailed form of the contact or of the relative hardness or softness of the two objects involved, the magnitudes of the forces that each object exerts on the other are always exactly equal. Note particularly that, from the very way they are defined, these two forces cannot both act on the same object. This may seem a trivial and obvious remark, but many calculations in elementary mechanics have come to grief through a failure to recognize it.

The production of a force of reaction in response to an applied force always involves deformation to some extent. You push on a wall, for example, and that is a conscious muscular act; but how does the wall know to push back? The answer is that it yields, however imperceptibly, and it is as a result of such elastic deformations that a contact force exerted by the wall comes into existence. No matter how rigid a surface may seem, it always gives a little under a push or a pull and cannot supply a contrary force until it has done so.

Rotational equilibrium; torque

We shall now consider the second condition for static equilibrium of an object, assuming that the first condition, $\Sigma F = 0$, has been satisfied. Whether the object is, in fact, in equilibrium will now depend on whether or not the forces are applied in such a way as to produce a resultant twist. Figure 2.2 illustrates the problem with the simplest possible example. An object is acted on by two equal and opposite forces. If, as in Fig. 2.2(a), the forces are along the line joining the points A and B at which the forces are applied, the object is truly in equilibrium; it has no tendency to rotate. In any other circumstances (e.g. as shown in Fig. 2.2b) the object is bound to twist. If the directions of the forces remain unchanged as the object turns, an equilibrium orientation is finally reached, as shown in Fig. 2.2(c). How do we construct out of such familiar knowledge a quantitative criterion for rotational equilibrium?

The law of the lever provides the clue. Look at the situation shown in Fig. 2.3. The balancing of the forces F_1 and F_2 with respect to the pivot at O requires the condition $F_1 l_1 = F_2 l_2$. The product of the force and its lever arm describes its 'leverage', or twisting ability; the technical term for this is

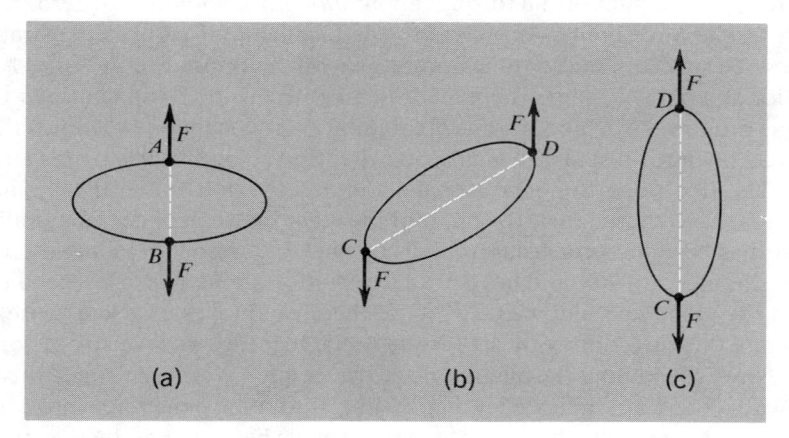

Fig. 2.2 (a) Rotational equilibrium with equal, opposite forces applied at different points of an extended object. (b) Equal, opposite forces applied in a way that does not give rotational equilibrium. (c) If free to rotate, the object moves from orientation (b) to an equilibrium orientation.

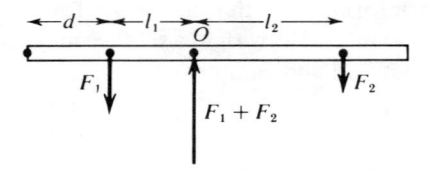

Fig. 2.3 All the forces acting on a pivoted bar (of negligible weight) in rotational equilibrium with one force applied on each side of the pivot.

torque or *moment*. The torques of F_1 and F_2 with respect to O are equal in magnitude but opposite in direction—that due to F_2 is clockwise and that due to F_1 is counterclockwise. Let us call one of them positive and the other one negative; then the condition of balance can be expressed in another way: *The total torque is equal to zero.*

Although the situation as described above is extremely simple, there is more to it than meets the eye, because a further force is exerted on the bar at the position of the pivot; it must be of magnitude $F_1 + F_2$ if the first condition of equilibrium is to be satisfied. Now, to be sure, this third force exerts no torque about the pivot point O itself. However, what if we choose to consider the torques about, let us say, the left-hand end of the bar, some distance d to the left of the point of application of F_1? Then clearly, with respect to this new origin, the force at O is supplying a counterclockwise torque—but it turns out that this is exactly balanced by the sum of the torques due to F_1 and F_2 (both clockwise, notice, with respect to the new, hypothetical pivot), provided that the condition $F_1 l_1 = F_2 l_2$ is satisfied. Thus if the vector sum of the forces on an object is zero, and if the sum of the torques about any one point is zero, then the sum of the torques about *any* other point is also zero.

So far we have limited ourselves to the balancing of torques of parallel forces. Now let us make things more general. Suppose that a force F is applied at a point P, somewhere on or in an object (Fig. 2.4a). Consider the torque produced by F about some other point O, which might be the position of a real pivot, or just an arbitrary point. Let the vector distance from O to P be r. The first thing to notice is that r and F between them define a plane, which we have chosen to be the plane of the diagram. Experience, so familiar that it has become second nature, tells us that if O were indeed a real pivot point, the effect of F would be to produce rotation about an axis perpendicular to the plane in which r and F lie. It therefore makes excellent sense to associate this direction with the torque itself, regarded as a vector of some sort. Now, what about the magnitude of the torque? We can calculate this in two ways. The first, indicated in Fig. 2.4(b), is to resolve the force into components along and perpendicular to r. If the angle between r and F is φ, these components are $F \cos \varphi$ and $F \sin \varphi$, respectively. The radial component represents a force directed straight through O and hence contributes nothing to the torque. The transverse component, perpendicular to r, gives a torque

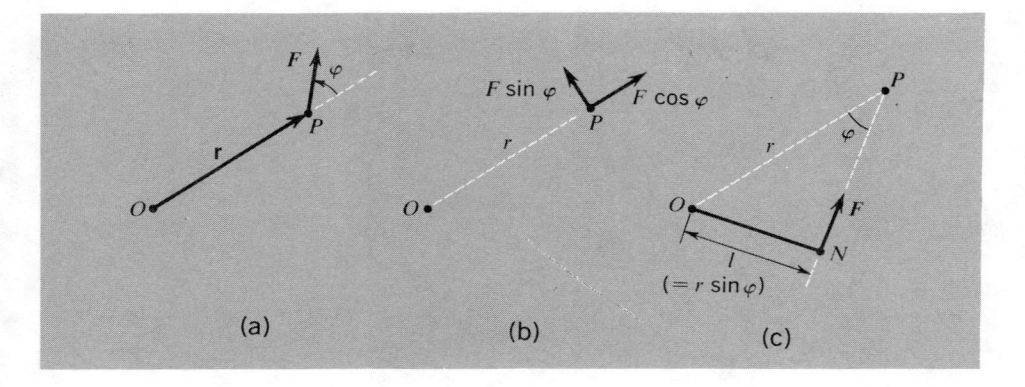

Fig. 2.4 (a) Force F applied at a vector distance r from a pivot point. (b) Resolution of F into components along and perpendicular to r. (c) Evaluation of torque of F by finding its effective lever arm, l.

of magnitude $rF \sin \varphi$. Another way of seeing this result is, as suggested in Fig. 2.4(c), to recognize that the effective lever arm of the total force F can be formed by drawing the perpendicular ON from O to the line along which F acts. Then the torque due to F is just the same as if it were actually applied at the point N, at right angles to a lever arm of length l equal to $r \sin \varphi$.

We shall introduce the single symbol M for the magnitude of the torque. Then we have

$$M = rF \sin \varphi \tag{2.2a}$$

This equation does not contain the necessary information about the *direction* of the torque, but a compact statement in vector algebra, invented specifically for such purposes, is at hand. This is the so-called cross product or vector product of two vectors.

Given two vectors A and B, the *cross product* C is defined to be a vector perpendicular to the plane of A and B and of magnitude given by

$$C = AB \sin \theta$$

where θ is the lesser of the two angles between A and B. There are, of course, two opposite vector directions normal to the plane of A and B. To establish a unique convention we proceed as follows. Imagine rotating the vector A through the (smaller) angle θ until it lies along the direction of B (see Fig. 2.5a). This establishes a sense of rotation. If the fingers of the *right* hand are curled around in the sense of rotation, keeping the thumb extended, the direction of the cross product is along the direction of the pointing thumb. The shorthand mathematical statement, which is understood to embody all these properties, is then written as

$$C = A \times B$$

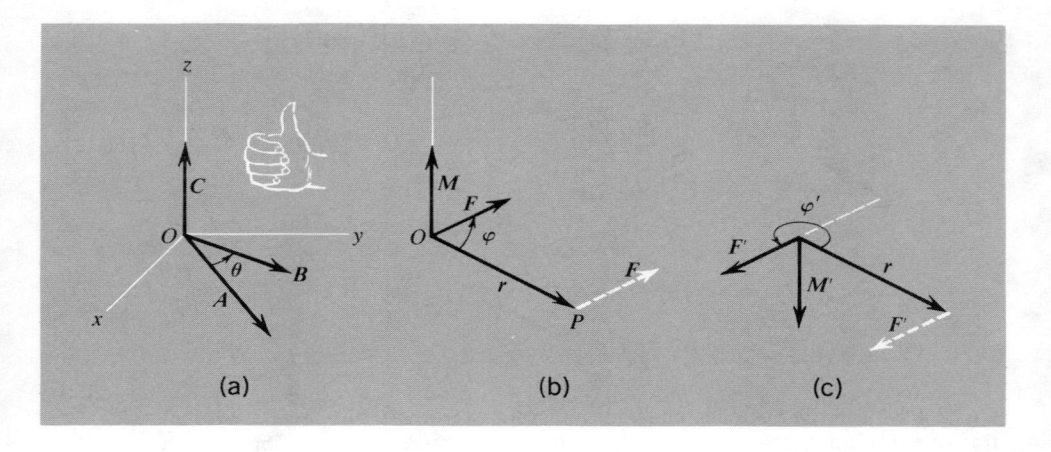

Fig. 2.5 (a) Cross product, **C**, of two arbitrary vectors, **A** and **B**. (b) Torque vector, **M**, as the cross product of **r** and **F**. (c) Situation resulting in a torque vector opposite in direction to that in (b).

Note carefully that the order of the factors is crucial; reversing the order reverses the sign:

$$\bm{B} \times \bm{A} = -(\bm{A} \times \bm{B})$$

Using this vector notation, the torque as a vector quantity is completely specified by the following equation:

$$\bm{M} = \bm{r} \times \bm{F} \tag{2.2b}$$

Figure 2.5(b) and (c) illustrates this for two different values of φ; in each case a right-handed rotation about the direction in which **M** points represents, as you can verify, the direction in which **F** would cause rotation to occur. Then, finally, we can write down the vector sum of all the torques acting on an object, and the second condition of equilibrium—equilibrium with respect to rotation—can be written as follows:

$$\Sigma \bm{M} = \bm{r}_1 \times \bm{F}_1 + \bm{r}_2 \times \bm{F}_2 + \bm{r}_3 \times \bm{F}_3 + \ldots = 0 \tag{2.3}$$

If the object on which a set of forces acts can be regarded as an ideal particle (i.e. a point object), then the condition of rotational equilibrium becomes superfluous. Since all the forces are applied at the same point, they cannot exert a net torque about this point; and if the condition $\Sigma \bm{F} = 0$ is also satisfied, they cannot exert a net torque about any other point either. If one wants to put this in more formal terms, one can say that the same value of **r** applies to every term in eqn (4.3), so that the condition $\Sigma \bm{M} = 0$ reduces to the condition

$$\bm{r} \times (\Sigma \bm{F}) = 0$$

and so embodies the condition $\Sigma F = 0$ for translational equilibrium; the equation for rotational equilibrium adds no new information.

Inertia

We now come to the central problem of Newtonian dynamics: how are motions of material objects affected by forces? Isaac Newton stated the principle of *inertia* in a generalized form in his 'first law' of motion as presented in the *Principia*: 'Every body perseveres in its state of rest, or of uniform motion in a right line, unless it is compelled to change that state by forces impressed upon it.' It is a familiar statement, which we probably all learn in our first encounter with mechanics. But what does it really say? The first thing we must recognize is that every statement about the motion of a given object involves a physical frame of reference; we can only measure displacements and velocities with respect to other objects. Thus the principle of inertia is not just a clear-cut statement about the behaviour of individual objects; it goes much deeper than that. We can, in fact, turn it around and make a statement that goes roughly as follows:

There exist certain frames of reference with respect to which the motion of an object, free of all external forces, is a motion in a straight line at constant velocity (including zero).

A reference frame in which the law of inertia holds good is called an *inertial frame*, and the question as to whether a given frame of reference is inertial then becomes a matter for observation and experiment. Most observations made within the confines of a laboratory on the earth's surface suggest that a frame of reference attached to that laboratory is suitable. A more critical scrutiny shows that this is not quite good enough, and we need to look further afield—but we shall do that later. For the moment we shall limit ourselves to introducing the main principle, which is not affected by the later refinements. Any deviation from a straight-line path is taken to imply the existence of a force. No deviation, no force—and vice versa. It must be recognized that we cannot 'prove' the principle of inertia by an experimental test, because we can never be sure that the object under test is truly free of all external interactions, such as those due to extremely massive objects at very large distances. Moreover, there is the far from trivial question of defining a straight line in a real physical sense: it is certainly not intuitively obvious, nor is it an abstract mathematical question. (How would *you* define a straight line for this purpose?) Nevertheless, it can be claimed that the principle of inertia is a valid generalization from experience; it is a possible interpretation of observed motions, and our belief in its validity grows with the number of phenomena one can correlate successfully with its help.

Force and inertial mass: Newton's second law

The law of inertia implies that the 'natural' state of motion of an object is a state of constant velocity. Closely linked to this is the recognition that the effect of an interaction between an object and an external physical system is to change the state of motion. For example, we have no doubt that the motion of a tennis ball is affected by the racket, that the motion of a compass needle is affected by a magnet, and that the motion of the earth is affected by the sun. *Inertial mass* is the technical phrase for that property which determines how difficult it is for a given applied force to change the state of motion of an object. Newton's second law gives us a quantitative relationship:

$$F = ma = m\frac{dv}{dt} \qquad (2.4)$$

where the proportionality factor m is called the *inertial mass* of the object and F is the net force acting on it. Embodied in this basic statement of Newton's law is the feature that force and acceleration are vector quantities and that the acceleration is always in the same direction as the net force.

An interesting historical fact, often overlooked, is that Newton's own statement of the basic law of mechanics was *not* in the form of eqn (2.4); the equation $F = ma$ appears nowhere in the *Principia*. Instead, Newton spoke of the change of 'motion' (by which he meant momentum) and related this to the value of force × time. In other words, Newton's version of the second law of motion was essentially the following:

$$F\Delta t = m\Delta v \qquad (2.5)$$

Some comments on Newton's second law

Simple and familiar as eqn (2.4) is, it nevertheless contains an enormous wealth of physical concepts—indeed, almost the whole basis of classical dynamics. First comes the assumption that quantitative measurements of displacements and time intervals lead us to a unique value of the acceleration of an object at a given instant. If we remind ourselves that displacements can only be measured with respect to other physical objects, we see that this, like the principle of inertia, cannot be separated from the choice of reference frame. In fact, we tacitly assume that the frame in which the acceleration is measured is an inertial frame.

Next comes the feature, already emphasized, that the acceleration vector is in the direction of the net force vector. This is an important result; it is an expression of the fact that the accelerative effects of several different forces combine in a linear way. It tells us that the instantaneous acceleration of an object is the consequence of a *linear superposition* of the applied forces or of the accelerations that they would individually produce.

Let us add a word of explanation and caution here. The linear superposition of *instantaneous* components of acceleration does not mean that we can always automatically proceed to calculate, let us say, the whole course of development of the *y* component of an object's motion without reference to what is happening in the *x* direction. To take an example that we shall consider in more detail later, if a charged particle is moving in a magnetic field, the component of force in a given direction depends on the component of velocity *perpendicular* to that direction. In such a case, we have to keep track of the way in which that perpendicular velocity component changes as time goes on.

In the case of an object subjected to a single force, one may be tempted to think that it is intuitively obvious that the acceleration is in the same direction as the force. It may be worth pointing out, therefore, that this is *not* in general true if high-velocity particles are involved—sufficiently fast to require the modified kinematics and dynamics of special relativity.

Thirdly comes the assertion that a given force, applied to a particular object, causes the velocity of that object to change at a certain rate *a*, the magnitude of which depends *only* on the magnitude and direction of *F* and on a single scalar quantity characteristic of the object—its inertial mass *m*. This is a very remarkable result; let us consider it further.

Newton's law asserts that the acceleration produced by any constant force, for example by a stretched spring, has the same value under all conditions. Thus, according to this statement, it does not matter whether the object is initially stationary or is travelling at high speed. Is this always, and universally true? No! It turns out that for extremely high speeds—speeds that are a significant fraction of the speed of light—the acceleration produced by a given force on a given object *does* depend on v. Under these high-speed conditions Newton's mechanics gives way to Einstein's, as described by the special theory of relativity: the inertia of a given object increases systematically with speed according to the formula

$$m(v) = \frac{m_0}{(1 - v^2/c^2)^{1/2}} \tag{2.6}$$

The quantity m_0, which is called the 'rest mass' of the object, represents what we can simply call *the* inertial mass in all situations to which classical mechanics applies, because for any $v \ll c$ the value of m according to eqn (2.6) is inappreciably different from m_0.

One last implication of Newton's law, as expressed by eqn (2.4), is that the basic dynamics of an object subjected to a given force does not depend on d^2v/dt^2 or on any of the higher time derivatives of the velocity. The absence of any such complication is in itself a remarkable result, which as far as we know continues to hold good even in the 'relativistic' region of very high velocities. It has, however, been pointed out that if one considers physiological effects, not just the basic physics, the existence and magnitude of d^2v/dt^2 ($= da/dt$) can be important. We all know the good feeling of a 'smooth acceleration' in a car, and what we mean by that phrase is an acceleration that is close to being constant. A rapid rate of change of acceleration produces great discomfort, and it has even been suggested that a unit of da/dt—to be called a 'jerk'—should be introduced as a quantitative measure of such effects!

The conclusion that we can draw from the above discussion is that Newton's law, although ultimately limited in its application, does express with insignificantly small error the relation between the acceleration of an object and the force acting on it for almost everything outside the realm of high-speed atomic particles.

The invariance of Newton's second law; relativity

We have emphasized how the experimental basis of Newton's second law involves the observation of motions with respect to an inertial frame of reference. The actual appearance of a given motion will vary from one such frame to another. It is worth seeing, therefore, how the dynamical conclusions are independent of the particular choice of frame—which means that Newton's mechanics embodies a principle of relativity.

The first point to establish is that, if we have identified any one inertial frame, S (i.e. a frame in which an object under no forces moves uniformly in a straight line), then any other frame, S', having a constant velocity relative to the first is also an inertial frame. This follows directly from the fact that if an object has the instantaneous velocity u in S, and if the velocity of S'

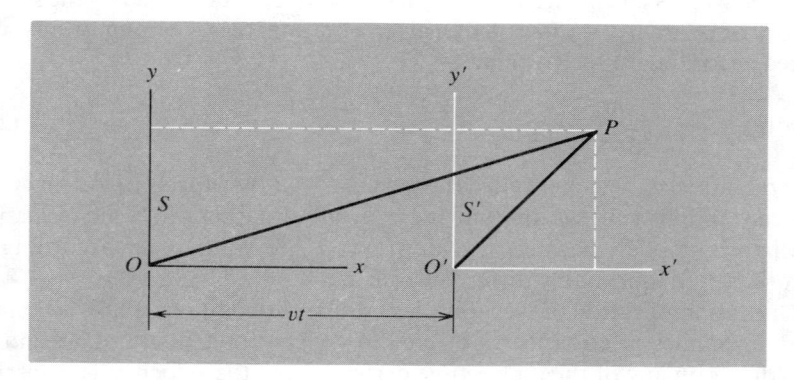

Fig. 2.6 Motion of a particle P referred to two frames that have a relative velocity v.

relative to S is v, then the instantaneous velocity of the object relative to S' is given (see Chapter 1) by

$$u' = u - v$$

Thus if u and v are constant velocities, so also is u', and the object will obey the law of inertia as observed in S'.

To discuss the problem further, let us set up rectangular coordinate systems in both frames, with their x axes along the direction of the velocity v (see Fig. 2.6). Let the origins O and O' of the two systems be chosen to coincide at $t = 0$, at which instant, also, the y and z axes of S' coincide with those of S. Let a moving object be at the point P at a later time, t, when the origin O' has moved a distance vt along the x axis of S. Then the coordinates of P in the two systems are related by the following equations. (It is appropriate, in view of Galileo's pioneer work in kinematics and especially of his clear statement of the law of inertia, that they should have become known as the Galilean transformations.)

(Galilean transformation: S' moves relative to S with a constant speed v in the $+x$ direction)
$$\begin{cases} x' = x - vt \quad (v = \text{const.}) \\ y' = y \\ z' = z \\ t' = t \end{cases} \quad (2.7)$$

The last of these equations expresses the Newtonian assumption of a universal, absolute flow of time, but it also embodies the specific convention that the zero of time is taken to be the same instant in both frames of reference, so that all the clocks in both frames agree with one another.

We can then proceed to obtain relationships between the components of an instantaneous velocity as measured in the two frames. Thus for the x components we have

$$u'_x = \frac{dx'}{dt'}; \quad u_x = \frac{dx}{dt}$$

Putting $x' = x - vt$, and $dt' = dt$, we have

$$u'_x = \frac{d}{dt}(x - vt) = u_x - v$$

The transformations of all three components of velocity are as follows:

$$\begin{aligned} u'_x &= u_x - v \\ u'_y &= u_y \\ u'_z &= u_z \end{aligned} \quad (2.8)$$

Finally, differentiating these velocity components with respect to time, we have (for $v = $ const.) three equalities involving the components of acceleration:

$$\frac{du'_x}{dt'} = \frac{du_x}{dt}; \quad \frac{du'_y}{dt'} = \frac{du_y}{dt}; \quad \frac{du'_z}{dt'} = \frac{du_z}{dt}$$

Thus the measure of any acceleration is the *same* in both frames:

$$\boldsymbol{a'} = \boldsymbol{a} \tag{2.9}$$

Since this identity holds for any two inertial frames, whatever their relative velocity, we say that the acceleration is an *invariant* in classical mechanics. This result is the central feature of relativity in Newtonian dynamics (and it ceases to hold good in the description of motion according to special relativity).

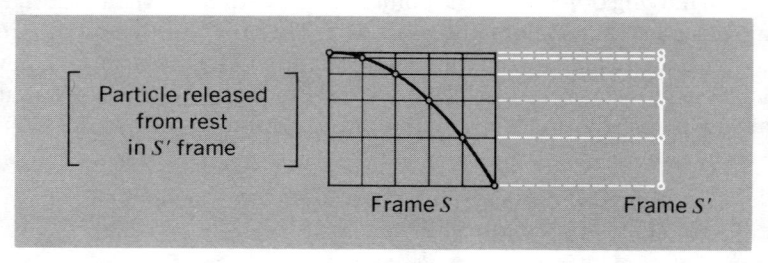

Fig. 2.7 Two different views of the trajectory of an object after it has been released from rest with respect to a moving frame, S'.

To illustrate the application of these ideas, consider the simple and familiar example of a particle falling freely under gravity. Suppose that at $t = 0$, when the axes of the systems S and S' are coincident, an experimenter in S' drops a particle from rest in this frame. The trajectories of the particle, as seen in S and S', are plotted in Fig. 2.7. In each frame the particle is observed to follow the expected trajectory according to the kinematic equations with a vertical acceleration g. In the S frame, the particle has an initial horizontal velocity and therefore it follows a parabolic path, whereas in S' the particle, under the action of gravity, falls straight down. Observers in these two different frames would agree that the equation $\boldsymbol{F} = m\boldsymbol{a}$, where they use the same $\boldsymbol{F}$, accounts properly for the trajectories for any particle launched in any manner in either frame. The frames are thus equivalent as far as dynamical experiments are concerned—either frame may be assumed stationary and the other frame in motion, with the same laws of mechanics providing correct explanations from the observed motions. This is a simple example of the *invariance* of Newton's law itself.

Concluding remarks

It will probably have become apparent to you during the course of this chapter that the foundation of classical mechanics, as represented by Newton's second law, is a complex and in many respects subtle matter. The precise content of the law is still a matter for debate, nearly three centuries after Newton stated the first version of it. In a fine discussion entitled 'The Origin and Nature of Newton's Laws of Motion', one author (Brian Ellis) has said: 'But what of Newton's second law of motion? What is the logical status of this law? Is it a definition of force? Of mass? Or is it an empirical proposition relating force, mass, and acceleration?' Ellis argues that it is something of all of these:

> Consider how Newton's second law is actually used. In some fields it is unquestionably true that Newton's second law is used to define a scale of force. How else, for example, can we measure interplanetary gravitational forces? But it is also unquestionably true that Newton's second law is sometimes used to define a scale of mass. Consider, for example, the use of the mass spectrograph. And in yet other fields, where force, mass, and acceleration are all easily and independently measurable, Newton's second law of motion functions as an empirical correlation between these three quantities. Consider, for example, the application of Newton's second law in ballistics and rocketry. . . . To suppose that Newton's second law of motion, or *any* law for that matter, must have a unique role that we can describe generally and call the logical status is an unfounded and unjustifiable supposition.

Since force and mass are both abstract concepts and not objective realities, we might conceive of a description of nature in which we dispensed with both of them. But, as one physicist (D.H. Frisch) has remarked, 'Whatever we think about ultimate reality it is convenient to follow Newton and split the description of our observations into "forces", which are what make masses accelerate, and "masses", which are what forces make accelerate. This would be just tautology were it not that the observed phenomena can best be classified as the result of *different forces* acting on the *same* set of masses.' Ellis spells out this same idea in more detail:

> Now there are, in fact, many and various procedures by which the magnitudes of the individual forces acting on a given system may be determined—electrostatic forces by charge and distance measurements, elastic forces by measurement of strain, magnetic forces by current and distance determinations, gravitational forces by mass and distance measurements, and so on. And it is an empirical fact that when all such force measurements are made and the magnitude of the resultant force determined, then the rate of change of momentum of the system under consideration is found to be proportional to the magnitude of this resultant force.

And so it is that we obtain an immensely fruitful and accurate description of a very large part of our whole experience of objects in motion, through the simple and compact statement of Newton's second law.

Problems

2.1 The ends of a rope are held by two men who pull on it with equal and opposite forces of magnitude F. Construct a clear argument to show why the tension in the rope is F, not $2F$.

2.2 It is a well-known fact that the total gravitational force on an object may be represented as a single force acting through a uniquely defined point—the 'centre of gravity'—regardless of the orientation of the object.

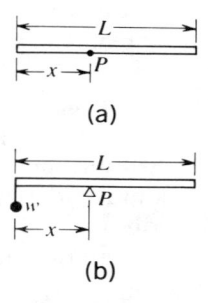

(a) For a uniform bar, the centre of gravity (CG) coincides with the geometrical centre. Use this fact to show that the total gravitational torque about the point P (see part (a) of the figure) may be considered as arising from a single force W at the bar's centre, or from two individual forces of magnitudes Wx/L and $W(L - x)/L$ acting at the midpoints of the two segments defined by P.

(b) If a bar or rod has a weight W, and a small weight w is hung at one end (see part (b) of the figure), use the simpler of the above two methods to show that the system balances on a fulcrum placed at P if $x = LW/2(W + w)$.

2.3 Diagram (a) represents a rectangular board, of negligible weight, with individual concentrated weights mounted at its corners.

(a) To find the position of the CG of this system, one can proceed as follows: choose an origin at the corner O, and introduce x and y axes as shown. Imagine the board to be pivoted about a horizontal axis along y, and calculate the distance $\bar{x}$ from this axis at which an upward force $W (= w_1 + w_2 + w_3 + w_4)$ will keep the system in rotational equilibrium. Next imagine the board to be pivoted about a horizontal axis along x, and calculate the corresponding distance $\bar{y}$. Then the centre of gravity, C, is at the point $(\bar{x}, \bar{y})$.

(b) An experimental method of locating the CG is to hang the board from two corners in succession (or any other two points, for that matter) and mark the direction of a plumbline across the board in each case. To verify that this is consistent with (a), imagine the board to be suspended from O in a vertical plane (diagram b) and show by direct consideration of the balancing of torques due to w_2, w_3 and w_4 that the board hangs in such a way that the vertical line from O passes through C (so that $\tan \theta = \bar{y}/\bar{x}$).

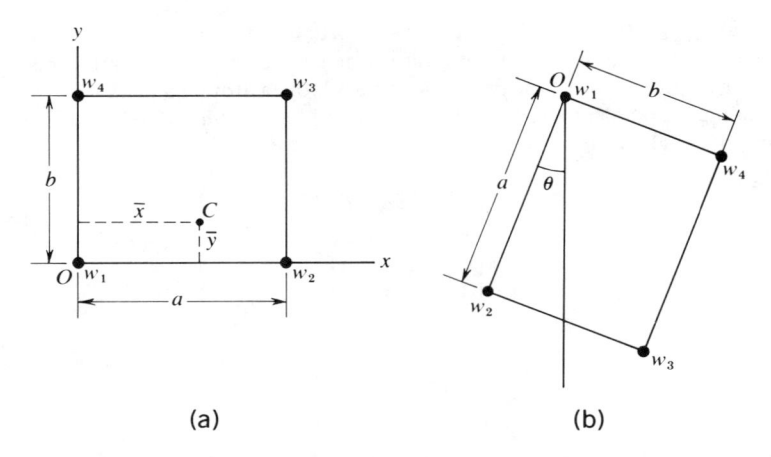

2.4 (a) Over 2000 years ago, Archimedes gave what he believed to be a theoretical proof of the law of the lever. Starting from the necessity that equal forces, F, at equal distances, l, from a fulcrum must balance (by symmetry), he argued that one of these forces could, again by symmetry, be replaced by a force $F/2$ at the fulcrum and another force $F/2$ at $2l$. Show that this argument depends on the truth of what it is purporting to prove.

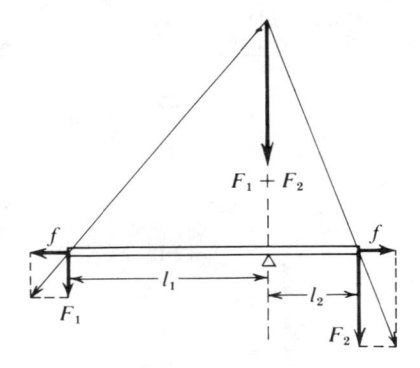

(b) A less vulnerable argument is based on the experimental knowledge that forces combine as vectors. Suppose that parallel forces F_1 and F_2 are applied to a bar as shown. Imagine that equal and opposite forces of magnitude f are introduced as shown. This gives us two resultant force vectors that intersect at a point that defines the line of action of their resultant (of magnitude $F_1 + F_2$). Show that this resultant intersects the bar at the pivot point for which $F_1 l_1 = F_2 l_2$.

2.5 Prove that if three forces act on an object in equilibrium, they must be coplanar and their lines of action must meet at one point (unless all three forces are parallel).

2.6 Painters sometimes work on a plank supported at its ends by long ropes that pass over fixed pulleys, as shown in the figure. One end of each rope is attached to the plank. On the other side of the pulley the rope is looped around a hook on the plank, thus holding the plank at any desired height. A painter weighing 175 lb works on such a plank, of weight 50 lb.

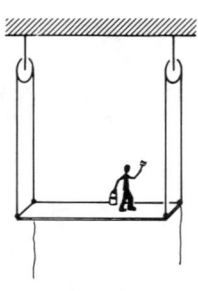

(a) Keeping in mind that he must be able to move from side to side, what is the maximum tension in the ropes?

(b) Suppose that he uses a rope that supports no more than 150 lb. One day he finds a firm nail on the wall and loops the rope around this instead of around the hook on the plank. But as soon as he lets go of the rope, it breaks and he falls to the ground. Why?

2.7 A man begins to climb up a 12-ft ladder (see the figure). The man weighs 180 lb, the ladder 20 lb. The wall against which the ladder rests is very smooth, which means that the tangential (vertical) component of force at the contact between ladder and wall is negligible. The foot of the ladder is placed 6 ft from the wall. The ladder, with the man's weight on it, will slip if the tangential (horizontal) force at the contact

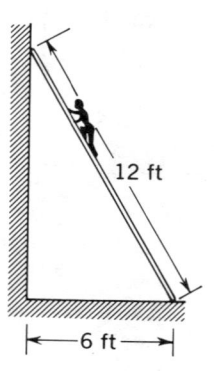

between ladder and ground exceeds 80 lb. How far up the ladder can the man safely climb?

2.8 A yo-yo rests on a table (see the figure) and the free end of its string is gently pulled at an angle θ to the horizontal as shown.

(a) What is the critical value of θ such that the yo-yo remains stationary, even though it is free to roll? (This problem may be solved geometrically if you consider the torques about P, the point of contact with the table.)

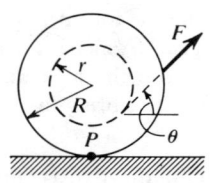

(b) What happens for greater or lesser values of θ? (If you have a yo-yo, test your conclusions experimentally.)

2.9 A simple and widely used chain hoist is based on what is called a differential pulley. In this arrangement two pulleys of slightly different diameter are rigidly connected with a common, fixed axis of rotation. An endless chain passes over these pulleys and around a free pulley from which the load W is suspended (see the figure). If the components of the differential pulley have radii a and $0.9a$, respectively, what downward pull applied to one side of the freely hanging part of the chain will (ignoring friction) suffice to prevent the load from descending if:

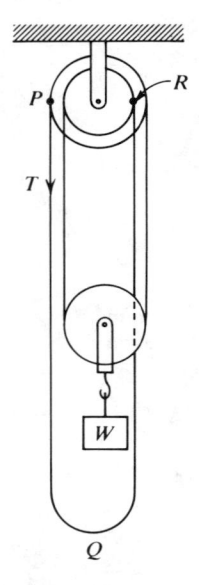

(a) the weight of the chain itself can be neglected?

(b) (more realistic) one takes account of the fact that the freely hanging portion of the chain (PQR) has a total weight w?

2.10 Analyse in qualitative but careful terms how the act of pushing vertically downward on the pedal of a bicycle results in the production of a horizontal force that can accelerate the bicycle forward. (Clearly the contact of the rear wheel with the ground plays an essential role in this situation.)

2.11 A cabin cruiser of mass 15 metric tons drifts in towards a dock at a speed of 0.3 m/s after its engines have been cut. (A metric ton is 10^3 kg.) A man on the dock is able to touch the boat when it is 1 m from the dock, and thereafter he pushes on it with a force of 700 N to try to stop it. Can he bring the boat to rest before it touches the dock?

2.12 (a) A man of mass 80 kg jumps down to a concrete patio from a window ledge only 0.5 m above the ground. He neglects to bend his knees on landing, so that his motion is arrested in a distance of about 2 cm. With what average force does this jar his bone structure?

(b) If the man jumps from a ledge 1.5 m above the ground but bends his knees so that his centre of gravity descends an additional distance h after his feet touch the ground, what must h be so that the average force exerted on him by the ground is only five times his normal weight?

2.13 The graphs shown give information regarding the motion in the xy plane of three different particles. In diagrams (a) and (b) the small dots indicate the positions at equal intervals of time. For each case, write equations that describe the force components F_x and F_y.

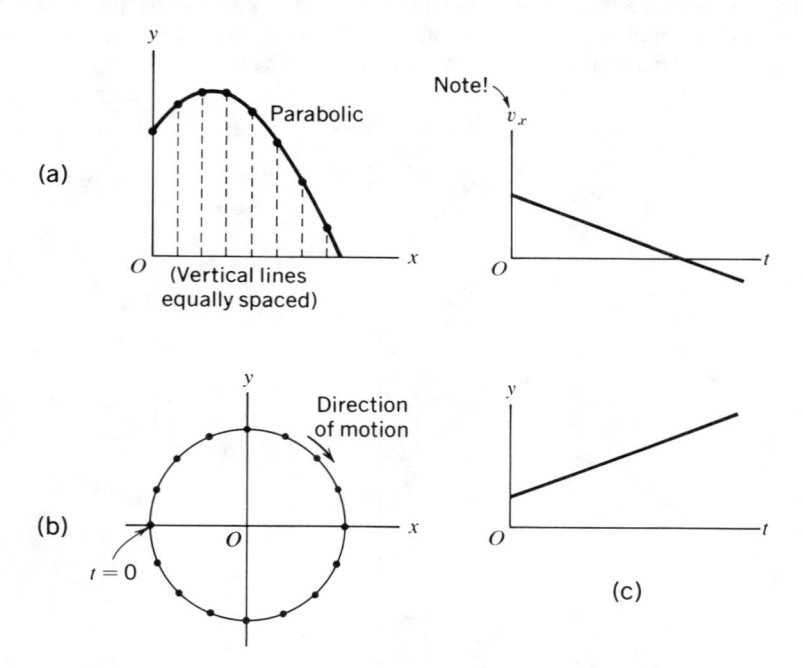

2.14 A particle of mass 2 kg oscillates along the x axis according to the equation

$$x = 0.2 \sin\left(5t - \frac{\pi}{6}\right)$$

where x is in metres and t in seconds.

(a) What is the force acting on the particle at $t = 0$?
(b) What is the maximum force that acts on the particle?

2.15 A particle of mass m follows a path in the xy plane that is described by the following equations:

$$x = A(\alpha t - \sin \alpha t)$$
$$y = A(1 - \cos \alpha t)$$

(a) Sketch this path.
(b) Find the time-dependent force vector that causes this motion. Can you suggest a way of producing such a situation in practice?

2.16 A piece of string of length l, which can support a maximum tension T, is used to whirl a particle of mass m in a circular path. What is the maximum speed with which the particle may be whirled if the circle is (a) horizontal; (b) vertical?

3 Using Newton's laws

It is worth re-emphasizing the fact that Newton's second law may be used in two primary ways:

1. Given a knowledge of all the forces acting on a body, we can calculate its motion.
2. Given a knowledge of the motion, we can infer what force or forces must be acting.

This may seem like a very obvious and quite trivial separation, but it is not. The first category represents a purely *deductive* activity—using known laws of force and making clearly defined predictions therefrom. The second category includes the *inductive*, exploratory use of mechanics—making use of observed motions to learn about hitherto unknown features of the interactions between objects. Skill in the deductive use of Newton's laws is of course basic to successful analytical and design work in physics and engineering and can bring great intellectual satisfaction. But, for the physicist, the real thrill comes from the inductive process of probing the forces of nature through the study of motions. It was in this way that Newton discovered the law of universal gravitation, that Rutherford discovered the atomic nucleus, and that the particle physicists explore the structure of nucleons (although, to be sure, this last field requires analysis in terms of quantum mechanics rather than Newtonian mechanics).

Some examples of $F = ma$

In Chapter 4 we shall have something to say about the way in which Newton arrived at his insight into gravitational forces from a study of planetary motions. But first we shall discuss how one goes about calculating motions from given forces by analysing a few simple examples. The general procedure is first to identify *all* the forces acting on the object and (providing that the mass remains constant) use $F_{net} = ma$ to calculate the resultant acceleration. We can then use the kinematic equations (page 14) to describe the subsequent motion.

Example 1: Two connected masses. This example illustrates the impor-

SOME EXAMPLES OF $F = ma$

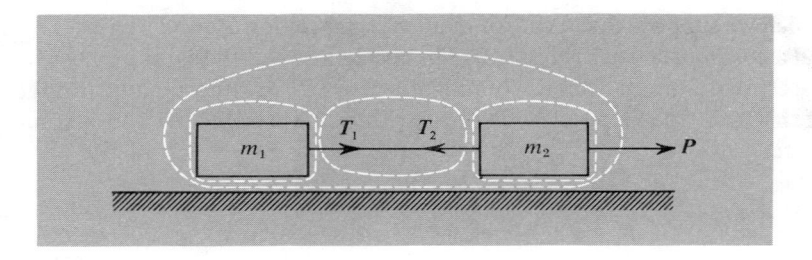

Fig. 3.1 Two connected blocks pulled horizontally on a perfectly smooth surface. Newton's second law must apply to any part of the system that one chooses to consider.

tant point that one is free to isolate, in one's imagination, *any part* of a complete system, and apply $F = ma$ to it alone. Figure 3.1 shows two masses connected by a light (massless) string on a smooth (frictionless) surface. A horizontal force P pulls at the right-hand mass. What can we deduce about the situation?

First, we can imagine an isolation boundary drawn around both m_1 and m_2 and the string that connects them. The only external horizontal force applied to this system is P, and the total mass is m_1 plus m_2. Hence we have

$$P = (m_1 + m_2)a$$

This at once tells us the acceleration that is common to both masses. Next, we can imagine an isolation boundary surrounding the connecting string alone. In Fig. 3.1 we indicate the forces T_1 and T_2 with which the string pulls on the masses; by the equality of action and reaction, the string has forces equal to $-T_1$ and $-T_2$ applied to its ends. The sum of these forces must equal the mass of the string times its acceleration a. Assuming the mass of the string to be negligible, this means that T_1 and T_2 would have the same magnitude, T, which we call *the* tension in the string. (This idealized result is, of course, rather obvious; what *is* worth noting is that, in any real situation, there would have to be enough difference of tensions at the ends of the string to supply the requisite accelerative force to the mass of the string itself.)

Finally, we can imagine drawing isolation boundaries around m_1 and m_2 separately, and applying Newton's law to the horizontal motion of each:

$$T = m_1 a$$
$$P - T = m_2 a$$

Adding these equations, we arrive at the equation of motion of the total system once again. But if we take either equation alone, substitution of the already determined value of a will give the value of the tension T in terms of P and the masses.

Example 2: Car travelling round a curve. Consider a simple case, in which we suppose that a car, of mass m, is travelling around a circle of radius r at a *constant* speed v (Fig. 3.2). The acceleration of the car is purely towards the centre of the circle and is of magnitude v^2/r. We know that the production of this 'centripetal' acceleration necessitates the existence of a corresponding force:

$$F = ma = \frac{mv^2}{r} \tag{3.1}$$

The *banking* of a curve makes it possible for the car, travelling at some reasonable speed, to be held in its curved path by a force exerted on it purely normal to the road surface, i.e. there would be no tangential force, as there would have to be if the road surface were horizontal.

Consider the ideal case as shown in Fig. 3.2(b). Resolving the forces vertically and horizontally, and applying $F = ma$, we have

$$\begin{aligned} N \sin \alpha &= \frac{mv^2}{r} \\ N \cos \alpha - F_g &= 0 \end{aligned} \tag{3.2}$$

Replacing F_g by mg, and solving for α, we find that

$$\tan \alpha = \frac{v^2}{gr} \tag{3.3}$$

which defines the correct angle of banking for given values of v and r. Alternatively, given r and α, eqn (3.3) defines the speed at which the curve should be taken. The situations that arise if greater or lesser values of v are

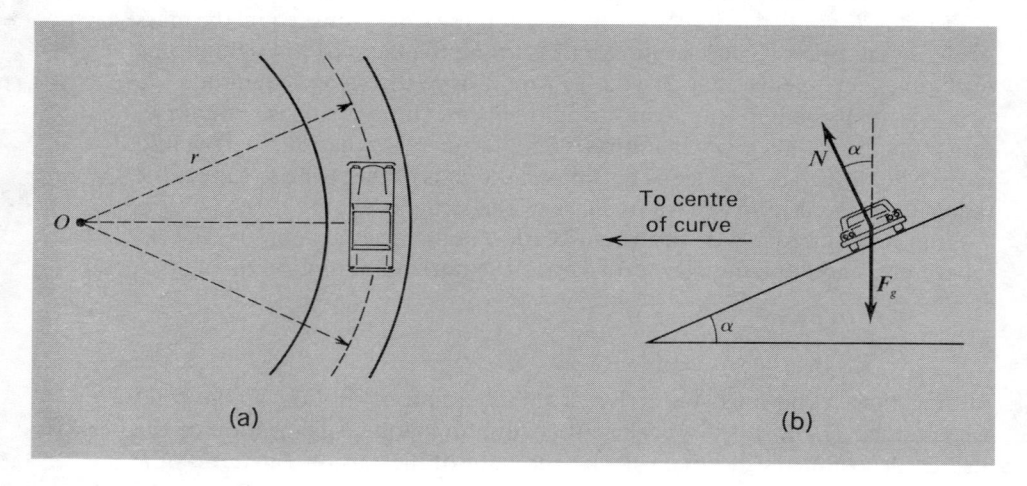

Fig. 3.2 (a) View of a curve in a road, as seen from vertically overhead. (b) Car on the banked curve, as seen from directly behind or in front.

used will require the introduction of a frictional force perpendicular to N (and of limiting magnitude μN) acting inward or outward along the slope of the banked surface.

Example 3: Curvilinear motion with changing speed. If a particle changes its speed as it travels along a circular path, it has, in addition to the centripetal acceleration toward the centre of the curvature, a component of acceleration tangential to the path. This tangential acceleration component represents the rate of change of the *magnitude* of the velocity vector. (This is in contrast to the centripetal acceleration component, which depends upon the rate of change of the *direction* of the velocity vector.) We derived the relevant results in Chapter 1, eqn (1.21).

The situation is most easily handled by considering the two components separately:

$$\left.\begin{array}{l}\textit{radial} \text{ component of} \\ \text{acceleration (at right} \\ \text{angles to the path)}\end{array}\right\} = a_r = -\frac{v^2}{r} \qquad (r \text{ is the radius of curvature})$$

$$\left.\begin{array}{l}\textit{transverse} \text{ acceleration} \\ \text{(tangent to the path)}\end{array}\right\} = a_\theta = \lim_{\Delta t \to 0} \frac{\Delta v}{\Delta t} = \frac{dv}{dt} \quad \begin{array}{l}\text{(note that this is the change of magnitude [only] of the velocity vector)}\end{array}$$

(3.4)

With only slight reinterpretation, we may apply these results to the case of motion along any arbitrary curvilinear path. For every point along such a path, there is a centre of curvature and an associated radius of curvature (both of which change as one moves along the path). Provided that we interpret the symbols r and v to mean the *instantaneous* values of the radius of curvature and the speed, the above expressions are perfectly general. They

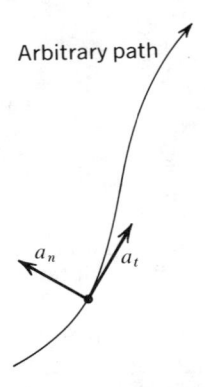

Fig. 3.3 Acceleration components at a point on an arbitrarily curved path.

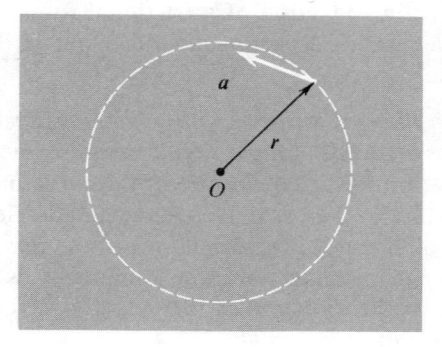

Fig. 3.4 Net acceleration vector of a particle attached to a disc if the angular velocity of the disc is changing.

give the instantaneous acceleration components—tangent to the path, and normal to the path—for the general curvilinear motion. In this case Fig. 3.3 indicates a more appropriate notation for these acceleration components.

A particle of dust that rides, without slipping, on a gramophone turntable as it starts up provides a simple and familiar example of a particle possessing both radial and tangential acceleration components. Its total acceleration a (Fig. 3.4) is the combination of a_r and a_θ at right angles: $a = (a_r^2 + a_\theta^2)^{1/2}$. The net horizontal force applied to the particle by the turntable must be in the direction of a and of magnitude ma. If the contact between the turntable and the particle cannot supply a force of this magnitude, such circular motion is not possible and the particle will slip relative to the surface.

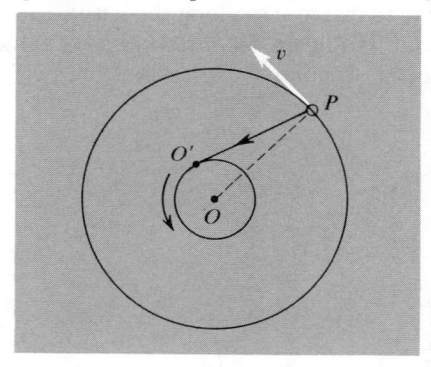

Fig. 3.5 A particle travels in a circular path about O. To increase its speed, the string PO' must provide a force component tangential to the circle.

If we analyse the process of whirling an object at the end of a string, so as to bring it from rest up to some high speed of circular motion (as, for example, in the athletic event 'throwing the hammer') we see that the string must perform two functions: (1) supply the tangential force to increase the speed of the object, and (2) supply the force of the appropriate magnitude mv^2/r towards the centre of the circle. To fill this dual role, the pull supplied by the string must 'lead' the object, so that the tension supplied by it has a tangential component, as shown in Fig. 3.5. This can be maintained if the inner end, O', of the string is continually moved around in a circular path, as indicated. This is the kind of thing we do more or less instinctively in practice.

Circular paths of charged particles in uniform magnetic fields

One of the most important examples of circular motion is the behaviour of electrically charged particles in magnetic fields. It is a matter of experimental fact that an electrically charged particle may, at a given point in space, experience a force when it is moving which is absent if it is at the same point but stationary. The existence of such a force depends on the presence somewhere in the neighbourhood (although perhaps quite far away) of magnets or electric currents. Detailed observations reveal the following features (cf. Fig. 3.6):

1. The force is always exerted at right angles to the direction of the velocity v of the particle. The force is reversed if the direction of the velocity is reversed.

2. The force is proportional to the amount of charge, Q, carried by the particle. The force reverses if the sign of the charge is reversed.

3. For a given value of Q and a given direction of v, the size of the force is proportional to the magnitude of v.

4. For motion parallel to one certain direction at a given point, the force is zero. This direction coincides with the direction in which a compass needle placed there would align itself. It is called the direction of the magnetic *field* at that point.

5. For any other direction of motion, the direction of the force is perpendicular to the plane formed by v and the field direction.

6. The magnitude of the force is proportional, for given values of Q and v and for a fixed magnetic field arrangement, to the sine of the angle between v and the field direction.

All the above results can be summarized in a very compact mathematical statement. We are going to introduce a quantitative measure of the magnetic field strength and denote it by the vector symbol B, in such a way that Fig. 3.6 shows the relation of the directions of Qv, B and F for a charged particle. (Note that Q may be of either sign and that F is normal to the plane containing v and B.) Then the value of F, in both magnitude and direction, is given by the following equation, involving the cross product of vectors (see p. 29).

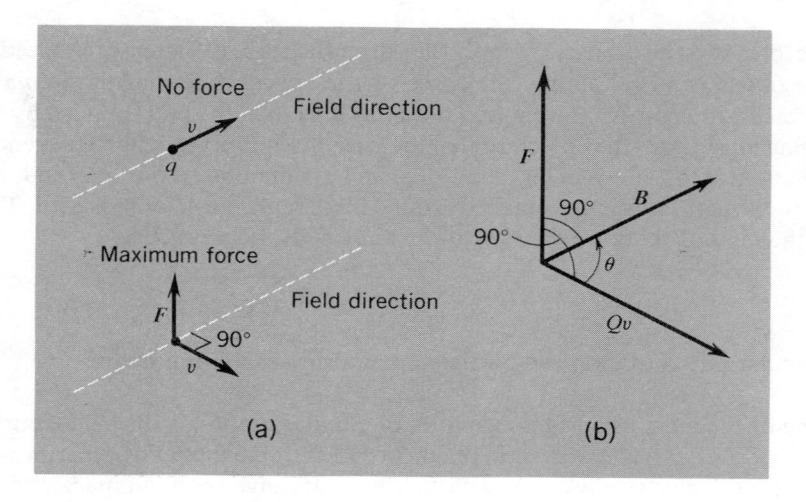

Fig. 3.6 (a) Situations of zero force and maximum force for a charged particle moving in a magnetic field. (b) General vector relationship of velocity, magnetic field and magnetic force.

$$F = Qv \times B \tag{3.5}$$

We can then use this equation to define the quantitative measure of the magnetic field strength, such that a field of unit strength, applied at right angles ($\sin \theta = 1$) to a charge of 1 C moving with a speed of 1 m s^{-1}, would produce a force of 1 N.

Let us now imagine a charged particle of (positive) charge Q, moving in the plane of this page, in a region in which the magnetic field B points perpendicularly down into the paper. Then from eqn (3.5) we have

$$F = QvB \tag{3.6}$$

This is a pure deflecting force, always at right angles to the particle's motion. Hence the magnitude of the velocity v cannot change, but its direction changes uniformly with time. Thus, although there is no centre of force in the usual sense, the particle describes a circular path of radius r. The centripetal acceleration is v^2/r, and so by Newton's law we have

$$QvB = \frac{mv^2}{r}$$

whence

$$r = \frac{mv}{QB} \tag{3.7}$$

Thus the radius of the circle is a measure of the momentum mv of a particle of given charge in a given magnetic field. This fact underlies the nuclear physicist's method for determining the momenta of charged particles. From the radius of the circular track which a charged particle generates in a cloud chamber, bubble chamber, etc., placed between the poles of a magnet the momentum can be readily found if the charge of the particle is known. To obtain the period T (or the angular velocity), we write

$$QvB = m\omega v$$

or

$$\omega = \frac{2\pi}{T} = \frac{Q}{m} B \qquad (3.8)$$

which is independent of the particle's speed. The angular frequency given by eqn (3.8) is called the 'cyclotron' frequency and depends, for a fixed magnetic field, only on the ratio of charge to mass of the particle. It was the recognition of this by E.O. Lawrence that led him in 1929 to design the first cyclotron, in which protons could be raised to high energies by the application of an alternating electric field of constant frequency. The holding of charged particles in an orbit of a given radius by means of a magnetic field is an essential feature of most high-energy particle accelerators.

Having the velocity in a plane perpendicular to B is, of course, a very special case. But if this condition is not satisfied, we can imagine v to be resolved into one component perpendicular to B and a second component parallel to B. The latter, by eqn (3.5), has no magnetic force associated with it; the former changes with time in precisely the way described above. Thus the resultant motion is a helix, whose projection on the plane perpendicular to B is a fixed circle, as shown in Fig. 3.7.

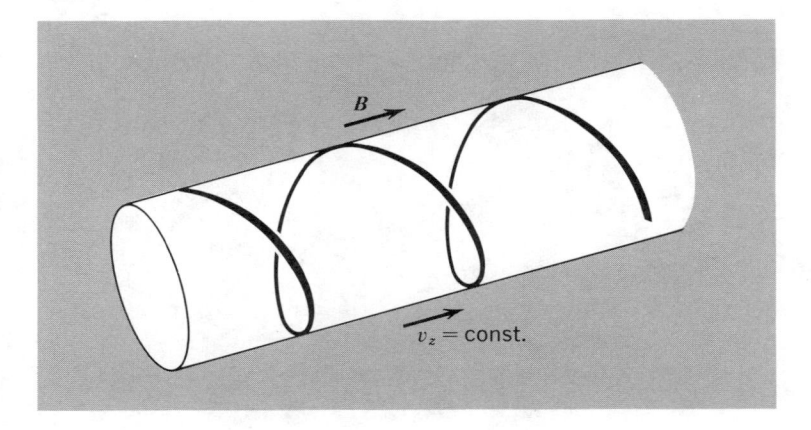

Fig. 3.7 Helical path of charged particle having a velocity component parallel to a magnetic field.

The fracture of rapidly rotating objects

The question of the stresses set up in a rotating object, and the possibility of fracture if they become excessive, provides another good example of Newton's law applied to uniform circular motion. Whenever an object such as a wheel is rotating, every portion of it has an acceleration towards the axis of rotation, and a corresponding accelerative force is required. Suppose, for example, that a thin wheel or hoop of radius r is rotating about its axis at n revolutions per second (rps) (see Fig. 3.8a). Then any small section of the hoop, such as the one shown shaded, must be supplied with a force equal to its mass, m, multiplied by its centripetal acceleration v^2/r. In this case the magnitude of the angular velocity ω is defined by the equation

$$\omega = 2\pi n$$

Thus the instantaneous speed is given by

$$v = \omega r = 2\pi r n \tag{3.9}$$

If we make an isolation diagram (Fig. 3.8b) for the small portion of material shown shaded in Fig. 3.8(a), it is clear that the forces acting on it must be supplied via its contact with adjacent material of the rim. These forces must, by symmetry, be tangential to the rim at each point. (For example, if the force exerted on Δm at one end had a component radially outward, then by the equality of action and reaction the portion of material with which it was in contact would be subjected to a force with a component radially inward. But in a uniform hoop, all portions such as Δm are equivalent; there is no basis for any asymmetries of this kind.) Thus we can picture the small portion of the rim being acted on by a force of magnitude T at each end. If the length of arc represented by this portion is Δs, it subtends

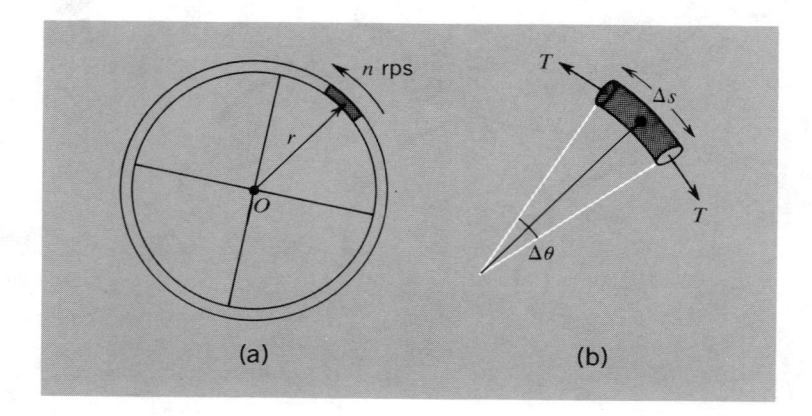

Fig. 3.8 (a) Rotating hoop. (b) Forces acting on a small element of the hoop.

an angle $\Delta\theta$, equal to $\Delta s/r$, at the centre O, and each force has a component equal to $T\sin(\Delta\theta/2)$ along the bisector of $\Delta\theta$. Thus by Newton's law we have

$$2T\sin(\Delta\theta/2) = \Delta m \frac{v^2}{r}$$

Putting $\sin(\Delta\theta/2) \approx \Delta\theta/2$, this then gives us

$$T\Delta\theta \approx \Delta m \frac{v^2}{r}$$

i.e.

$$T\frac{\Delta s}{r} \approx \Delta m \frac{v^2}{r}$$

or

$$T\Delta s = v^2 \Delta m \tag{3.10}$$

Let us now express the mass Δm in terms of the density, ρ, of the material and the volume of the piece of the rim. If the cross-sectional area of the rim is A, we have

$$\Delta m = \rho A \, \Delta s$$

Substituting this into eqn (3.10) and substituting $v = 2\pi rn$ from eqn (3.9), we obtain the following result:

$$T = 4\pi^2 n^2 r^2 \rho A \tag{3.11}$$

Now, it is an experimental fact that a bar or rod of a given material will fracture under tension if the ratio of the applied force to the cross-sectional area exceeds a certain critical value—the ultimate tensile strength, S. Thus we can infer from eqn (3.11) that a hoop of the kind we have been discussing has a critical rate of rotation, above which it will burst. We have, in fact,

$$n_{\max} = \frac{1}{2\pi r}\left(\frac{S}{\rho}\right)^{1/2} \tag{3.12}$$

Suppose, for example, that we have a steel hoop of radius 1 ft (i.e., about 0.3 m). The density of steel is about 7600 kg m^{-3}, and its ultimate tensile strength is about 10^9 N m^{-2}. These values lead to a value of $n_{\max}$ of about 500 rps, or 30,000 rpm—much faster than any such wheel would normally be driven. However, the rotors of ultracentrifuges are, in fact, driven at speeds of this order, up to a significant fraction of the bursting speed for their particular radius.

Motion against resistive forces

We shall now consider an important class of problems in which an object is subjected to a constant driving force, F_0, but has its motion opposed by a resistive force, R, that always acts in a direction opposite to the instantaneous velocity. Typical of such forces are the frictional resistance as an object is pulled along a solid surface, or the air resistance to falling raindrops, moving cars, and so on. In general this resistive force is a function of speed, so that the statement of Newton's law must be written as follows:

$$F_0 - R(v) = m \frac{dv}{dt} \tag{3.13}$$

The resistive force of dry friction is in fact nearly independent of v, as indicated in Fig. 3.9(a), so that we can put

$$R(v) = \mathscr{F} \approx \text{const.}$$

Equation (3.13) then reduces to the simple case of acceleration under a constant net force. The situation is very different in the case of fluid resistance, for which $R(v)$ increases monotonically with v, as indicated in Fig. 3.9(b) and as described by the relation

$$R(v) = Av + Bv^2 \tag{3.14}$$

In this case, if we consider an object starting out from rest under the force F_0, the initial acceleration is F_0/m, but the net driving force is immediately reduced to a value below F_0, because as soon as the object has any appreciable velocity it is exposed, in its own frame of reference, to a wind or flow of fluid past it at the speed v. The statement of Newton's law as it applies to this problem must now be written

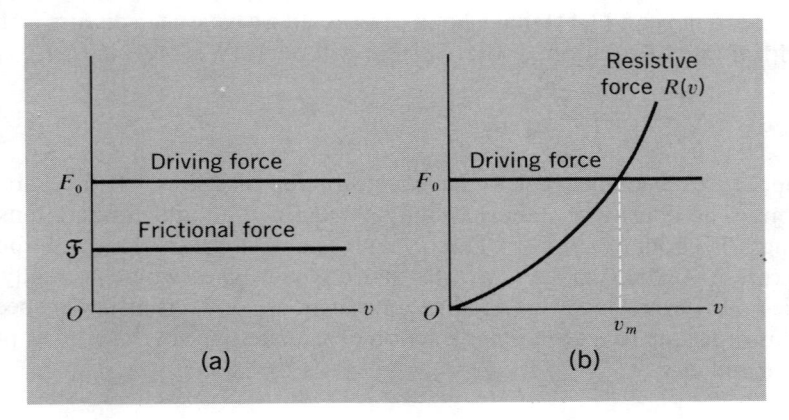

Fig. 3.9 (a) Driving and resistive forces for an object resisted by dry friction. (b) Driving and resistive forces for an object resisted by fluid friction.

$$m\frac{dv}{dt} = F_0 - Av - Bv^2 \tag{3.15}$$

(In this equation F_0, A, B and v are all taken to be positive. One must be careful to consider what is the appropriate statement of Newton's law for this system if the direction of v is reversed.)

The solving of eqn (3.15) is not at all such a simple matter as our familiar problems involving constant forces or forces perpendicular to the velocity. We are now faced with finding the solution to an awkward differential equation. We do not intend to plunge into all the formal mathematics of this problem. Let us, however, consider some individual features of the solution:

1. For some range of small values of v, the acceleration will be almost constant and v will start off as a linear function of t with slope F_0/m.

2. As v increases, a decreases monotonically, giving a steadily decreasing slope in the graph of v versus t (Fig. 3.10).

3. There is a *limiting speed*, v_m, under any given applied force. It is the speed at which the graph of $R(v)$ versus v is intersected by a horizontal line at the ordinate equal to F_0 (see Fig. 3.9b). Algebraically, it is the positive root of the quadratic equation

$$Bv^2 + Av - F_0 = 0$$

Notice the contrast between the sharply defined value of v_m, in the graph of $R(v)$ versus v, and the gradual manner in which this velocity is approached (and in principle never quite reached) if one considers v as a function of time (Fig. 3.10).

It is well worth taking a moment to consider the dynamical situation represented by $v = v_m$. It is a motion with zero acceleration under zero net force, but it seems a far cry from the unaccelerated motion of objects moving under no force at all; and it is certainly not an application of what we understand by the principle of inertia. But let us emphasize that, like static equilibrium, it *is*

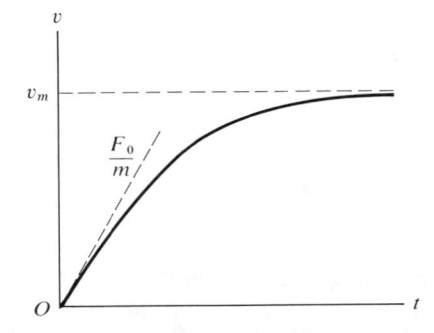

Fig. 3.10 Asymptotic approach to terminal speed for object in a fluid resistive medium.

a case of $\Sigma F = 0$. Every time we see a car hurtling along a straight road at a steady 80 mph, or a jet plane racing through the air at a constant 600 mph, we are seeing objects travelling under zero *net* force. This basic dynamical fact tends to be obscured because what matters in practical terms is the large value of the driving force F_0 needed to maintain the steady motion once it has been established.

Detailed analysis of resisted motion

In order to see how Newton's law for resisted motion (eqn 3.15) works in practice we need to introduce the additional feature that the two terms in the resistive force differ not only in their dependence on v but also in their dependence on the linear dimensions of the object (see Fig. 3.11). Specifically, for a sphere of radius r we have

$$A = C_1 r$$
$$B = C_2 r^2$$

and thus

$$R(v) = C_1 r v + C_2 r^2 v^2 \qquad (3.16)$$

The two terms in the resistance thus become equal at a critical speed, v_c, defined by the formula

$$v_c = \frac{c_1}{c_2 r} \qquad (3.17)$$

We know that the term proportional to v will dominate the picture if v is sufficiently small (since $Bv^2/Av \sim v$), but we know equally that for speeds much in excess of v_c (say $v \gtrsim 10 v_c$) the quadratic term takes over. If the resistive medium is air, the coefficients c_1 and c_2 have the following approximate values:

$$c_1 = 3.1 \times 10^{-4} \text{ kg m}^{-1} \text{ s}^{-1}$$
$$c_2 = 0.87 \text{ kg m}^{-3}$$

Thus, if r is expressed in metres, we have

$$v_c (\text{m s}^{-1}) = \frac{3.6 \times 10^{-4}}{r} \qquad (3.18)$$

This means that for an object such as a small pebble, with $r \approx 1$ cm, the value of v_c is only a few centimetres per second; a speed equal to 10 times this (say 0.5 m s^{-1}) would be acquired in free fall under gravity within a time of about 0.05 s and a distance of about 0.5 in. (Check these numbers!) Thus

DETAILED ANALYSIS OF RESISTED MOTION

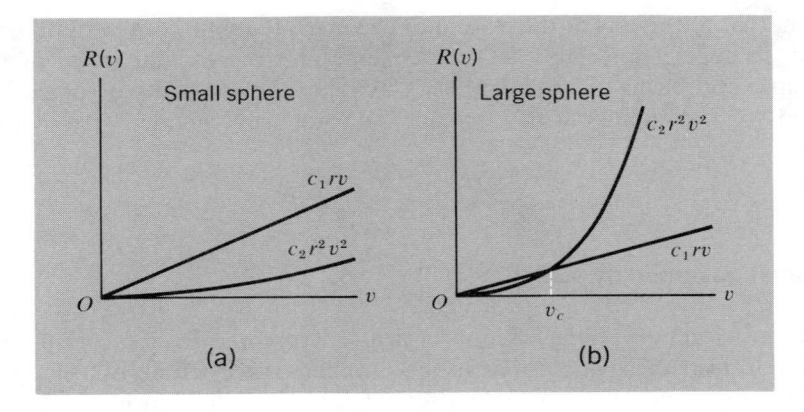

Fig. 3.11 (a) Linear and quadratic terms in fluid resistance for a small object, with viscous resistance predominant. (b) Similar graph for a large object, for which the v^2 term predominates at all except low speeds.

for most problems of practical interest (we shall consider an exception in the next section) the motion under a constant applied force in a resisting medium can be extremely well described with the help of the following simplified version of eqn (3.15):

$$m\frac{dv}{dt} \approx F_0 - Bv^2 \qquad (3.19)$$

The resistive term Bv^2 is quite important in the motion of ordinary objects falling through the air under gravity. This becomes very apparent if we calculate the *terminal speed*, v_t, by putting $dv/dt = 0$ in eqn (3.19), with F_0 set equal to the gravitational force mg. Take, for example, the case of our pebble of radius 1 cm. The density of stone or rock is about 2.5 times that of water, i.e. about 2500 kg m^{-3}, so we have

$$m = \frac{4\pi}{3}\rho r^3 \approx 4(2.5 \times 10^3) \times (10^{-2})^3 \approx 10^{-2} \text{ kg}$$

The value of F_0 is thus about 0.1 N. The value of the coefficient B ($= c_2 r^2$) for an object of this size is about 10^{-4} kg m^{-3}. Substituting these values in the equation

$$F_0 - Bv_t^2 = 0$$

we find

$$v_t \approx 30 \text{ m s}^{-1}$$

Under the assumptions of genuinely free fall, this speed would be attained in a time of about 3 s and a vertical distance of about 150 ft. We can be sure, then, that the effects of the resistance become quite significant within times and distances appreciably less than these. This means that many of the idealized problems of free fall under gravity, of the sort that everyone meets in his first encounter with mechanics, do not correspond very well with reality.

Motion governed by viscosity

If we are dealing with microscopic or near-microscopic objects, such as particles of dust, then, in contrast to the situations discussed above, the resistance is due almost entirely to the viscous term, Av, up to quite high values of v. If, for example, we consider a tiny particle of radius 1 μm ($= 10^{-6}$ m), then eqn (3.18) tells us that the critical speed, v_c, at which the contributions Av and Bv^2 become equal, is 360 m s^{-1}. This implies a wide range of lower speeds for which the motion is controlled by viscous resistance alone, and the statement of Newton's law can be written, without any appreciable error, in the following form:

$$m\frac{dv}{dt} = F_0 - Av \tag{3.20}$$

with

$$A = c_1 r$$

Motion under these conditions played the central role in R. A. Millikan's celebrated 'oil-drop' experiment to determine the elementary electric charge. The basic idea was to measure the electric force exerted on a small charged object by finding the terminal speed of the object in air. If the radius of the particle is known, the resistive force is completely determined, and the driving force must be equal and opposite to it at $v = v_t$.

In Millikan's original experiments the charged particles were tiny droplets of oil in the mist of vapour from an 'atomizer'. Such droplets have a high probability of carrying a net electric charge of either sign when they are produced. In order to apply electric forces to them, Millikan used the arrangement shown schematically in Fig. 3.12. Two parallel metal plates, spaced by

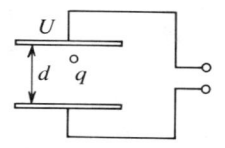

Fig. 3.12 Basic parallel-plate arrangement in Millikan experiment.

MOTION GOVERNED BY VISCOSITY

a small fraction of their diameter, are connected to the terminals of a battery. The force on a particle of charge Q anywhere between these plates is given by

$$F_{el} = Q \frac{U}{d} \tag{3.21}$$

where U is the voltage difference applied to the plates and d is their separation in metres. If Q is measured in coulombs, F is given in newtons by this equation. Thus if a dynamic balance were set up between this electric driving force and the resistive force Av, we should have

$$Q \frac{U}{d} = Av_t = c_1 r v_t$$

The droplets randomly produced in a mist of oil vapour are of various sizes. The ones that Millikan found most suitable for his experiments were the smallest (partially because they had the lowest terminal speed under their own weight). But these droplets were so tiny that even through a medium-power microscope they appeared against a dark background merely as points of light; no direct measurement could be made of their size. Millikan, however, used the clever trick of exploiting the law of viscous resistance a second time by applying it to the fall of a droplet under the gravitational force alone, with no voltage between the plates. Under these conditions we have

$$F_0 = mg = \frac{4\pi}{3} \rho r^3 g$$

where ρ is the density of oil (about 800 kg m^{-3}). The terminal speed of fall under gravity is then given by

$$\frac{4\pi}{3} \rho r^3 g = c_1 r v_g$$

(Gravitational) $\quad v_g = \dfrac{4\pi}{3} \dfrac{\rho g}{c_1} r^2 \tag{3.22}$

Putting in the approximate numerical values we find

(Gravitational) $\quad v_g \approx 10^8 r^2 \quad$ (v in m s^{-1}, r in m)

Putting $r \approx 1~\mu\text{m} = 10^{-6}$ m, we have

$$v \approx 10^{-4} \text{ m s}^{-1} = 0.1 \text{ mm s}^{-1}$$

Such a droplet would take over 1 min to fall 1 cm in air under its own weight, thus allowing precision measurements of its speed. (It is clear, incidentally, that for such motions as this the resistive term Bv^2 is utterly negligible.)

It is worth noting the dynamical stability of this system, and indeed of any situation involving a constant driving force and a resistive force that increases monotonically with speed. If by chance the droplet should slow down a little, there is a net force that will speed it up. Conversely, if it should speed up, it is subjected to a net retarding force. If one could observe the motion of a falling droplet in sufficient detail, this behaviour would doubtless be found, because the air, being made up of individual molecules, does not behave as a perfectly homogeneous fluid. In other words, the speed of the droplet would fluctuate about some average value, although the fluctuations would be exceedingly small.

In the Millikan experiment proper, the vertical motion of the charged droplets is studied with the electric force either aiding or opposing the gravitational force. Thus if we measure velocities as positive downward, the terminal velocity in both magnitude and sign will be defined by the equation

$$mg + \frac{QU}{d} = c_1 r v_t \tag{3.23}$$

where U is the voltage of the upper plate relative to the lower and Q is the net charge on the drop (positive or negative). Although in principle the terminal velocity is approached but never quite reached (see Fig. 3.10), the small droplets under the conditions of the Millikan experiment do, in effect, reach this speed within a very short time—much less than 1 ms in most cases.

Millikan was able to follow the motion of a given droplet for many hours on end, using its electric charge as a handle by which to pull it up or down at will. In the course of such protracted observations the charge on the drop would often change spontaneously, and several different values of v_t would be obtained. The crucial observation was that in any such experiment, with a given value of the voltage U, the speed v_t was limited to a set of sharp and distinct values, implying that the electric charge itself comes in discrete units. But Millikan went further and obtained the first precise value of the absolute magnitude of the elementary charge.

You might wonder why Millikan used such a roundabout method to measure the electric force exerted on a charged particle. After all, it would in principle be possible to hang such an electrified particle on a balance and measure the force in a static arrangement. However, in practice, when only a few elementary charges are involved, the forces are extremely weak and such a method is not feasible. For example, the force on a particle with a net charge of 10 elementary units, between plates 1 cm apart with 500 V between them, is only about 10^{-13} N.

Growth and decay of resisted motion

We have seen how the velocity under a constant force in a resistive fluid medium rises asymptotically toward the terminal value. What happens if the driving force is suddenly removed? We can guess that the velocity will decay away towards zero in a similar asymptotic way, as indicated in Fig. 3.13. If the initial speed is small enough, the whole decay depends on viscous resistance alone and is governed by a special, simplified form of eqn (3.20) in which F_0 is set equal to zero:

$$m\frac{dv}{dt} = -Av$$

or (3.24)

$$\frac{dv}{dt} = -\alpha v$$

where $\alpha = A/m$.

Equation (3.24) is the basic differential equation for exponential decay with the following solution for $v(t)$:

$$v(t) = v_0 e^{-\alpha t} \qquad (3.25)$$

The recipocal of α in eqn (3.25) is of the dimension of time and represents a characteristic *time constant*, τ, for the exponential decay process. In the particular case of small spheres moving through the air, τ is defined by the equation

$$\tau = \frac{m}{A} = \frac{m}{c_1 r}$$

We can express this in much more vivid terms by introducing the terminal velocity of fall under gravity, v_g. For from eqn (3.22) we have

$$v_g = \frac{mg}{c_1 r}$$

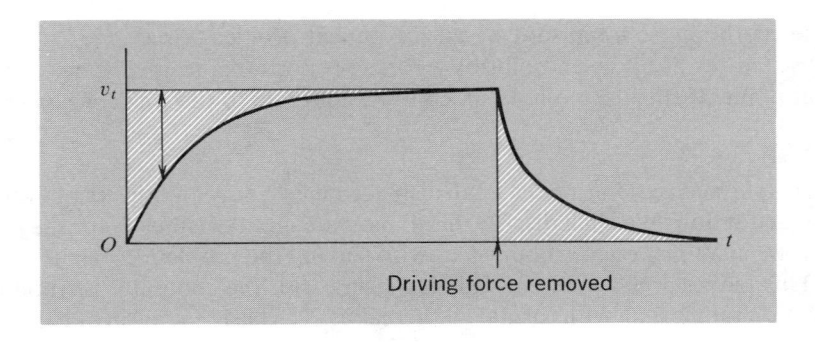

Fig. 3.13 Growth and decay of the velocity of a particle controlled by a viscous resistive force proportional to v.

It follows that we have

$$\tau = \frac{v_g}{g} \tag{3.26}$$

Thus τ is equal to the time that a particle would take to reach a velocity equal to the terminal velocity under conditions of free fall. For an oil drop of radius 1 μm, with $v_g \approx 0.1$ mm s^{-1}, this time would be only about 10^{-5} s ($= 10$ μs).

In a time equal to any substantial multiple of τ, the value of $v(t)$ as given by eqn (3.25) falls to a very small fraction of v_0. For example, if we take the basic equation

$$v(t) = v_0 e^{-t/\tau} \tag{3.27}$$

and put $t = 10\tau$, then the value of v becomes less than 10^{-4} of v_0, which for many purposes can be taken to be effectively zero.

It is more or less intuitively clear that the *growth* of velocity toward its limiting value, after a particle starts from rest under the action of a suddenly applied driving force, must be a kind of upside-down version of the decay curve (see Fig 3.13). In fact, if the terminal velocity is v_t, the approach to it, under these conditions of viscous resistance, is described by

$$v(t) = v_t(1 - e^{-t/\tau}) \tag{3.28}$$

If one chooses to write this in the form

$$v_t - v(t) = v_t e^{-t/\tau}$$

one can see explicitly how closely the growth and decay curves are related. Indeed, one could almost deduce eqn (3.28) from eqn (3.27) plus the recognition of this symmetry.

Simple harmonic motion

One of the most important of all dynamical problems is that of a mass attracted toward a given point by a force proportional to its distance from that point. If the motion is assumed to be confined to the x axis we have

$$F(x) = -kx \tag{3.29}$$

Many physical systems under small displacements from equilibrium obey this basic equation. We shall discuss this in more detail in Chapter 6; for the present we shall just concern ourselves with solving the problem as such.

This makes a good case in which to begin with the computer method of solution rather than with formal mathematics. The basic equation of motion, expressed in the form $ma = F$, is as follows:

$$m \frac{d^2 x}{dt^2} = -kx \tag{3.30}$$

SIMPLE HARMONIC MOTION

Rewriting this as

$$\frac{d^2x}{dt^2} = -\frac{k}{m}x$$

we recognize that k/m is a constant, of dimension (time)$^{-2}$. Denoting this by ω^2, our basic equation thus becomes

$$\frac{d^2x}{dt^2} = -\omega^2 x \tag{3.31}$$

We can read this as a direct statement of the way in which dx/dt is changing with time and can proceed to calculate the approximate change of dx/dt in a small interval of time Δt:

$$\frac{d}{dt}\left(\frac{dx}{dt}\right) = -\omega^2 x$$

and so

$$\Delta\left(\frac{dx}{dt}\right) \approx -\omega^2 x \, \Delta t \tag{3.32}$$

Suppose, to be specific, that we start out at $t = 0$ with $x = x_0$ and v ($= dx/dt$) $= v_0$ both positive, as shown in Fig. 3.14. Then at time $t = \Delta t$ we have

$$x \approx x_0 + v_0 \Delta t$$

$$\frac{dx}{dt} \approx v_0 - \omega^2 x_0 \Delta t$$

The displacement is a little bigger, the slope a little less. Using these new values, we take another step of Δt, and so on. Several features can be read from eqn (3.32).

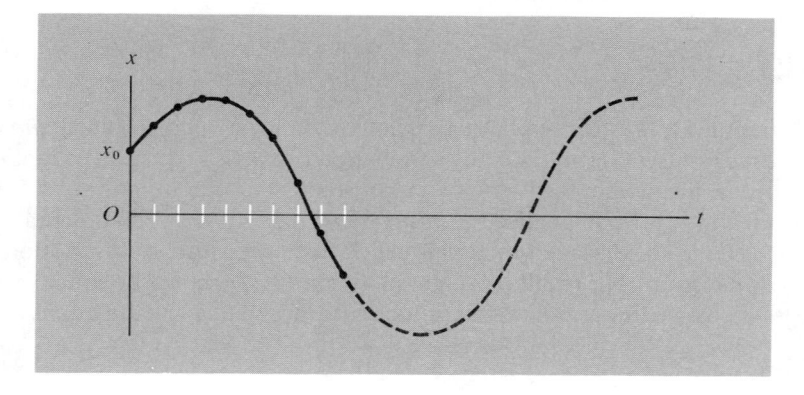

Fig. 3.14 Displacement versus time in simple harmonic motion.

1. As long as x is positive, the slope dx/dt decreases when we go from t to $t + \Delta t$. That is, if dx/dt is positive it gets smaller; if dx/dt is negative it becomes more negative.
2. The rate of change of dx/dt is proportional to x. The graph has its greatest curvature at the largest x, and as $x \to 0$ it becomes almost a straight line.
3. As soon as x becomes negative, dx/dt becomes less negative or more positive with each time increment Δt.

Using these considerations, we can construct the picture of a curve that is always curving toward the line $x = 0$ (i.e. the t axis), necessarily forming a repetitive wavy pattern.

Now anyone who has ever drawn graphs of trigonometric functions will recognize that Fig. 3.14 looks remarkably like a sine or cosine curve. More specifically, it suggests a comparison with the following analytic expression for the distance x as a function of time:

$$x = A \sin(\alpha t + \varphi_0)$$

where A is the maximum value attained by x during the motion, α a constant with the dimension (time)$^{-1}$, and φ_0 an adjustable angle that allows us to fit the value of x at $t = 0$.

Testing this trial function against the original differential equation of motion (eqn 3.31) requires differentiating x twice with respect to t:

$$v = \frac{dx}{dt} = \alpha A \cos(\alpha t + \varphi_0)$$

$$a = \frac{d^2 x}{dt^2} = -\alpha^2 A \sin(\alpha t + \varphi_0) = -\alpha^2 x$$

We see that the solution does indeed fit, provided that we put $\alpha = \omega$. This then brings us to the following final result:

$$x(t) = A \sin(\omega t + \varphi_0) \quad \text{where} \quad \omega = \left(\frac{k}{m}\right)^{1/2} \tag{3.33a}$$

Equation (3.33a) is the characteristic equation of what is called *simple harmonic motion* (SHM), and any system that obeys this equation of motion is called a harmonic oscillator. The constant A is the *amplitude* of the motion, and φ_0 is what is called the *initial phase* (at $t = 0$). The complete argument, $\omega t + \varphi_0$, of the sine function is called just '*the phase*' of the motion at any given instant. The result represented by eqn (3.33a) could be equally well expressed by writing x as a cosine function, rather than a sine function, of t:

$$x(t) = A \cos(\omega t + \varphi_0') \tag{3.33b}$$

with an appropriate value of the constant φ_0'. This form of the solution is found more convenient for some purposes.

SIMPLE HARMONIC MOTION

The harmonic motion is characterized by its *period*, T, which defines successive equal intervals of time at the end of which the state of the motion reproduces itself exactly in both displacement and velocity. The value of T is readily obtained from eqns (3.33) by noting that each time the phase angle $(\omega t + \varphi_0)$ changes by 2π, both x and v have passed through a complete cycle of variation. Thus we can put

$$\varphi_1 = \omega t_1 + \varphi_0$$
$$\varphi_1 + 2\pi = \omega(t_1 + T) + \varphi_0$$

Therefore, by subtraction,

$$2\pi = \omega T$$

or

$$T = \frac{2\pi}{\omega} = 2\pi \left(\frac{m}{k}\right)^{1/2} \tag{3.34}$$

Fitting the initial conditions

It can be recognized, both as a general principle and through various examples, that the complete solution of any problem in the use of Newton's second law requires not only a knowledge of the force law but also the specification of two independent quantities that correspond to the constants of integration introduced as we go from $a\ (=\ d^2x/dt^2)$ to x. Most commonly we have talked of giving the initial position, x_0, and the initial velocity, v_0. Here, in our analysis of the motion of a harmonic oscillator, we also need initial conditions or their equivalent. Actually they appear as the two constants A and φ_0 in eqns (3.33). We have already identified A as the amplitude of the motion and φ_0 as the initial phase. If we are given the values of x_0 and v_0 we can readily solve for A and φ_0 as follows:

From eqn (3.33a),

$$x = A \sin(\omega t + \varphi_0)$$
$$v = \frac{dx}{dt} = \omega A \cos(\omega t + \varphi_0) \tag{3.35}$$

Therefore,

$$x_0 = A \sin \varphi_0$$
$$v_0 = \omega A \cos \varphi_0$$

It follows that

$$A = \left[x_0^2 + \left(\frac{v_0}{\omega}\right)^2\right]^{1/2}$$

(3.36)

$$\tan \varphi_0 = \frac{\omega x_0}{v_0}$$

Problems

3.1 Two blocks, of masses $M = 3$ kg and $m = 2$ kg, are in contact on a horizontal table. A constant horizontal force $F = 5$ N is applied to block M as shown. There is a *constant* frictional force of 2 N between the table and the block m but *no* frictional force between the table and the first block M.

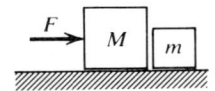

(a) Calculate the acceleration of the two blocks.
(b) Calculate the force of contact between the blocks.

3.2 A man is raising himself and the platform on which he stands with a uniform acceleration of 5 m/s² by means of the rope-and-pulley arrangement shown. The man has mass 100 kg and the platform is 50 kg. Assume that the pulley and rope are massless and move without friction, and neglect any tilting effects of the platform. Assume $g = 10$ m/s².

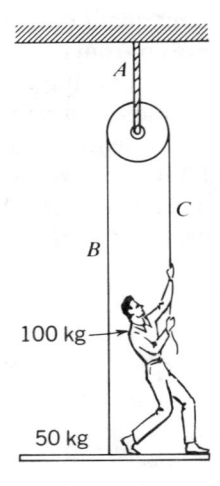

(a) What are the tensions in the ropes A, B and C?
(b) Draw isolation diagrams for the man and the platform and draw a separate

force diagram for each, showing all the forces acting on them. Label each force and clearly indicate its direction.

(c) What is the force of contact exerted on the man by the platform?

3.3 In an equal-arm arrangement, a mass $5m_0$ is balanced by the masses $3m_0$ and $2m_0$, which are connected by a string over a pulley of negligible mass and prevented from moving by the string A (see the figure). Analyse what happens if the string A is suddenly severed, e.g. by means of a lighted match.

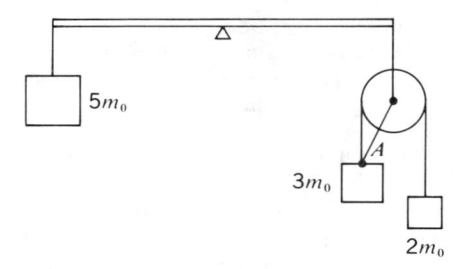

3.4 A prisoner in jail decides to escape by sliding to freedom down a rope provided by an accomplice. He attaches the top end of the rope to a hook outside his window; the bottom end of the rope hangs clear of the ground. The rope has a mass of 10 kg, and the prisoner has a mass of 70 kg. The hook can stand a pull of 600 N without giving way. If the prisoner's window is 15 m above the ground, what is the least velocity with which he can reach the ground, starting from rest at the top end of the rope?

3.5 (a) Suppose that a uniform rope of length L, resting on a frictionless horizontal surface, is accelerated along the direction of its length by means of a force F pulling it at one end. Derive an expression for the tension T in the rope as a function of position along its length. How is the expression for T changed if the rope is accelerated vertically in a constant gravitational field?

(b) A mass M is accelerated by the rope in part (a). Assuming the mass of the rope to be m, calculate the tension for the horizontal and vertical cases.

3.6 A certain loaded car is known to have its centre of gravity half-way between the front and rear axles. It is found that the drive wheels (at the rear) start slipping when the car is driven up a 20° incline. How far back must the load (weighing a quarter the weight of the empty car) be shifted for the car to get up a 25° slope? (The distance between the axles is 10 ft.)

3.7 A beam of electrons from an electron gun passes between two parallel plates, 3 mm apart and 2 cm long. After leaving the plates the electrons travel to form a spot on a fluorescent screen 25 cm further on. It is desired to get the spot to deflect vertically through 3 cm when 100 V are applied to the deflector plates. What must be the accelerating voltage V_0 on the electron gun? (Show first, in general, that if the linear displacement caused by the deflector plates can be neglected, the required voltage is given by $V_0 = V(lD/2Yd)$, where Y is the linear displacement of the spot on the screen, l and d are the length and spacing of the plates, and D is the distance from plates to screen.)

3.8 A ball of mass m is attached to one end of a string of length l. It is known that the string will break if pulled with a force equal to nine times the weight of the ball. The ball, supported by a frictionless table, is made to travel a horizontal circular path, the other end of the string being attached to a fixed point O. What is the largest number of revolutions per unit time that the mass can make without breaking the string?

3.9 A curve of 300 m radius on a level road is banked for a speed of 25 m/s ($\approx$ 55 mph) so that the force exerted on a car by the road is normal to the surface at this speed.

(a) What is the angle of bank?

(b) The friction between tyres and road can provide a maximum tangential force equal to 0.4 of the force normal to the road surface. What is the highest speed at which the car can take this curve without skidding?

3.10 A model rocket rests on a frictionless horizontal surface and is joined by a string of length l to a fixed point so that the rocket moves in a horizontal circular path of radius l. The string will break if its tension exceeds a value T. The rocket engine provides a thrust F of constant magnitude along the rocket's direction of motion. The rocket has a mass m that does not decrease appreciably with time.

(a) Starting from rest at $t = 0$, at what later time t_1 is the rocket travelling so fast that the string breaks? Ignore air resistance.

(b) What was the magnitude of the rocket's instantaneous net acceleration at time $t_1/2$? Obtain the answer in terms of F, T and m.

(c) What distance does the rocket travel between the time t_1 when the string breaks and the time $2t_1$? The rocket engine continues to operate after the string breaks.

3.11 A trick cyclist rides his bike around a 'wall of death' in the form of a vertical cylinder (see the figure). The maximum frictional force parallel to the surface of the cylinder is equal to a fraction μ of the normal force exerted on the bike by the wall.

(a) At what speed must the cyclist go to avoid slipping down?

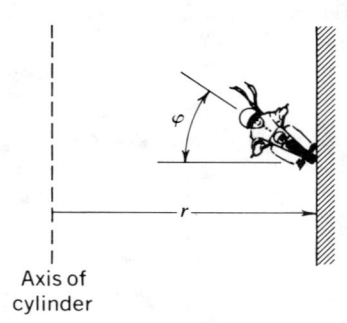

(b) At what angle (φ) to the horizontal must he be inclined?

(c) If $\mu \approx 0.6$ (typical of rubber tyres on dry roads) and the radius of the cylinder is 5 m, at what minimum speed must the cyclist ride, and what angle does he make with the horizontal?

3.12 The following expression gives the resistive force exerted on a sphere of radius r moving at speed v through air. It is valid over a very wide range of speeds.

$$R(v) = 3.1 \times 10^{-4} rv + 0.87 r^2 v^2$$

where R is in N, r in m and v in m/s. Consider water drops falling under their own weight and reaching a terminal speed.

(a) For what range of values of r is the terminal speed determined within 1% by the first term alone in the expression for $R(v)$?

(b) For what range of values of larger r is the terminal speed determined within 1% by the second term alone?

(c) Calculate the terminal speed of a raindrop of radius 2 mm. If there were no air resistance, from what height would it fall from rest before reaching this speed?

3.13 Analyse in retrospect the legendary Galilean experiment that took place at the leaning tower of Pisa. Imagine such an experiment done with two iron spheres, of radii 2 and 10 cm, respectively, dropped simultaneously from a height of about 15 m. Make calculations to determine, approximately, the difference in the times at which they hit the ground. Do you think this could be detected without special measuring devices? (Density of iron ≈ 7500 kg/m^3.)

3.14 Estimate the terminal speed of fall (in air) of an air-filled toy balloon, with a diameter of 30 cm and a mass (not counting the air inside) of about 0.5 g. About how long would it take for the balloon to come to within a few percent of this terminal speed? Try making some real observations of balloons inflated to different sizes.

3.15 (a) A small bead of mass m is attached to the midpoint of a string (itself of negligible mass). The string is of length L and is under constant tension T. Find the frequency of the SHM that the mass describes when given a slight transverse displacement.

(b) Find the frequency in the case where the mass is attached at a distance D from one end instead of the midpoint.

3.16 A block rests on a tabletop that is undergoing simple harmonic motion of amplitude A and period T.

(a) If the oscillation is vertical, what is the maximum value of A that will allow the block to remain always in contact with the table?

(b) If the oscillation is horizontal, and the coefficient of friction between block and tabletop is μ, what is the maximum value of A that will allow the block to remain on the surface without slipping?

4 Universal gravitation

The discovery of universal gravitation

In chapters 2 and 3 we have built up the kind of foundation in dynamics that Newton himself was the first to establish. In a nutshell, it is the quantitative identification of force as the cause of acceleration, coupled with the purely kinematic problem of relating accelerations to velocities and displacements. We shall now consider, as a topic in its own right, the first and most splendid example of how a *law of force* was deduced from the study of motions.

It is convenient, and historically not unreasonable, to consider separately three aspects of this great discovery:

1. The analysis of the data concerning the orbits of the planets around the sun, to the approximation that these orbits are circular with the sun at the centre.
2. The proof that gravitation is universal, in the sense that the law of force that governs the motion of objects near the earth's surface is also the law that controls the motion of celestial bodies.
3. The proof that the true planetary orbits, which are ellipses rather than circles, are explained by an inverse-square law of force.

In the present chapter we shall begin by discussing the first of these questions, using only our basic results in the kinematics and dynamics of particles. The second question requires us to learn how to analyse the gravitating properties of a body, like the earth, which is so obviously not a geometrical point when viewed from close to its surface. We shall present one approach to the problem here and complete the story in Chapter 7, where this special feature of the gravitational problem is discussed. The third question, concerning the exact mathematical description of the orbits, is something that we shall not go into at all at this stage; such orbit problems will be the concern of Chapter 9.

Kepler's third law

The precise form of the relationship between planetary periods and distances was first discovered by Johann Kepler in 1618 and published by him the

KEPLER'S THIRD LAW

Table 4.1 Kepler's Third Law

Planet	Radius r of orbit of planet, AU	Period T, days	r^3/T^2, $(AU)^3/(day)^2 \times 10^{-6}$
Mercury	0.389	87.77	7.64
Venus	0.724	224.70	7.52
Earth	1.000	365.25	7.50
Mars	1.524	686.98	7.50
Jupiter	5.200	4,332.62	7.49
Saturn	9.510	10,759.20	7.43

following year in his book *The Harmonies of the World*. In it he triumphantly wrote: 'I first believed I was dreaming . . . But it is absolutely certain and exact that the ratio which exists between the periodic times of any two planets is precisely the ratio of the $\frac{3}{2}$th powers of the mean distances. . . .' Table 4.1 shows the data used by Kepler and a test of the near constancy of the ratio r^3/T^2. Figure 4.1 is a different presentation of the planetary data plotted on log-log graph paper so as to show this relationship:

$$T \sim r^{3/2} \tag{4.1}$$

This is known as Kepler's third law, having been preceded, 10 years earlier, by the statement of his two great discoveries concerning the paths of the individual planets, namely (1) that the planets move in ellipses with the sun at one focus, and (2) that the line joining the sun to a given planet sweeps out equal areas in equal times.

The dynamical explanation of Kepler's third law had to await Newton's discussion of such problems in the *Principia*. A very simple analysis of it is possible if we use the simplified picture of the planetary orbits as circles with the sun at the centre. It then becomes apparent that eqn (4.1) implies that an inverse-square law of force is at work. For in a circular orbit of radius r we have

$$a_r = -\frac{v^2}{r}$$

Expressing v in terms of the known quantities r and T, we have

$$v = \frac{2\pi r}{T}$$

$$a_r = -\frac{4\pi^2 r}{T^2} \quad \text{(towards the centre)} \tag{4.2}$$

From Newton's law, then, we infer that the force on a mass in a circular orbit must be given by

$$F_r = ma_r = -\frac{4\pi^2 mr}{T^2} \tag{4.3}$$

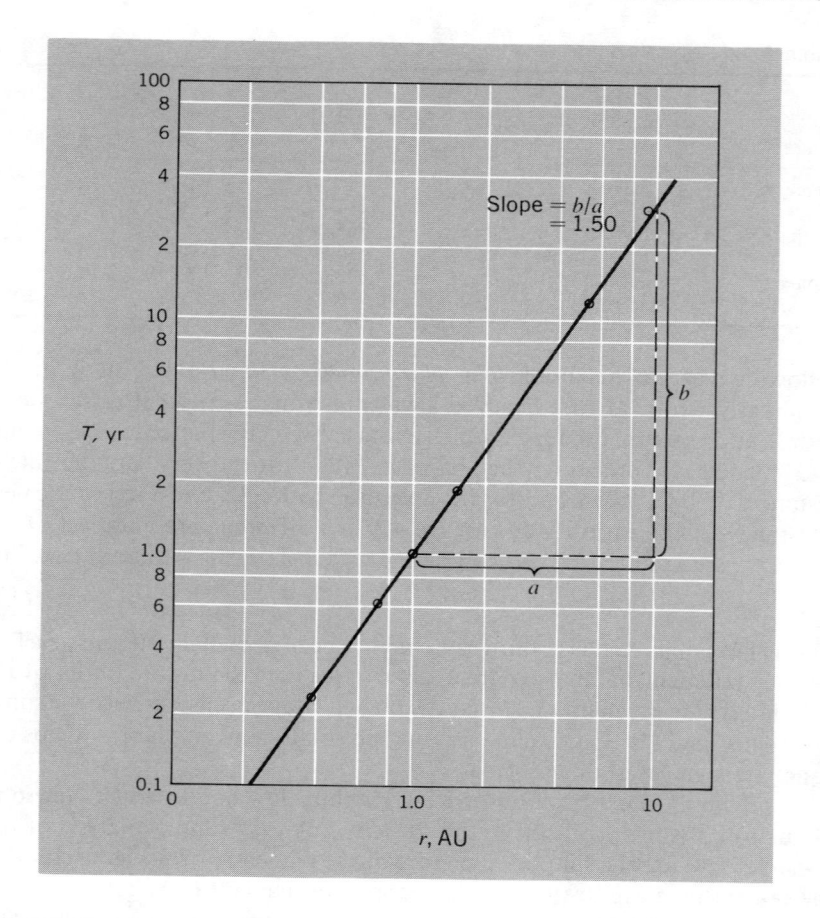

Fig. 4.1 Log-log plot of planetary period T versus orbit radius r. The graph shows that T is proportional to $r^{3/2}$ (Kepler's third law).

From Kepler's third law, however, we have the relation

$$\frac{r^3}{T^2} = K \tag{4.4}$$

where K might be called Kepler's constant—the same value of it applies to all the planets travelling around the sun. From eqn (4.4) we thus have $1/T^2 = K/r^3$, and substituting this in eqn (4.3) gives us

$$F_r = -\frac{4\pi^2 Km}{r^2} \tag{4.5}$$

The implication of Kepler's third law, therefore, when analysed in terms of Newton's dynamics, is that the force on a planet is proportional to its inertial mass m and inversely proportional to the square of its distance from the sun.

The proportionality of the force to the attracted mass, as required by eqn (4.5), was a feature that Newton fully appreciated. With his grasp of the concept of interactions exerted mutually between pairs of objects, he saw that the reciprocity in the gravitational interaction must mean that the force is proportional to the mass of the attracting object just as it is to the mass of the attracted. Each object is the attracting agent as far as the other one is concerned. Hence the magnitude of the force exerted on either one of a mutually gravitating pair of particles must be expressed in the famous mathematical statement of universal gravitation:

$$F = - \frac{Gm_1 m_2}{r^2} \qquad (4.6)$$

where G is a constant to be found by experiment, and m_1 and m_2 are the inertial masses of the particles.

The moon and the apple

It is an old story, but still an enthralling one, of how Newton, as a young man of 23, came to think about the motion of the moon in a way that nobody had ever done before. The path of the moon through space, as referred to the 'fixed' stars, is a line of varying curvature (always, however, bending toward the sun), which crosses and recrosses the earth's orbit. But of course there is a much more striking way of looking at it—the familiar earth-centred view, which shows the moon describing an approximately circular orbit around the earth. To this extent it is quite like the planetary-orbit problem that we have just been discussing. But Newton, with his extraordinary insight, constructed an intellectual bridge between this motion and the behaviour of falling objects—the latter being such a commonplace phenomenon that it needed a genius to recognize its relevance. He saw the moon as being just an object falling toward the earth like any other—as, for example, an apple dropping off a tree in his garden. A very special case, to be sure, because the moon was so much farther away than any other falling object in our experience. But perhaps it was all part of the same pattern.

As Newton himself described it, he began in 1665 or 1666 to think of the earth's gravity as extending out to the moon's orbit, with an inverse-square relationship already suggested by Kepler's third law. We could of course just restate the centripetal acceleration formula and apply it to the moon, but it is illuminating to trace the course of Newton's own way of discussing the problem. In effect he said this: imagine the moon at any point A in its orbit (Fig. 4.2). If freed of all forces, it would travel along a straight line AB, tangent to the orbit at A. Instead, it follows the arc AP. If O is the centre of

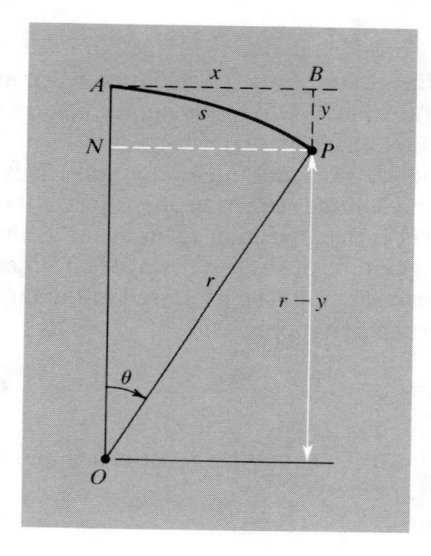

Fig. 4.2 Geometry of a small portion of a circular orbit, showing the deviation y from the tangential straight-line displacement AB ($= x$) that would be followed in the absence of gravity.

the earth, the moon has in effect 'fallen' the distance BP toward O, even though its radial distance r is unchanged. Let us calculate how far the moon falls, in this sense, in 1 s, and compare it with the distance of about 4.9 m that an object projected horizontally near the earth's surface would fall in that same time.

If we denote the distance AB as x, and the distance BP as y, it will be an exceedingly good approximation to put

$$y \approx \frac{x^2}{2r} \tag{4.7}$$

Since for small θ the arc length AP ($= s$) is almost equal to the distance AB, we can equally well put

$$y \approx \frac{s^2}{2r} \tag{4.8}$$

In order to put numbers into this formula we need to know both the radius and the period of the moon's orbital motion. The moon's orbit radius r is about 240,000 miles $\approx 3.8 \times 10^8$ m, and its period T is 27.3 days $\approx 2.4 \times 10^6$ s. Therefore, in 1 s it travels a distance along its orbit given by

$$\text{(in 1 s)} \quad s = \frac{2\pi \times 3.8 \times 10^8}{2.4 \times 10^6} \approx 1000 \text{ m}$$

During this same time it falls a vertical distance y_1 given (via eqn 4.8) by

(in 1 s) $y_1 \approx \dfrac{10^6}{7.6 \times 10^8} = 1.3 \times 10^{-3}$ m

In other words, in 1 s, while travelling 'horizontally' through a distance of 1 km, the moon falls vertically through just over 1 mm; its deviation from a straight-line path is indeed slight. On the other hand, for an object near the earth's surface, projected horizontally, the vertical displacement in 1 s is given by

$y_2 = \tfrac{1}{2}gt^2 = 4.9$ m

Thus

$\dfrac{y_1}{y_2} = 2.7 \times 10^{-4} \approx \dfrac{1}{3700}$

Newton knew that the radius of the moon's orbit was about 60 times the radius of the earth itself, as the ancient Greeks had first shown. And with an inverse-square law, if it applied equally well at all radial distances from the earth's centre, we would expect y_1/y_2 to be about 1/3600. It *must* be right! And yet, what an astounding result. Even granted an inverse-square law of attraction between objects separated by many times their diameters, one still has the task of proving that an object a few feet above the earth's surface is attracted as though the whole mass of the earth were concentrated at a point 4000 miles below the ground. Newton did not prove this result until 1685, nearly 20 years after his first great insight into the problem. He published nothing, either, until it all came out, perfect and complete, in the *Principia* in 1687.

The gravitational attraction of a large sphere

Suppose we have a large solid sphere, of radius R_0, as shown in Fig. 4.3(a), and wish to calculate the force with which it attracts a small object of mass m at an arbitrary point P. We shall assume that the density of the material of the sphere may vary with distance from the centre (as is the case for the earth, to a very marked degree) but is spherically symmetric. We can then consider the solid sphere to be built up of a very large number of thin uniform spherical shells, like the successive layers of an onion. The total gravitational effect of the sphere can be calculated as the superposition of the effects of all these individual shells. Thus the basic problem becomes that of calculating the force exerted by a thin spherical shell of arbitrary radius, assuming that the fundamental law of force is that of the inverse square between point masses.

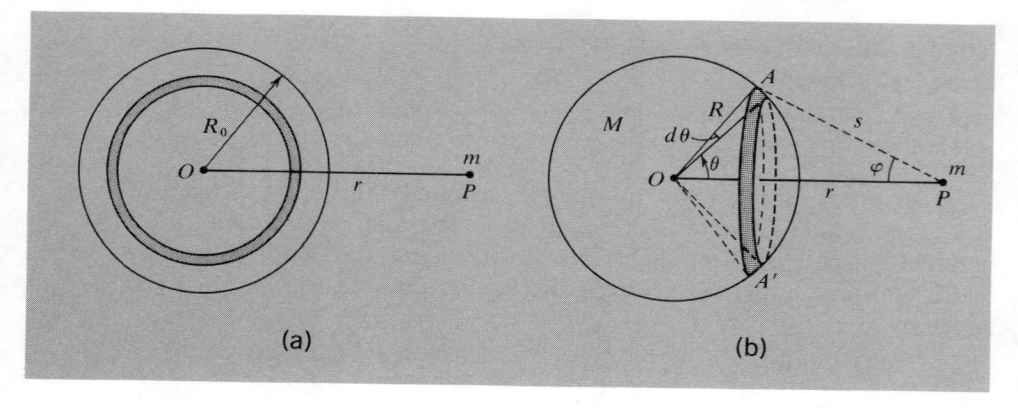

Fig. 4.3 (a) A solid sphere can be regarded as built up of a set of thin concentric spherical shells. (b) The gravitational effect of an individual shell can be found by treating it as an assemblage of circular zones.

In Fig. 4.3(b) we show a shell of mass M, radius R, of negligible thickness, with a particle of mass m at a distance r from the centre of the shell. If we consider a small piece of the shell, near point A, the force that it exerts on m is along the line AP. It is clear from the symmetry of the system, however, that the resultant force due to the whole shell must be along the line OP; any component of force transverse to OP due to material near A will be cancelled by an equal and opposite contribution from material near A'. Thus if we have an element of mass dM near A, we need only consider its contribution to the force along OP, i.e. the radial direction from the centre of the shell to m. Hence we have

$$dF_r = - \frac{Gm\, dM}{s^2} \cos \varphi \tag{4.9}$$

Let us now consider a complete belt or zone of the shell, shown shaded in the diagram. It represents the portion of the shell that is contained between the directions θ and $\theta + d\theta$ to the axis OP, and the same mean values of s and φ apply to every part of it. Thus, if we calculate its mass, we can substitute this value as dM in eqn (4.9) to obtain the contribution of the whole belt to the resultant gravitational force along OP. Now the width of the belt is $R\, d\theta$ and its circumference is $2\pi R \sin \theta$; thus its area is $2\pi R^2 \sin \theta\, d\theta$. The area of the whole shell is $4\pi R^2$; hence the mass of the belt is given by

$$dM = \frac{2\pi R^2 \sin \theta\, d\theta}{4\pi R^2} M = \frac{M}{2} \sin \theta\, d\theta$$

Thus eqn (4.9) gives us

$$dF_r = - \frac{GMm}{2} \frac{\cos \varphi \sin \theta\, d\theta}{s^2} \tag{4.10}$$

THE GRAVITATIONAL ATTRACTION OF A LARGE SPHERE

Our task now is to sum the contributions such as dF_r over the whole of the shell, i.e. over the whole range of values of s, φ and θ. This looks like a formidable task, but with the help of a little calculus (another of Newton's inventions!) the solution turns out to be surprisingly straightforward.

From the geometry of the situation (Fig. 4.3b), it is possible to express both of the angles θ and φ in terms of two fixed distances, r and R, and the variable distance s. By two separate applications of the cosine rule we have

$$\cos\theta = \frac{r^2 + R^2 - s^2}{2rR} \quad ; \quad \cos\varphi = \frac{r^2 + s^2 - R^2}{2rs}$$

From the first of these, by differentiation, we have

$$\sin\theta \, d\theta = \frac{s \, ds}{rR}$$

Hence, substituting the values of $\cos\varphi$ and $\sin\theta \, d\theta$ in eqn (4.10) we obtain

$$dF_r = -\frac{GMm}{4r^2R} \frac{(r^2 + s^2 - R^2) \, ds}{s^2}$$

The total force is found by integrating this expression from the minimum value of s ($= r - R$) to its maximum value ($r + R$). Thus we have

$$F_r = -\frac{GMm}{4r^2R} \int_{r-R}^{r+R} \frac{r^2 + s^2 - R^2}{s^2} \, ds \qquad (4.11)$$

The integral is just the sum of two elementary forms; we have

$$\int \frac{r^2 + s^2 - R^2}{s^2} \, ds = \int ds + (r^2 - R^2) \int \frac{ds}{s^2}$$

$$= s - \frac{r^2 - R^2}{s}$$

Inserting the limits, we then find that

$$\int_{r-R}^{r+R} \frac{r^2 + s^2 - R^2}{s^2} \, ds = [(r + R) - (r - R)]$$

$$- \left(\frac{r^2 - R^2}{r + R} - \frac{r^2 - R^2}{r - R} \right)$$

$$= 2R - (r - R) + (r + R)$$

$$= 4R$$

Substituting this value of the definite integral in eqn (4.11) we have

$$F_r = -\frac{GMm}{r^2} \qquad (4.12)$$

What a wonderful result! It is of extraordinary simplicity, and the radius R of the shell does not appear at all. It is uniquely a consequence of an inverse-square law of force between particles; no other force law would yield such a simple result for the net effect of an extended spherical object.

Once we have eqn (4.12), the total effect of a solid sphere follows at once. Regardless of the particular way in which the density varies between the centre and the surface (provided that it depends only on R) the complete sphere does indeed act as though its total mass were concentrated at its centre. It does not matter how close the attracted particle P is to the surface of the sphere, as long as it is in fact outside. Take a moment to consider what a truly remarkable result this is. Ask yourself: is it obvious that an object a few feet above the apparently flat ground should be attracted as though the whole mass of the earth (all 6 000 000 000 000 000 000 000 tons of it!) were concentrated at a point (the earth's centre) 4000 miles down? It is about as far from obvious as could be, and there can be little doubt that Newton had to convince himself of this result before he could establish, to his own satisfaction, the grand connection between terrestrial gravity and the motion of the moon and other celestial objects.

Other satellites of the earth

Let us derive the formulas for the required speed v and the period T of a satellite launched horizontally in a circular orbit at a distance r from the centre of the earth. The necessary force to maintain circular motion is provided by gravitational attraction:

$$\frac{mv^2}{r} = G\frac{M_E m}{r^2}$$

where M_E is the mass of the earth, m the mass of the satellite, and G the universal gravitational constant. Solving for v,

$$v = \left(\frac{GM_E}{r}\right)^{1/2} \tag{4.13}$$

It is often convenient to express this result in terms of more familiar quantities. We can do this by noticing that, for an object of mass m at the earth's surface, the gravitational force on it, by eqn (4.6), is

$$F_g = \frac{GM_E m}{R_E^2}$$

But this is the force that can be set equal to mg for the mass in question. Hence we have

$$mg = \frac{GM_E m}{R_E^2}$$

or

$$GM_E = gR_E^2$$

Substituting this in eqn (4.13), we get

$$v = \left(\frac{gR_E^2}{r}\right)^{1/2}$$

The period, T, of the satellite is then given by

$$T = \frac{2\pi r}{v} = \frac{2\pi r^{3/2}}{g^{1/2} R_E} \tag{4.14}$$

Putting $g = 9.8$ m s^{-2}, $R_E = 6.4 \times 10^6$ m, we have a numerical formula for the period of any satellite in a circular orbit of radius r around the earth:

(earth satellites) $\quad T \approx 3.14 \times 10^{-7} r^{3/2}$ \hfill (4.15)

where T is in seconds and r in metres.

For example, a satellite at minimum practicable altitude (about 200 km, say) has $r \approx 6.6 \times 10^6$ m, and hence

$$T \approx 5.3 \times 10^3 \text{ s} \approx 90 \text{ min}$$

The value of G, and the mass of the earth

The presently accepted value of the universal gravitation constant, G, as obtained from laboratory measurements of the force exerted between two known masses, is

$$G = 6.672 \times 10^{-11} \text{ N m}^2 \text{ kg}^{-2} \tag{4.16}$$

If we denote the mean density of the earth as ρ and its radius as R, the gravitational force exerted on a particle of mass m just at the earth's surface is given by

$$F = \frac{GMm}{R^2} = \frac{4\pi}{3}(G\rho R)m \tag{4.17}$$

But $F = mg$ so

$$g = \frac{4\pi}{3} G\rho R \tag{4.18}$$

Given the directly determined value of G (eqn 4.16) we substitute in eqns (4.17) and (4.18) to find the mass and the mean density of the earth. The result of these substitutions (with $R = 6.37 \times 10^6$ m) is

$M = 5.97 \times 10^{24}$ kg
$\rho = 5.52 \times 10^3$ kg m^{-3}

Local variations of *g*

If we take the idealization of a perfectly spherical, symmetrical earth, then the gravitational force on an object of mass *m* at a distance *h* above the surface is given by

$$F = \frac{GMm}{(R + h)^2}$$

If we identify *F* with *m* times the value of *g* at the point in question, we have

$$g(h) = \frac{GM}{(R + h)^2} \tag{4.19}$$

For $h \ll R$, this would imply an almost exactly linear decrease of *g* with height. Using the binomial theorem, we can rewrite eqn (4.19) as follows:

$$g(h) = \frac{GM}{R^2}\left(1 + \frac{h}{r}\right)^{-2}$$

Hence, for small *h*, we have

$$g(h) \approx g_0\left(1 - \frac{2h}{R}\right) \tag{4.20}$$

where $g_0 = GM/R^2$, the value of *g* at points extremely close to the earth's surface.

The variation of *g* with *h* described by eqn (4.20) is very small. For example, one would have to ascend to a point about 1000 ft above ground (to the top of the Empire State Building) before the decrease of *g* was even as great as 1 part in 10,000.

Superimposed on the systematic variations of the gravitational force with height are the variations produced by inhomogeneities in the material of the earth's crust. For example, if one is standing above a subterranean deposit of salt or sand, much lower in density than ordinary rocks, one would expect the value of *g* to be reduced. Such changes, although extremely small, can be measured with remarkable accuracy by modern instruments and have become a very valuable tool in geophysical prospecting.

Inertial and gravitational mass

Perhaps the most profound contribution that Newton made to science was the fundamental connection that he recognized between the inertial mass of an object and the earth's gravitational force on it.

It had been known since Galileo's time that all objects near the earth fall with about the same acceleration, g. Until Newton it was just a kinematic fact. But in terms of Newton's law it took on a much deeper significance. If an object is observed to have this acceleration, there must be a force W acting on it given by $F = ma$, i.e.

$$W = mg \qquad (4.21)$$

It then becomes a vitally significant *dynamical* fact that, since the acceleration g is the same for all objects, the force W causing it is strictly proportional to the inertial mass. To appreciate how remarkable this result is, imagine starting from scratch to investigate the force of attraction between two objects in a purely static experiment. One measures the force by balancing it with a springlike device—a torsion fibre. One finds a quantity, which might be called (by analogy with electrical interactions) a gravitational charge, Q_g. This 'charge' is characteristic of any object and has, as far as these experiments are concerned, nothing at all to do with the inertial mass, which is defined solely in terms of acceleration (under the action of forces produced, for example, by stretched springs). One experiments with objects of all sorts of materials, in different states of aggregation, and so on. It then turns out that, in each and every instance, the gravitational charge is strictly proportional to an independently established quantity, the inertial mass. This proportionality has been established by R. H. Dicke and his collaborators to 1 part in 10^{10} or better. Is this just a remarkable coincidence, or does it point to something very fundamental? For a long time this apparent coincidence was regarded as one of the unexplained mysteries of nature. It took the genius of an Einstein to suspect that gravitation may, in a sense, be *equivalent* to acceleration. Einstein's 'postulate of equivalence', that the gravitational charge Q_g and the inertial mass m are measures of the same quantity, provided the basis of his own theory of gravitation as embodied in the general theory of relativity. We shall come back to this in Chapter 8 when we discuss non-inertial frames of reference.

Weight

In eqn (4.21) the symbol W was introduced. What does this really represent? It is often described interchangeably as the gravitational force on an object or its *weight*. In this book we shall define *weight* as the downward force equal in

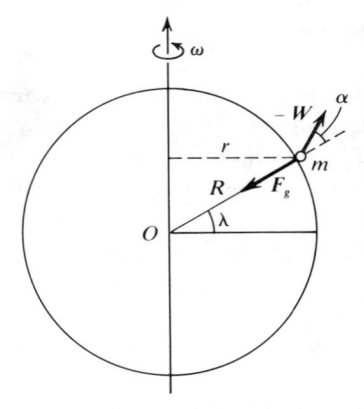

Fig. 4.4 The force needed to balance the weight of an object is different in both direction and magnitude from the true force of gravity.

magnitude to the upward force needed to hold an object at rest relative to the surface of the earth. With this definition the weight of an object will *not* be quite identical with the true gravitational force on it (F_g) because an object at rest relative to the surface of the earth must, in fact, have an unbalanced force acting on it. If we take the idealization of a perfectly spherical earth (Fig. 4.4) and if W is the weight of the object as we have defined it, then the equilibrium of the object is maintained by applying a force $-W$ at an angle α to the radius such that the following conditions are satisfied:

$$W \sin \alpha = m\omega^2 r \sin \lambda$$
$$F_g - W \cos \alpha = m\omega^2 r \cos \lambda$$

where $r = R \cos \lambda$. Since α is certainly a very small angle, it is justifiable to put $\cos \alpha \approx 1$ in the second equation, thus giving the result

$$W \approx F_g - m\omega^2 R \cos^2 \lambda$$

It follows that

$$W(\lambda) \approx W_0 + m\omega^2 R \sin^2 \lambda$$

where W_0 is the measured weight on the equator. Putting $W_0 = mg_0$, we can also obtain a corresponding expression for the latitude dependence of g:

$$g(\lambda) = g_0 + \omega^2 R \sin^2 \lambda$$

If in this expression we substitute $\omega = 2\pi/86{,}400$ s^{-1} and $R = 6.4 \times 10^6$ m, we find $\omega^2 R \approx 3.4 \times 10^{-2}$ m s^{-2}, which with $g_0 \approx 9.8$ ms^{-2} gives us

$$g(\lambda) \approx 9.8(1 + 0.0035 \sin^2 \lambda) \quad \text{m s}^{-2}$$

This formula is more successful than it deserves to be, for we have no right to ignore the significant flattening of the earth, due again to the rotation, which makes the equatorial radius of the earth about 1 part in 300 greater than the polar radius. The actual value of g at sea level is quite well described by the following formula:

$$g(\lambda) = 9.7805(1 + 0.00529 \sin^2 \lambda) \tag{4.22}$$

Thus our simple calculation has the correct form, but its value for the numerical coefficient of the latitude-dependent correction is only about two thirds of the true figure.

As a postscript to the above discussion, note that W, as we have defined it, *is* equal to m times the observed local acceleration of free fall (g) with respect to the earth, but that g is *not* equal to the true gravitational force divided by m, except at the poles (see p. 207).

Weightlessness

It is appropriate, after the detailed discussion of the relations between mass, gravitational force and weight, to devote a few words to the property that is called weightlessness. The very explicit distinction that we have drawn between the gravitational force on an object and its measured weight makes use of what is called an *operational* definition of the latter quantity. The weight, as we have defined it, is a force equal and opposite to the force that will hold an object at rest relative to the earth. Our definition of weightlessness derives very naturally from this: *An object is weightless whenever it is in a state of completely free fall.* In this state each part of the object undergoes the same acceleration, of whatever value corresponds to the strength of gravity at its location. (In saying this we assume that g does not change appreciably over the linear extent of the object.) An object that is prevented from falling, by being restrained or supported, inevitably has internal stresses and deformations in its equilibrium state. This may become very obvious, as when a drop of mercury flattens somewhat when it rests on a horizontal surface. All such stresses and deformations are removed in the weightless state of free fall. The mercury drop, for example, is free to take on a perfectly spherical shape.

The above definition of weightlessness can be applied in any gravitational environment, and this is the way it should be. The bizarre dynamical phenomena of life in a space capsule do not depend on getting into regions far from the earth, where the gravitational forces are much reduced, but simply on the fact that the capsule, and everything in it, is falling freely with the same acceleration, which in consequence goes undetected. For example, if a spacecraft is in orbit around the earth, 200 km above the earth's surface, the gravitational force on the spacecraft, and on everything in it, is still about 95% of what one would measure at sea level, but the phenomena associated with what we call weightlessness are just as pronounced under these conditions as they are in another spacecraft 200,000 miles from earth, where the earth's gravitational attraction is down to 1/2500 of that at the earth's surface. In both situations an object released inside the spacecraft will remain poised in midair. The same would be true in a spacecraft that was simply falling radially toward the earth's centre rather than pursuing a circular or elliptical orbit around the earth. When viewed in these terms the phenomena of weightlessness are not in the least mysterious, although they are still startling because they conflict so strongly with our normal experience.

The discovery of Neptune

Probably the most vivid illustration of the power of the gravitational theory has been the prediction and discovery of planets whose very existence had not previously been suspected. It is noteworthy that the number of known planets remained unchanged from the days of antiquity until long after Newton. Then, in 1781, William Herschel noticed the object that we now know as Uranus. He was engaged in a systematic survey of the stars, and his only clue to start with was that the object seemed slightly less pointlike than the neighbouring stars. Then, having a telescope with various degrees of magnification, he confirmed that the size of the image increased with magnification, which is not true for most stars—they remain below the limit of resolution of even the biggest telescopes, and always produce images indistinguishable from those due to ideal point sources.

Once his attention had been drawn to the object, Herschel returned to it night after night and confirmed that it was moving with respect to the other stars. Also, as has happened in various other cases, it was found that the existence of the object had in fact been recorded in earlier star maps (first by John Flamsteed in 1690). These old data suddenly became extremely valuable, because they were a ready-made record of the object's movements dating back through nearly a century. When combined with new measurements carried out over many months, they showed that the object (finally to be called Uranus) was indeed a member of our solar system, following an almost circular orbit with a mean radius of 19.2 AU and a period of 84 years.

This is where our main story begins. Once it was discovered, Uranus and its motions became the subject of a continuing study, and evidence began to accumulate that there were some extremely small irregularities in its motion that could not be ascribed to perturbing effects from any known source. Figure 4.5(a), a tribute to the wonderful precision of astronomical observation, shows the anomaly as a function of time since 1690. The suspicion began to grow that perhaps there was yet another planet beyond Uranus, unknown in mass, period or distance. Two men—J.C. Adams in England

THE DISCOVERY OF NEPTUNE

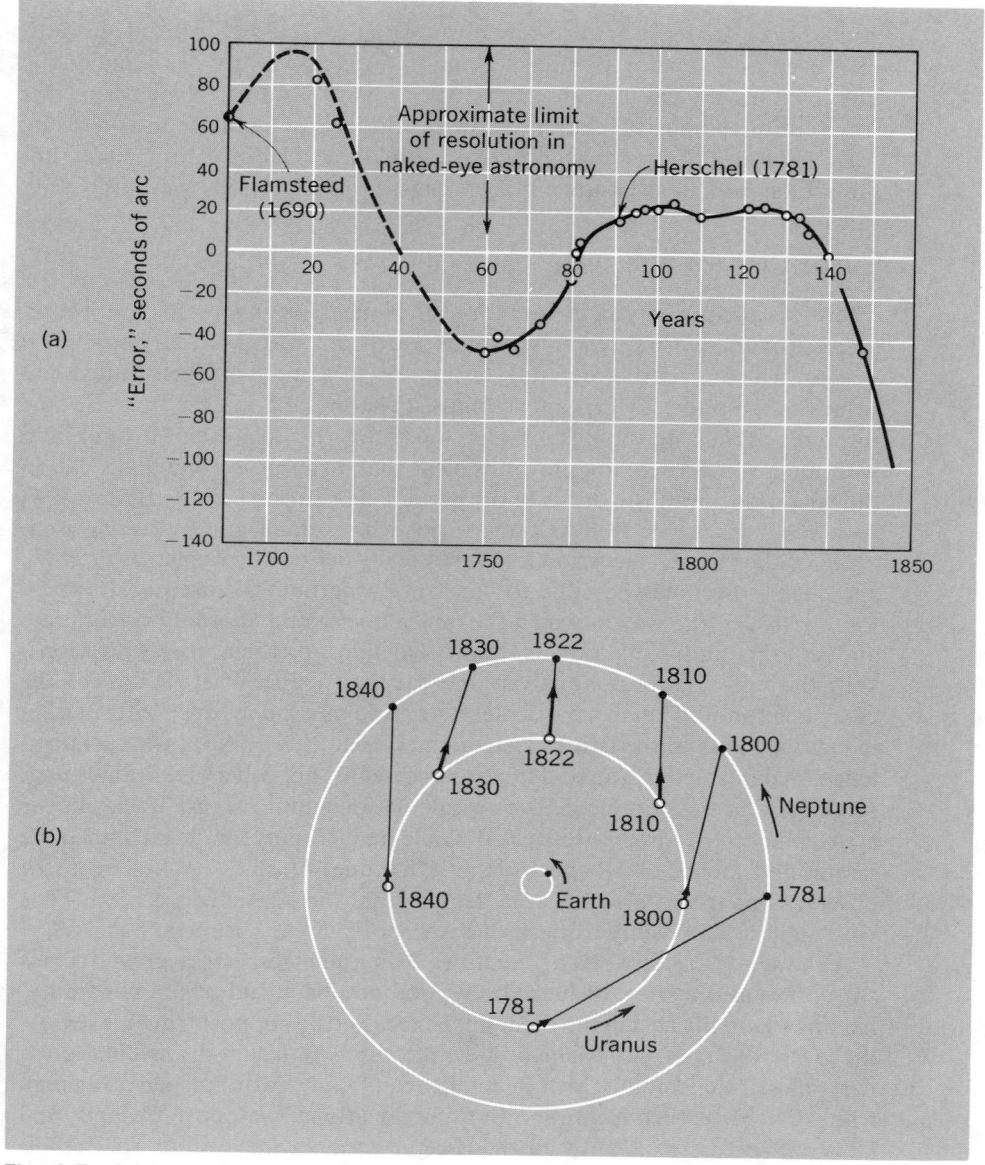

(a)

(b)

Fig. 4.5 (a) Unexplained residual deviations in the observed positions of Uranus between 1690 and 1840. (b) Basis of ascribing the deviations to the influence of an extra planet. The arrows indicate the relative magnitude of the perturbing force at different times.

and U. J. LeVerrier in France—independently worked on the problem. Both men used as a starting point the assumption that the orbital radius of the unknown planet was almost exactly twice that of Uranus. The basis of this was a curious empirical relation, known as Bode's law (actually discovered by J. D. Titius in 1772, but publicized by J. Bode), which expresses the fact that the orbital radii of the known planets can be roughly fitted by the following formula:

$$R_n (AU) = 0.4 + (0.3)(2^n)$$

where n is an appropriate integer for each planet. Putting $n = 0, 1, 2$, we get the approximate orbital radii for Venus, earth and Mars (Mercury requires $n = -\infty$, which is hard to defend). Using $n = 4, 5$ and 6 one gets quite good values for Jupiter, Saturn and Uranus. (The missing integer, $n = 3$, corresponds to the asteroid belt.) Figure 4.6 shows this relation of orbital radii with the help of a semilog plot; it is clear that a simple exponential relation (linear on this graph) works almost as well, but if one accepts Bode's law, then $n = 7$ gives $r = 38.8$ AU, and this is what Adams and LeVerrier used.

Given the radius, the period is automatically defined by Kepler's third law, and then it becomes possible to construct a definite picture, as shown in Fig. 4.5(b), of the way in which the new planet could alternately accelerate and retard Uranus in its orbital motion, depending on their relative positions. With the help of laborious analysis, one can then deduce where in its orbit the new planet should be on a particular date. Adams supplied such information in October 1845 to the British Astronomer Royal, G. B. Airy, who acknowledged Adams' letter, raised a question of detail, but otherwise did nothing. LeVerrier did not complete his own calculations until August 1846, but the astronomer to whom he wrote (J. G. Galle, in Germany) made an immediate search and identified the new planet (Neptune) on his very first night of observation. It was only about a degree from the predicted position. (The next night it had visibly shifted, thereby confirming its planetary status.)

Although the discovery of Neptune is in some respects a great success story, it is also a story of luck, both good and bad, and of human frailty. Adams was really first in the field, but he received no support from his seniors (he was fresh from his bachelor's degree when he began his calculations). Airy missed the credit, which he might readily have won, of being the man who first identified Neptune. But the locations that both Adams and LeVerrier predicted might well have been hopelessly misleading, for in their reliance on Bode's law they used an orbital radius (and hence a period) that was far from correct. The true value is about 30 AU instead of nearly 40 as they had assumed, which means that they overestimated the period by nearly 50%. It was therefore largely a lucky accident that the planet was so near to its predicted position on the particular date that Galle sought and found it. But let this not be taken as disparagement. A great discovery was made, with the help of the laws of motion and the gravitational force law, and it remains

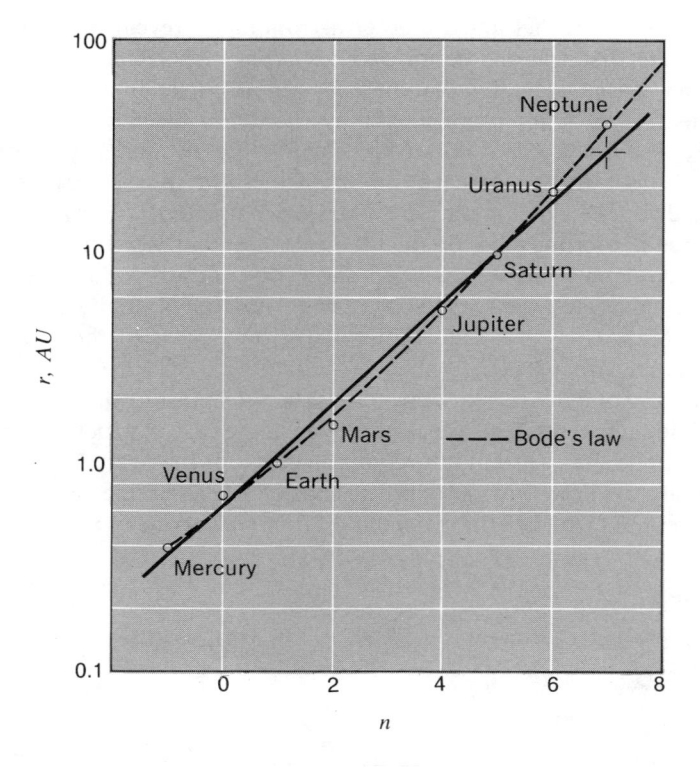

Fig. 4.6 Graph for predicting the orbital radius of the new planet with the help of Bode's law.

as the most triumphant confirmation of the dynamical model of the universe that Newton invented. The discovery of Pluto by C. Tombaugh in 1930, on the basis of a detailed record of the irregularities of Neptune's own motion, provided an echo of the original achievement.

Gravitation outside the solar system

When Newton wrote his *System of the World*, nothing was known about the distances or possible motions of the stars. They simply provided a seemingly fixed background against which the dynamics of the solar system proceeded. There were exceptions. A few prominent stars, e.g. Sirius, known since antiquity by naked-eye astronomy, were found to have shifted position within historic time. But the serious and systematic investigation of stellar motions was begun by William Herschel. His observations, continued and refined by his son John Herschel and by other astronomers, revealed two classes of results. The first was the continuing apparent displacement of individual stars in a way that suggested that the solar system is itself involved in a general movement of the stars in our neighbourhood, at a speed of the order of 10 miles per second (comparable to the earth's own orbital speed around the sun). This, as it stood, was just an empirical fact. But the second type of result pointed directly to the operation of Newton's dynamics. For the Herschels discovered numerous pairs of stars that were evidently orbiting around one another as *binary systems*. Figure 4.7 shows one of the best documented early examples, and the first to be subjected to a detailed analysis in terms of Kepler's laws. (It is ξ-Ursae, in one of the hind paws of the constellation known as the Great Bear.)

The period of a binary star depends on the *total* mass of the system, not on the individual masses. This is easily proved in the case in which the orbits are assumed to be circles around the common centre of mass (see Fig. 4.8a). The individual stars are always at opposite ends of a straight line passing through the centre, C. If we write the statement of $F = ma$ for one of the stars, say m_1, we have

$$G \frac{m_2 m_1}{(r_1 + r_2)^2} = \frac{m_1 v_1^2}{r_1} = m_1 \omega^2 r_1$$

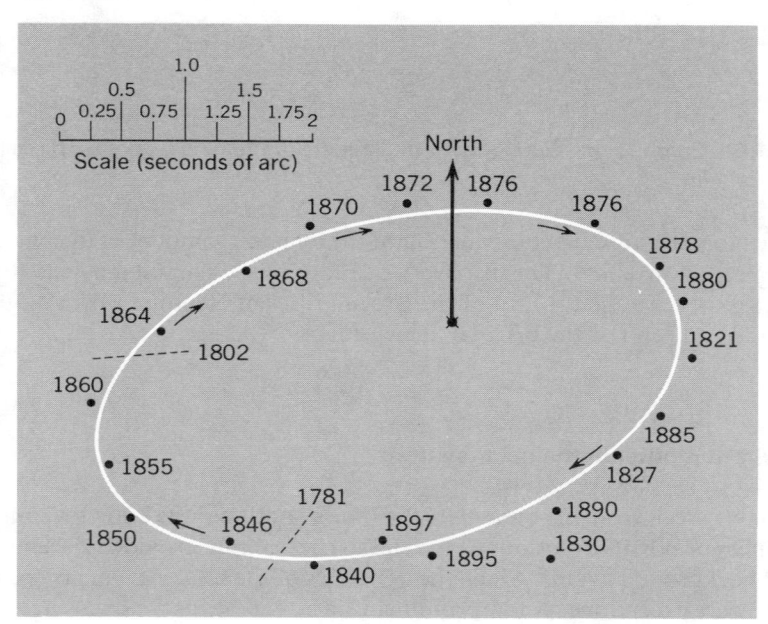

Fig. 4.7 Variation with time of the relative position vector of the members of a double-star system. (After Arthur Berry, *A Short History of Astronomy*, 1898; reprinted by Dover Publications, New York, 1961.)

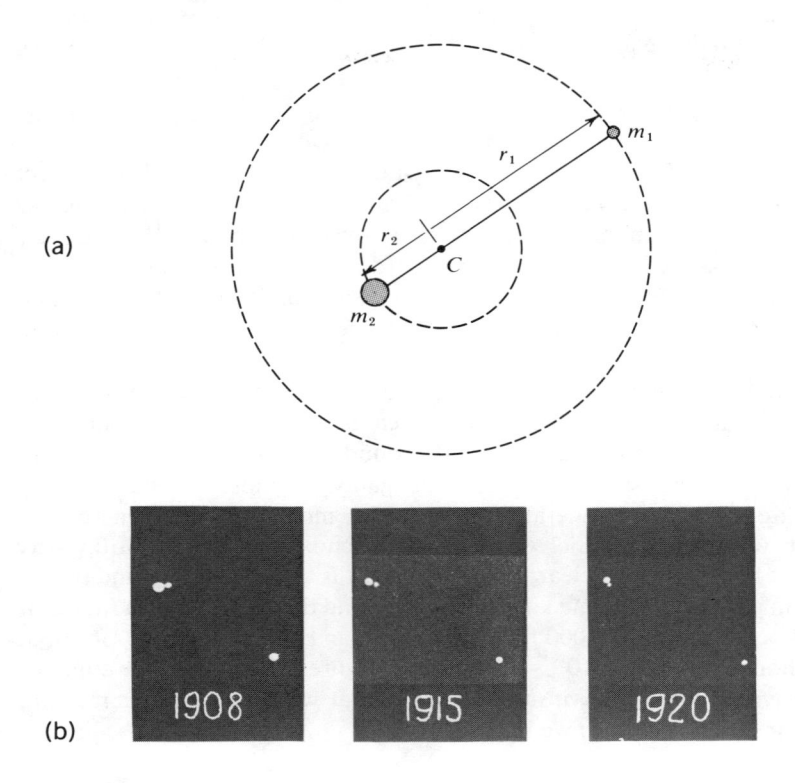

Fig. 4.8 (a) Motion of the members of a binary star system with respect to the centre of mass, C, for the case of circular orbits. (b) Direct visual evidence of the motion of a binary system—Krueger 60, photographed by E. E. Barnard. (Yerkes Observatory photograph.)

where $\omega \,(= 2\pi/T)$ is the angular velocity common to both stars. Hence

$$\omega^2 = \frac{Gm_2}{r_1(r_1 + r_2)^2}$$

However, by the definition of the centre of mass, we have

$$r_1 = \frac{m_2}{m_1 + m_2}\, r$$

where

$$r = r_1 + r_2$$

It follows at once that

$$\omega^2 = \frac{G(m_1 + m_2)}{r^3}$$

Thus if the distance r between the stars can be obtained by direct astronomical observation (e.g. starting from a knowledge of their angular separation), the sum of the masses is at once determined. Finding the individual masses entails the somewhat harder job of measuring the motion of each star in absolute terms against the background of the 'fixed' stars. Figure 4.8(b) shows convincing direct evidence of the orbital motion of an actual binary system.

With the development of modern astronomy, the systematic motions of our sun and its neighbours came to be seen as part of a greater scheme of movements controlled by gravity. All around us throughout the universe were the immense stellar systems—the galaxies—most of them vividly suggesting a state of general rotation. The most difficult structure to elucidate was the one in which we ourselves are embedded, i.e. the Milky Way galaxy. It finally became clear, however, that it is a spiral galaxy, and that in it our sun is describing some kind of orbit around the centre, with a radius of about 3×10^{20} m ($\approx 30\ 000$ light-years) and an estimated period of about 250 million years ($\approx 8 \times 10^{15}$ s). Using these figures we can infer the approximate gravitating mass M, inside the orbit, that would define this motion. Using eqns (4.3) and (4.6) we have

$$M = \frac{4\pi^2}{G} \frac{r^3}{T^2}$$

With $G \approx 7 \times 10^{-11}$ N m² kg⁻², we find that

$$M \approx \frac{40}{7 \times 10^{-11}} \times \frac{3 \times 10^{61}}{6 \times 10^{31}} \approx 3 \times 10^{41} \text{ kg}$$

Since the mass of the sun (a typical star) is about 2×10^{30} kg, we see that a core of about 10^{11} stars is implied. This is not really a figure that can be independently checked. It is a kind of ultimate tribute to our belief in the universality of the gravitational law that it is confidently used to draw conclusions like those above concerning masses of galactic systems.

Einstein's theory of gravitation

We have described earlier how Newton recognized that the proportionality of weight to inertial mass is a fact of fundamental significance; it played a central role in leading him to the conclusion that his law of gravitation must be a general law of nature. For Newton this was a strictly dynamical result, expressing the basic properties of the force law. But Albert Einstein, in 1915, looked at the situation through new eyes. For him the fact that all objects fall towards the earth with the same acceleration g, whatever their size or physical state or composition, implied that this must be in some truly profound way a kinematic or geometrical result, not a dynamical one. He regarded it as being on a par with Galileo's law of inertia, which expressed the tendency of objects to persist in straight-line motion.

Building on these ideas, Einstein developed the theory that a planet (for example) follows its characteristic path around the sun because in so doing it is travelling along what is called a *geodesic* line—that is to say, the most economical way of getting from one point to another. His proposition was that although in the absence of massive objects the geodesic path is a straight line in the Euclidean sense, the presence of an extremely massive object such as the sun modifies the geometry locally so that the geodesics become curved lines. The state of affairs in the vicinity of a massive object is, in this view, to be interpreted not in terms of a gravitational field of force but in terms of a 'curvature of space'—a facile phrase that covers an abstract and mathematically complex description of non-Euclidean geometries.

For the most part the Einstein theory of gravitation gives results indistinguishable from Newton's; the grounds for preferring it might seem to be conceptual rather than practical. But in one celebrated instance of planetary motions there is a real discrepancy that favours Einstein's theory. This is in the so-called 'precession of the perihelion' of Mercury. The phenomenon is that the orbit of Mercury, which is distinctly elliptical in shape, very gradually rotates or precesses in its own plane, so that the major axis is along a slightly different direction at the end of each complete revolution. Most of this precession (amounting to about 10 minutes of arc per century) can be understood in terms of the disturbing effects of the other planets according to Newton's law of gravitation. But there remains a tiny, obstinate residual rotation equal to 43 seconds of arc per century. The attempts to explain it on Newtonian theory—for example by postulating an unobserved planet inside Mercury's own orbit—all came to grief by conflicting with other facts of observation concerning the solar system. Einstein's theory, on the other hand, without the use of any adjustable parameters, led to a calculated precession rate that agreed exactly with observation. It corresponded, in effect, to the existence of a very small force with a different dependence on distance than the dominant $1/r^2$ force of Newton's theory. The way in which a disturbing effect of this kind causes the orbit to precess is discussed in Chapter 9. Other empirical modifications of the basic law of gravitation—small departures from the inverse-square law—had been tried before Einstein developed his theory, but apart from their arbitrary character they also led to false predictions for the other planets. In Einstein's theory, however, it emerged automatically that the size of the disturbing term was proportional to the square of the angular velocity of the planet and hence was much more important for Mercury, with its short period, than for any of the other planets.

Problems

4.1 Given a knowledge of Kepler's third law as it applies to the solar system, together with the knowledge that the disc of the sun subtends an angle of about $\frac{1}{2}°$ at the earth, deduce the period of a hypothetical planet in a circular orbit that skims the surface of the sun.

4.2 It is well known that the gap between the four inner planets and the five outer planets is occupied by the asteroid belt instead of by a tenth planet. This asteroid belt extends over a range of orbital radii from about 2.5 to 3.0 AU. Calculate the corresponding range of periods, expressed as multiples of the earth's year.

4.3 It is proposed to put up an earth satellite in a circular orbit with a period of 2 h.
   (a) How high above the earth's surface would it have to be?
   (b) If its orbit were in the plane of the earth's equator and in the same direction as the earth's rotation, for how long would it be continuously visible from a given place on the equator at sea level?

4.4 A satellite is to be placed in synchronous circular orbit around the planet Jupiter to study the famous 'red spot' in Jupiter's lower atmosphere. How high above the surface of Jupiter will the satellite be? The rotation period of Jupiter is 9.9 h, its mass M_J is about 320 times the earth's mass, and its radius R_J is about 11 times that of the earth. You may find it convenient to calculate first the gravitational acceleration g_J at Jupiter's surface as a multiple of g, using the above values of M_J and R_J, and then use a relationship analogous to that developed in the text for earth satellites (eqn 4.14 or 4.15).

4.5 A satellite is to be placed in synchronous circular earth orbit. The satellite's power supply is expected to last 10 years. If the maximum acceptable eastward or westward drift in the longitude of the satellite during its lifetime is 10°, what is the margin of error in the radius of its orbit?

4.6 Imagine that in a certain region of the ocean floor there is a roughly cone-shaped mound of granite 250 m high and 1 km in diameter. The surrounding floor is relatively flat for tens of kilometres in all directions. The ocean depth in the region is 5 km and the density of the granite is 3000 kg/m³. Could the mound's presence be detected by a surface vessel equipped with a gravity meter that can detect a change in g of 0.1 mgal? (1 gal = 1 cm/s², so 1 mgal ≈ 10^{-6} g.)
   (*Hint:* Assume that the field produced by the mound at the surface can be approximated by the field of a mass point of the same total mass located at the level of the surrounding floor. Note that in calculating the change in g you must keep in mind that the mound has displaced its own volume of water. The density of water, even at such depths, can be taken as about equal to its surface value of about 1000 kg/m³.)

4.7 Show that the period of a particle that moves in a circular orbit close to the surface of a sphere depends only on G and the mean density of the sphere. Deduce what this period would be for any sphere having a mean density equal to the density of water. (Jupiter almost corresponds to this case.)

4.8 Calculate the mean density of the sun, given a knowledge of G, the length of the earth's year, and the fact that the sun's diameter subtends an angle of about $0.55°$ at the earth.

4.9 An astronaut who can lift 50 kg on earth is exploring a planetoid (roughly spherical) of 10 km diameter and density 3500 kg/m^3.

(a) How large a rock can he pick up from the planetoid's surface, assuming that he finds a well-placed handle?

(b) The astronaut observes a rock falling from a cliff. The rock's radius is only 1 m and as it approaches the surface its velocity is 1 m/s. Should he try to catch it? (This is obviously a fanciful problem. One would not expect a planetoid to have cliffs or loose rocks, even if an astronaut were to get there in the first place.)

(c) The astronaut is pleased to discover that, by running horizontally off the edge of the cliff, he can put himself into a circular orbit around the planetoid. How fast does he need to run?

4.10 A dedicated scientist performs the following experiment. After installing a huge spring at the bottom of a 20-storey-high lift shaft, he takes the lift to the top, positions himself on a bathroom scale inside the airtight car with a stopwatch and with pad and pencil to record the scale reading, and directs an assistant to cut the car's support cable at $t = 0$. Presuming that the scientist survives the first encounter with the spring, sketch a graph of his measured weight versus time from $t = 0$ up to the beginning of the second bounce. (*Note:* Twenty stories is ample distance for the elevator to acquire terminal velocity.)

4.11 A planet of mass M and a single satellite of mass $M/10$ revolve in circular orbits about their stationary centre of mass, being held together by their gravitational attraction. The distance between their centres is D.

(a) What is the period of this orbital motion?

(b) What fraction of the total kinetic energy resides in the satellite?

(Ignore any spin of planet and satellite about their own axes.)

4.12 The sun appears to be moving at a speed of about 250 km/s in a circular orbit of radius about 25 000 light-years around the centre of our galaxy. (One light-year $\simeq$ 10^{16} m.) The earth takes 1 year to describe an almost circular orbit of radius about 1.5 $\times$ 10^{11} m around the sun. What do these facts imply about the total mass responsible for keeping the sun in its orbit? Obtain this mass as a multiple of the sun's mass M. (Note that you do not need to introduce the numerical value of G to obtain the answer.)

4.13 The continuous output of energy by the sun corresponds (through Einstein's relation $E = Mc^2$) to a steady decrease in its mass M, at the rate of about 4×10^6 tons/s. This implies a progressive increase in the orbital periods of the planets, because for an orbit of a given radius we have $T \sim M^{-1/2}$.

A precise analysis of the effect must take into account the fact that as M decreases the orbital radius itself increases—the planets gradually spiral away from the sun. However, one can get an order-of-magnitude estimate of the size of the effect, albeit a little bit on the low side, by assuming that r remains constant.

Using the simplifying assumption of constant r, estimate the approximate increase in the length of the year resulting from the sun's decrease in mass over the time span of accurate astronomical observations—about 2500 years.

4.14 It is mentioned at the end of the chapter how Einstein's theory of gravitation leads to a small correction term on top of the basic Newtonian force of gravitation. For a planet of mass m, travelling at speed v in a circular orbit of radius r, the gravitational force becomes in effect the following:

$$F = \frac{GMm}{r^2}\left(1 + 6\frac{v^2}{c^2}\right)$$

where c is the speed of light. (Correction terms of the order of v^2/c^2 are typical of relativistic effects.)

(a) Show that, if the period under a pure Newtonian force GMm/r^2 is denoted by T_0, the modified period T is given approximately by

$$T \approx T_0\left(1 - \frac{12\pi^2 r^2}{c^2 T_0^2}\right)$$

(Treat the relativistic corrections as representing, in effect, a small fractional increase in the value of G, and use the value of v corresponding to the Newtonian orbit.)

(b) Hence show that, in each revolution, a planet in a circular orbit would travel through an angle greater by about $24\pi^3 r^2/c^2 T_0^2$ than under the pure Newtonian force, and that this is also expressible as $6\pi GM/c^2 r$, where M is the mass of the sun.

(c) Apply these results to the planet Mercury and verify that the accumulated advance in angle amounts to about 43 seconds of arc per century. This corresponds to what is called the precession of its orbit.

5 Collisions and conservation laws

The conservation of linear momentum

Some of the most powerful aspects of our physical description of the world are embodied in statements about conserved quantities. In mechanics the law of conservation of linear momentum is such a statement—one might even claim that it is the most important single principle in dynamics. It is based directly on the results of collision experiments. If, for a given particle, we introduce the word *momentum* to describe the product mv, then we have a compact statement:

> **The total momentum of a system of two colliding particles remains unchanged by the collision, i.e. the total linear momentum is a conserved quantity:**

$$p_{1i} + p_{2i} = p_{1f} + p_{2f} \tag{5.1}$$

where the subscripts i and f are used to denote initial and final values, respectively (i.e. pre-collision and post-collision).

Underlying this generalization concerning two-particle collisions is the tacit assumption that the system is effectively *isolated*—the particles interact with each other but not with anything else.

The vector equation (5.1) defines the magnitude and the direction of any one of the momentum vectors if the other three are known. It will very often be convenient to separate eqn (5.1) into three equations in terms of the resolved parts of the vectors along three mutually orthogonal axes (x, y, z). Each of these component equations must then be separately satisfied. Thus, for example, if two bodies, of masses m_1 and m_2, have initial and final velocities u_1, u_2 and v_1, v_2, eqn (5.1) becomes

$$m_1 u_1 + m_2 u_2 = m_1 v_1 + m_2 v_2 \tag{5.2}$$

which contains the following three *independent* statements:

$$\begin{aligned} m_1 u_{1x} + m_2 u_{2x} &= m_1 v_{1x} + m_2 v_{2x} \\ m_1 u_{1y} + m_2 u_{2y} &= m_1 v_{1y} + m_2 v_{2y} \\ m_1 u_{1z} + m_2 u_{2z} &= m_1 v_{1z} + m_2 v_{2z} \end{aligned} \tag{5.3}$$

In carrying out actual numerical calculations this resolution of the vectors will often be necessary. But manipulations involving unspecified masses and velocities are best made in terms of the unresolved equations (5.1) or (5.2), without reference to any particular coordinate system. This is, indeed, one of the main strengths (and economies) of using vector notation.

Action, reaction, and impulse

A collision is a process involving two objects, each of which exerts a force on the other. Object 1 exerts a force F_{12} on object 2; object 2 exerts a force F_{21} on object 1 (Fig. 5.1). We make no assumptions about the relationship between the two forces, *except* that they act for equal times. This last assumption is certainly reasonable, because we recognize that the forces come into being as a result of the collision, and surely the duration of the collision must be the same for both objects. We can then take the statement of $F = ma$ as it applies to each object separately:

$$F_{21} = m_1 a_1 \;; \qquad F_{12} = m_2 a_2$$

i.e.

$$a_1 = \frac{F_{21}}{m_1} \;; \qquad a_2 = \frac{F_{12}}{m_2} \tag{5.4}$$

Suppose, to simplify this present argument, that each force remains constant throughout a collision that lasts for a time Δt. Then we have

$$v_1 = u_1 + \frac{F_{21}}{m_1} \Delta t \;; \qquad v_2 = u_2 + \frac{F_{12}}{m_2} \Delta t \tag{5.5}$$

where u_1 and u_2 are the initial velocities of the two objects and v_1 and v_2 are their final velocities. From these equations we therefore have

$$m_1 v_1 = m_1 u_1 + F_{21} \Delta t$$

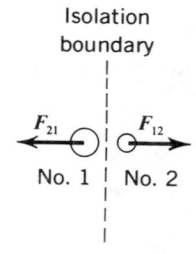

Fig. 5.1 Inferring the equality of action and reaction forces from the fact of momentum conservation.

$$m_2 v_2 = m_2 u_2 + F_{12} \Delta t$$

Adding these two, we thus get

$$m_1 v_1 + m_2 v_2 = m_1 u_1 + m_2 u_2 + (F_{21} + F_{12}) \Delta t \tag{5.6}$$

Experimentally, however, we have eqn (5.2). We *deduce*, therefore, that

$$F_{21} + F_{12} = 0$$

i.e.

$$F_{21} = -F_{12} \tag{5.7}$$

This is, of course, the famous statement known usually as Newton's third law, that 'action and reaction are equal and opposite'.

Having introduced these forces of interaction, we can now relate them to the changes of momentum of the individual objects in a collision. Thus, if Δt is the duration of the collision and F_{21} is a constant force exerted by particle 2 on particle 1, the change of momentum Δp_1 of particle 1 in the collision is given by

$$F_{21} \Delta t = \Delta p_1$$

More generally, if any constant force F acts on a particle for some short time interval Δt, the change of momentum that it generates is called the *impulse* and is given by

$$F \Delta t = \Delta p$$

If F is varying in magnitude and/or direction during the time span Δt, one can proceed to the limit of vanishingly small time and so obtain the following equation:

$$(\text{Newton's law reformulated}) \quad F = \frac{dp}{dt} \tag{5.8}$$

This equation, written on the assumption that F is the net force acting on a particle, then becomes the basic statement of Newtonian dynamics. It is, in a sense, broader in scope than $F = ma$, or at least a more efficient statement of it. For example, a given force applied in turn to a number of different masses causes the same rate of change of momentum in each but not the same acceleration. In short, we come to recognize that momentum is a valuable *single* quantity to be accepted in its own right and most importantly for its property of being *conserved* in an isolated system of interacting particles.

In our earlier discussion of the collision problem, we took the forces of interaction as being constant in time. In most cases that would be quite unrealistic. It is easy to see, however, that through the integration of eqn (5.8) we obtain the net momentum change caused by a varying force. Thus, for example, if a force has some arbitrary variation between $t = 0$ and $t = \Delta t$, the momentum change that it generates is given by

$$\Delta p = \int_0^{\Delta t} F\,dt \tag{5.9}$$

In a two-body collision, the duration of which is Δt, all that we actually observe is that the total momentum after the collision is equal to the total momentum before the collision. In terms of the impulses Δp_1 and Δp_2 given to the separate objects this result is expressed by the condition

$$\Delta p_1 + \Delta p_2 = 0$$

From this we infer that

$$\int_0^{\Delta t} F_{21}\,dt = -\int_0^{\Delta t} F_{12}\,dt$$

In principle, F_{12} and F_{21} could have quite unrelated values at any particular instant, as long as the above integrals are equal. However, failing any evidence to the contrary we assume that they are equal and opposite at each and every instant. Thus in a one-dimensional collision the graphs of these forces as a function of time, whatever their exact form, are taken to be mirror images of each other as shown in Fig. 5.2. It is important to realize, however, that this *is* a postulate. And it is not always true! There is no difficulty as far as 'contact' collisions between ordinary objects are concerned. But in situations in which objects influence one another at a distance, as for example through the long-range forces of electricity or gravitation, Newton's third law *may* cease to apply. For no interaction is transmitted instantaneously, and if the propagation time cannot be ignored in comparison with the time-scale of the motion, the concept of instantaneous action and reaction can no longer be used.

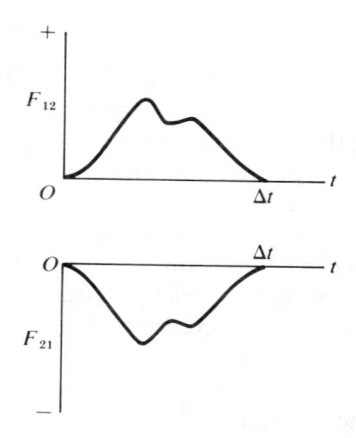

Fig. 5.2 Corresponding variations of action and reaction forces during the course of a collision.

In electromagnetism we have examples of important *delayed* interactions. We know that the interaction between two separated charges takes place via the electromagnetic field, and the propagation of such a field takes place at a speed which, although extremely large, is still finite—the speed of light. The transfer of momentum from one charge to another (resulting, let us say, from a sudden movement of the first charge) involves a time lag equal to the distance between them divided by c. If we looked at the charged particles alone, we would see a sudden change in the momentum of the first charge without an equal and opposite change in the momentum of the other at the same time. There would thus appear to be a failure of momentum conservation, instant by instant, unless we associate some momentum with the electromagnetic field that carries the interaction. And that is precisely what electromagnetic theory suggests. The picture becomes even more vivid when we introduce the quantization of the electromagnetic field and recognize that radiation is carried in the form of photons, or light quanta. There is even a well-developed theoretical description of the static interaction between electric charges in terms of a continual exchange of so-called 'virtual photons'. In this case, however, since the forces are constant in time, the equality of action and reaction holds good at every instant, and the existence of a finite time for propagating the interaction ceases to be evident.

Extending the principle of momentum conservation

It is perhaps worth amplifying the remarks of the last section a little. In Newton's view, the law of conservation of momentum was closely tied in with the action-reaction idea. There is, however, an alternative approach to simple collisions which loosens this connection and eases the transition to non-Newtonian mechanics.

This approach can be defined in terms of a question: what do we actually *observe* in a collision experiment? The answer is that our observations are purely kinematic ones—measurements of the velocities of the two objects before and after impact. Suppose that two objects, A and B, with initial velocities u_1 and u_2, respectively, collide with one another and afterwards have velocities v_1 and v_2. In any individual collision of this type, it is always possible to find a set of four scalar multiples (α) that permit one to write an equation of the following form:

$$\alpha_1 u_1 + \alpha_2 u_2 = \alpha_3 v_1 + \alpha_4 v_2 \tag{5.10}$$

This as it stands is a quite uninteresting statement. But experiments for all sorts of values of u_1 and u_2 reveal the remarkable result that in every such collision, for two given objects, we can obtain a vector identity by putting $\alpha_1 = \alpha_3 = \alpha_A$ (a scalar property of A) and $\alpha_2 = \alpha_4 = \alpha_B$ (a corresponding scalar property of B). In other words, the purely kinematic observations on a collision process permit us to introduce a unique dynamical property of each object. Notice that this simple situation ceases to hold if any of the velocities involved become comparable to that of light. In that case it is still possible to construct a vector balance equation in the form of eqn (5.10), but only if the parameters α are made explicit functions of speed. In fact, we arrive at the relativistic formula for the variation of mass with speed (eqn 2.6).

Let us return now to the results of experiments on collisions at low velocities. The basic statement of these results is that, in the impact of any two given objects, the velocity change of one always bears a constant, negative ratio to the velocity change of the other:

$$v_2 - u_2 = -\text{const}(v_1 - u_1) \tag{5.11}$$

It is precisely because this ratio of velocity changes is found to come out to the same value, whatever the type, strength, or duration of the interaction between the objects, that we can infer that it provides a measure of some intrinsic property of the objects themselves. We *define* this property as the inertial mass.

One can set up an inertial mass scale for a number of objects 1, 2, 3, . . . by finding the velocity changes pairwise in such interaction processes and defining the inertial mass ratio, let us say of objects 1 and 2, by

$$\frac{m_2}{m_1} = \frac{|\Delta v|_1}{|\Delta v|_2}$$

Similarly for objects 1 and 3,

$$\frac{m_3}{m_1} = \frac{|\Delta v'|_1}{|\Delta v'|_3}$$

and so on. If m_1 is the standard kilogram, we then have an operation to determine the inertial masses of any other objects. We must, however, do more. If we let objects 2 and 3 interact, then the ratio

$$\frac{m_3}{m_2} = \frac{|\Delta v''|_2}{|\Delta v''|_3}$$

must be consistent with the same ratio obtained from the first two measurements. In fact it is, and this internal experimental consistency then allows us to use the values m_2, m_3, . . . as measures of the inertial masses of the respective objects.

Having set up a consistent measure of inertial mass in this way, we can then rewrite eqn (5.11) in the form

$$\frac{v_1 - u_1}{v_2 - u_2} = -\frac{m_2}{m_1} \tag{5.12}$$

which when rearranged gives us

$$m_1 u_1 + m_2 u_2 = m_1 v_1 + m_2 v_2 \tag{5.13}$$

Thus an equation identical in appearance with the usual momentum-conservation relation emerges, but notice how this contrasts with the Newtonian analysis. We have used the collision processes themselves to define mass ratios through eqn (5.11). Once this has been done, the terms in eqn (5.13) *automatically* add up to the same total before and after the collision.

What we have done here, in effect, is to give primacy of place to momentum conservation. The question as to whether or not action equals reaction does not arise. And this can be very valuable. For when one is confronted with non-Newtonian interactions (i.e. those for which action and reaction are *not* equal opposite forces at each instant) one faces the problem of how to incorporate them into physics—whether to abandon the law of momentum conservation in its limited form or to extend the idea of momentum and retain the conservation law. The momentum-conservation principle has proved so extremely powerful that the latter course has been chosen, and conservation of linear momentum is a central feature of relativistic dynamics.

This way of analysing the basic results of collision processes exposes the very intimate relation that exists between kinematics and dynamics. If we change our description of space, time and motion, then we can expect that our dynamics must be changed also. This is, in fact, precisely the situation as we make the transition from the kinematics of Galileo and Newton to the kinematics of Einstein's special theory of relativity.

Jet propulsion

An important application of Newton's third law in its generalized form (eqn 5.7) is the force due to a jet or stream of fluid. Such a stream can be characterized by its speed v_0 and its rate of flow, the latter being described by the rate of mass flow, μ.

If such a stream is stopped by an obstacle, or if it is set in motion from fluid at rest, there is an associated rate of destruction or creation of momentum in the jet equal to μv_0. This momentum change requires a force given by

$$F = \mu v_0$$

The stream exerts an equal and opposite force on the other partner in this interaction. This result is, of course, the basis of rocket propulsion, in which the thrust is the reaction to the force that generates backward momentum in the ejected gases.

A jet engine of an aircraft presents a somewhat more complicated application of these dynamical results. In this case the air that enters at the front of the engine, and leaves as part of the exhaust gases at the rear, plays an important role in the overall process of momentum transfer. The main function of the fuel that is carried with the plane is to give the ejected gases a high speed with respect to the plane, and most of the moving mass is supplied by the air. It is very convenient to analyse the dynamics of this system from the standpoint of a reference frame in which the engine is instantaneously at rest. If the plane is travelling forward at a velocity v, the air is seen as entering the engine at the front with an equal and opposite velocity. Then, at the rear, all the ejected material has a backward velocity v_0 in this frame. Thus, if air is being carried through the engine at the rate μ_{air} (kg/s) and fuel is being burnt at the rate μ_{fuel}, the total rate of change of momentum defines a total forward force on the engine according to the following equation:

$$P = \mu_{fuel} v_0 + \mu_{air}(v_0 - v) \tag{5.14}$$

Example. A jet aircraft is travelling at a speed of 250 m s^{-1}. Each of its engines takes in 100 m^3 of air per second, corresponding to a mass of 50 kg of air per second at the plane's flying altitude. The air is used to burn 3 kg of fuel per second, and all the gases coming from the combustion chamber are ejected with a speed of 500 m s^{-1} relative to the aircraft. What is the thrust of each engine? Substituting directly in eqn (5.14) we have

$$P = 3 \times 500 + 50(500 - 250)$$
$$= 1.4 \times 10^4 \text{ N}$$

Four engines of this type would thus give a total driving force of about 12 000 lbf, a more or less realistic figure.

Rockets

This has become such a very large subject in recent years that it is clear that we shall do no more than touch upon the underlying dynamical principles. For simplicity, let us consider the motion of a rocket out in a region of space where the effects of gravity are sufficiently small to be ignored in the first approximation. Under this assumption, the only force acting on the body of the rocket is the thrust from the ejected fuel. Suppose that the burnt fuel has a speed v_0 relative to the rocket. Between time t and time $t + \Delta t$, a mass Δm of fuel is burnt and becomes separated from the rocket. The situations before and after ejection are shown in Fig. 5.3, where m is the total mass of the rocket plus its remaining fuel at $t + \Delta t$, and v is its velocity at time t.

The ejection of the fuel is a kind of inelastic collision in reverse, since

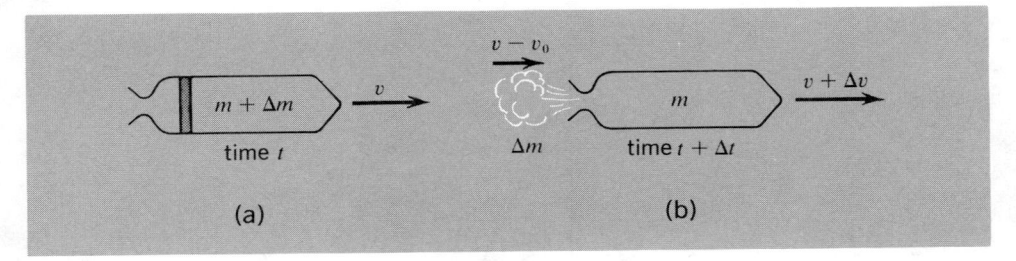

Fig. 5.3 Situations just before and just after the ejection of an element Δm of mass by a rocket.

initially the masses m and Δm have the same velocity. By conservation of linear momentum we have

$$(m + \Delta m)v = m(v + \Delta v) + \Delta m(v - v_0)$$

Therefore,

$$\Delta v = v_0 \frac{\Delta m}{m} \tag{5.15}$$

This equation is not quite exact. (Why?) But as we let Δt approach zero, the error approaches zero. As long as $\Delta v/v_0$ is much less than unity, eqn (5.15) is an excellent approximation. Given the initial total mass m_i of the rocket plus fuel, and the final mass m_f of the rocket at burnout, the velocity gained by the rocket can be evaluated. If the initial and final velocities of the rocket are v_i and v_f, eqn (5.15) tells us that

$$v_f - v_i = v_0 \sum \frac{1}{m} \Delta m$$

The answer could be obtained numerically, by drawing a graph (Fig. 5.4) of $1/m$ against m, and finding the area (shaded) between the limits m_f and m_i. This is a pure number, which when multiplied by v_0 gives the increase of velocity. Analytically, if m is used to denote the mass of the rocket (plus its remaining fuel) at any instant, it is more satisfactory to interpret dm as the change of mass *of the rocket* in a time dt. Defined in this way, dm is actually a negative quantity, which when integrated from the beginning to the end of the burning process gives the value of $m_f - m_i$ (< 0). In these terms the change of velocity can be written as the following integral in closed form:

$$v_f - v_i = -v_0 \int_{m_i}^{m_f} \frac{dm}{m} = v_0 \ln\left(\frac{m_i}{m_f}\right) \tag{5.16}$$

You should satisfy yourself that this is indeed equivalent to the final result of the numerical-graphical method described above.

Notice that the time does not enter into the calculation at all, although of

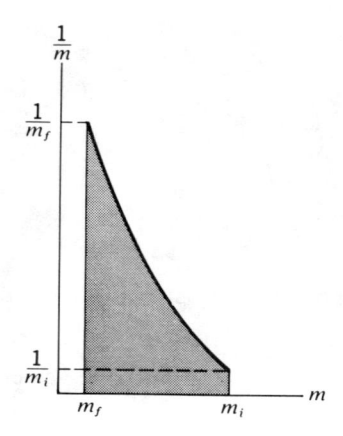

Fig. 5.4 Graph of $1/m$ versus m. The shaded area gives a relative measure of the gain of velocity resulting from a given mass change.

course it would do so if we wanted to consider the *rate* of increase of v or the magnitude of the thrust.

Notice also that we would be entitled, at each and every instant, to look at the situation from the frame of reference of the rocket itself. In this frame the velocity of the ejected fuel is always just $-v_0$, and eqn (5.15) follows immediately.

It is worth examining some of the implications of eqn (5.16). The first thing to notice is that the gain of velocity is directly proportional to the speed of the ejected gases. Thus it pays to make v_0 as great as possible. The highest values of v_0 attainable through chemical burning processes are of the order of 5000 m s^{-1}, and in practice, because of incomplete burning and other losses, it is hard to do better than about 50% of the ideal theoretical value for a given fuel. These velocities are, of course, very high in ordinary terms, but they are small compared to the velocities that can be given to charged particles by electrical acceleration. Hence the interest in developing ion-gun engines, or even using the highest available speed (that of light) by making an exhaust jet of radiation. The trouble with both of these, however, is the very small rate of ejection of mass, which makes the attainable thrust very small.

The other main feature of eqn (5.16) is the way in which the increase of velocity varies logarithmically with the mass ratio. This places rapidly increasing demands on the amount of fuel needed to confer larger and larger final velocities on a given payload. Suppose, for example, that we wanted to attain a velocity equal to the exhaust velocity v_0, starting from rest. Then, by eqn (5.16) we have

$$v_f - v_i = v_0 = v_0 \ln\left(\frac{m_i}{m_f}\right)$$

Therefore,

$$\frac{m_i}{m_f} = e = 2.718\ldots$$

But to attain twice this velocity, we need to have

$$\ln\left(\frac{m_i}{m_f}\right) = 2$$

i.e.

$$\frac{m_i}{m_f} = e^2 \approx 7.4$$

Table 5.1 presents the results of such calculations in more convenient form. The last column represents the extra mass needed, as a multiple of the payload. The practical problems of producing very large mass ratios are prohibitive, but the use of multistage rockets (which also have other advantages) avoids this difficulty.

Table 5.1

$v_f - v_i$	m_i/m_f	$(m_i - m_f)/m_f$
v_0	2.7	1.7
$2v_0$	7.4	6.4
$3v_0$	20.1	19.1
$4v_0$	54.5	53.5

A result that may seem surprising at first sight is that there is nothing, in principle, to prevent us from giving a rocket a forward velocity that is considerably greater than the speed v_0 of the exhaust gases. Thus at a late stage in the motion one would see both rocket and the ejected material moving forward with respect to the frame in which the rocket started out from rest. No violation of dynamical principles would be involved, and if one made a detailed accounting of the motion of all the material that was in the rocket initially, one would find that the total momentum of the system had remained at zero (as long as the effects of external forces, including gravity, could be ignored). It should of course be emphasized that our whole analysis would hold good, as it stands, only in the absence of gravity and of resistive forces due to the air.

The zero-momentum frame

The momentum of a particle or of a system of particles is not an invariant; it depends on the frame of reference in which one observes the motion. If, however, one compares the descriptions of the motion in two different inertial frames, related by a constant velocity, the difference between the measured values of the momentum of a particle is always a constant vector of magnitude mv, where m is the mass of the particle and v is the velocity of one frame relative to the other. One can always find a reference frame in which the total momentum of any particle vanishes; it is evidently a frame in which the particle is at rest at the instant of time one determines its momentum. One can likewise find a reference frame in which the total momentum of any system of particles is zero, and this zero-momentum frame of reference is of great importance, not only as a convenience for looking at collisions and interactions in general but also, as we shall show shortly, for its *dynamical* implications. To see how we identify this zero-momentum frame consider two particles m_1 and m_2 with velocities v_1 and v_2 in a frame of reference S. Our task is to find the velocity v_c of a reference frame S' relative to S such that in S' the total momentum $m_1 v'_1 + m_2 v'_2$ is equal to zero.

Let O' be the origin of S' moving with velocity v_c relative to O. The velocities of m_1 and m_2 relative to O' are given by

$$v'_1 = v_1 - v_c$$
$$v'_2 = v_2 - v_c$$

Hence the momentum of the two particles in S' is

$$m_1 v'_1 + m_2 v'_2 = m_1 v_1 - m_1 v_c + m_2 v_2 - m_2 v_c$$

and if this is to be zero, we must have

$$(m_1 + m_2) v_c = m_1 v_1 + m_2 v_2 \tag{5.17}$$

Equation (5.17) fixes the velocity of the reference frame S' relative to S but leaves the choice of the position (r_c) of O' arbitrary. We use this freedom to choose the location of O' relative to the positions of m_1 and m_2 as simply as possible. If we rewrite eqn (5.17) in the equivalent form

$$(m_1 + m_2) \frac{dr_c}{dt} = m_1 \frac{dr_1}{dt} + m_2 \frac{dr_2}{dt}$$

or better as

$$\frac{d}{dt}[(m_1 + m_2)r_c] = \frac{d}{dt}[m_1 r_1 + m_2 r_2]$$

evidently the simplest way of satisfying this is to equate the two sets of square brackets. The difference between them must be a constant and we choose the constant equal to zero. We then have for the position of O' in S,

$$r_c = \frac{m_1 r_1 + m_2 r_2}{m_1 + m_2} \tag{5.18}$$

THE ZERO-MOMENTUM FRAME

The origin of a zero-momentum reference frame chosen in this way is called the *centre of mass* of the two-particle system made up of m_1 and m_2. In general the velocity vectors of two colliding objects define a plane. Actually one must stand ready to go all the way into three dimensions, because the plane defined by the velocity vectors of two particles after a collision may not be the same as that defined by the initial velocities. Many collision processes are, however, purely two-dimensional, and we shall limit ourselves to such cases in discussing specific examples (even though the theory applies equally well to three-dimensional problems).

Whenever possible, two-dimensional collisions should be analysed from the standpoint of the zero-momentum frame. Since the total momentum in this frame is zero, the particles separate, as well as approach, with equal and opposite momenta, as shown in Fig. 5.5(a). Also the magnitudes of the final momentum vectors are independent of their direction in the CM frame. Thus, as shown in Fig. 5.5(b), the vectors p'_{1f} and p'_{2f} can be represented with their tips lying at opposite ends of the diameter of a circle.

The relative directions of p'_{1i} and p'_{1f} can be anything, depending on the details of the interaction. The relative lengths of these vectors also depend, of course, on the detailed mechanism. In a perfectly elastic collision we have $p'_f = p'_i$. In a completely inelastic collision we have $p'_f = 0$. (In an explosive process, for example rocket propulsion, we have the converse situation, where $p'_i = 0$, $p'_f > 0$.) And in atomic, chemical or nuclear reaction processes we may find p'_f less than, equal to, or greater than p'_i.

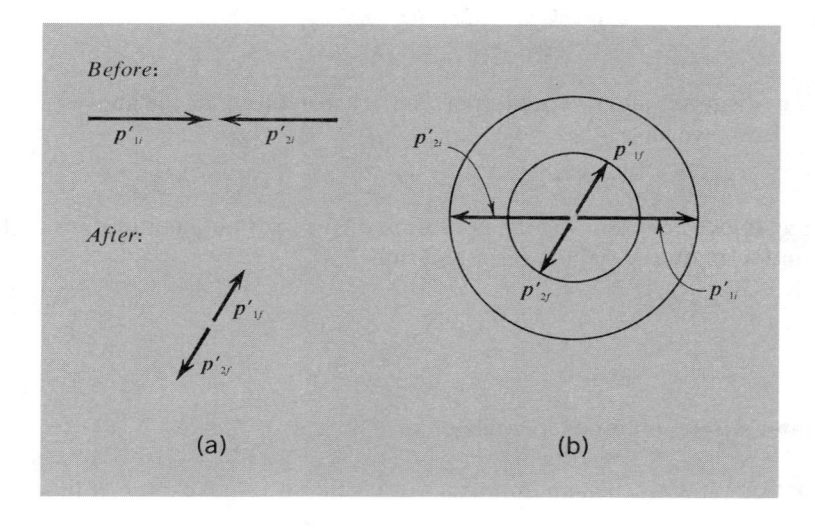

Fig. 5.5 (a) An arbitrary collision as described by equal and opposite momentum vectors in the CM frame. (b) In the CM frame, the end points of the momentum vectors lie on circles.

Kinetic energy of a two-body system

We shall now consider the kinetic energy of a two-body system from the standpoints of the laboratory frame and of the zero-momentum frame and the relationship between them.

Let two particles have masses m_1 and m_2 and velocities v_1 and v_2 in the laboratory frame (S). Then the velocity, v_c, of the CM frame (S') is defined by the equation

$$v_c = \frac{m_1 v_1 + m_2 v_2}{m_1 + m_2} \tag{5.19}$$

If the velocities of the particles as measured in S' are v_1' and v_2', we then have

$$v_1 = v_1' + v_c$$
$$v_2 = v_2' + v_c$$

We now write down the total kinetic energy in S, using the fact that the kinetic energy $\frac{1}{2}mv^2$ of a particle can also be expressed as $\frac{1}{2}m(v \cdot v)$, i.e. in terms of the scalar product of the vector v with itself. Thus we have

$$K = \tfrac{1}{2}m_1(v_1 \cdot v_1) + \tfrac{1}{2}m_2(v_2 \cdot v_2)$$
$$= \tfrac{1}{2}m_1(v_1' + v_c) \cdot (v_1' + v_c) + \tfrac{1}{2}m_2(v_2' + v_c) \cdot (v_2' + v_c)$$

Now consider one of these scalar products, using the distributive and commutative laws that apply to the dot products of vectors:

$$(v_1' + v_c) \cdot (v_1' + v_c) = v_1' \cdot v_1' + 2v_1' \cdot v_c + v_c \cdot v_c$$
$$= v_1'^2 + 2v_1' \cdot v_c + v_c^2$$

Using this result and its counterpart in the second term of the above expression for K, we have

$$K = (\tfrac{1}{2}m_1 v_1'^2 + \tfrac{1}{2}m_2 v_2'^2) + (m_1 v_1' + m_2 v_2') \cdot v_c + \tfrac{1}{2}(m_1 + m_2)v_c^2$$

But again we note that, by the definition of the zero-momentum frame, the second term on the right is zero, and thus

$$K = K' + \tfrac{1}{2}Mv_c^2 \tag{5.20}$$

Kinetic energy changes in collisions

Equation (5.20) is such an important result that we shall express it in words also:

> The kinetic energy of a system of two particles is equal to the kinetic energy of motion relative to the centre of mass of the system (the internal

kinetic energy), **plus the kinetic energy of a single particle of mass equal to the total mass of the system moving with the centre of mass.**

The great importance of this separation of the kinetic energy into two parts—and note well that it works *only* if the new reference frame is the zero-momentum frame—is that it opens the way to a very powerful and simplifying procedure. *We have the possibility of analysing the internal motion of a system (relative to the CM) without reference to the bodily motion of the system as a whole.*

One of the implications of eqn (5.20) is that a certain amount of kinetic energy is locked up, as it were, in the motion of the centre of mass. In the absence of external forces the velocity v_c remains constant throughout the course of a collision process, and the kinetic energy $\frac{1}{2}Mv_c^2$ must likewise remain unchanged. This means that in the collision of two objects, only a certain fraction of their total kinetic energy, as measured in the laboratory, is available for conversion to other purposes. The amount of this available kinetic energy is, in fact, identical with K'. We can calculate it with the help of eqn (5.20):

$$K' = K - \tfrac{1}{2}Mv_c^2$$
$$= (\tfrac{1}{2}m_1v_1^2 + \tfrac{1}{2}m_2v_2^2) - \tfrac{1}{2}(m_1 + m_2)v_c^2$$

Substituting for v_c from eqn (5.19), this leads by simple algebra to the following result:

$$K' = \frac{1}{2}\frac{m_1m_2}{m_1 + m_2}(v_2 - v_1)^2 \qquad (5.21\mathrm{a})$$

The value of K', as expressed by this equation, is thus the kinetic energy of what can be regarded as being effectively a single mass, of magnitude $m_1m_2/(m_1 + m_2)$, moving at a velocity equal to the *relative* velocity of the colliding particles. The effective mass is called the *reduced mass* of the system and is given the symbol μ. Thus we may write eqn (5.21a) in the more compact form

$$K' = \tfrac{1}{2}\mu v_{\mathrm{rel}}^2 \qquad (5.21\mathrm{b})$$

where

$$\mu = \frac{m_1m_2}{m_1 + m_2}$$
$$v_{\mathrm{rel}} = v_2 - v_1 = v_2' - v_1'$$

For example, if a moving object of mass 2 units strikes a stationary object of mass 1 unit, two thirds of the initial kinetic energy is locked up in centre-of-mass motion, and only one third is available for the purpose of producing deformations, and so on, when the objects collide.

Let us apply eqn (5.20) to the states of a two-particle system before and after an interaction of any kind. We have

$$K_i = K_i' + \tfrac{1}{2}Mv_c^2$$
$$K_f = K_f' + \tfrac{1}{2}Mv_c^2$$

We assume that the total mass remains unchanged and, in the absence of external forces, so does v_c. Thus we arrive at the result

$$K_f - K_i = K_f' - K_i' = Q$$

where Q is the amount by which the final kinetic energy exceeds (algebraically) the initial kinetic energy. The actual value of Q may, of course, be positive, negative or zero.

This result can be extended, with virtually no modification, to such processes as nuclear reactions, in which the actual identity of the particles in the final state may be quite different from what one has at the beginning. Suppose, for example, that there is a collision between two nuclei, of masses m_1 and m_2, which react to produce two different nuclei, of masses m_3 and m_4. Then we can write the following statements of conservation:

Mass: $m_1 + m_2 = m_3 + m_4 = M$

Momentum $m_1v_1 + m_2v_2 = m_3v_3 + m_4v_4 = Mv_c$
$$m_1v_1' + m_2v_2' = m_3v_3' + m_4v_4' = 0$$

Thus the initial and final kinetic energies can be written as follows:

$$K_i = \tfrac{1}{2}m_1(v_1')^2 + \tfrac{1}{2}m_2(v_2')^2 + \tfrac{1}{2}Mv_c^2$$
$$K_f = \tfrac{1}{2}m_3(v_3')^2 + \tfrac{1}{2}m_4(v_4')^2 + \tfrac{1}{2}Mv_c^2$$

with $K_f - K_i = Q$.

Example 1. Proton-proton elastic collisions. Figure 5.6 shows a collision between two protons, as recorded in a photographic emulsion. One of the protons belonged to a hydrogen atom in the emulsion and was effectively stationary before the collision took place; the other entered the emulsion with a kinetic energy of about 5 MeV (8×10^{-13} J).

The most notable feature of the collision is that the paths of the two protons after collision make an angle of 90° with each other. This is true for all such proton-proton collisions, until we get up to energies so high that Newtonian mechanics is no longer adequate to describe the situation. By first looking into the centre-of-mass frame we can readily understand this. Let the velocity of the incident proton as observed in the laboratory be v. Then the zero-momentum frame has velocity $v/2$, and in this frame the protons approach and recede with equal and opposite velocities as shown in Fig. 5.7(a). Suppose one proton emerges from the collision in the direction

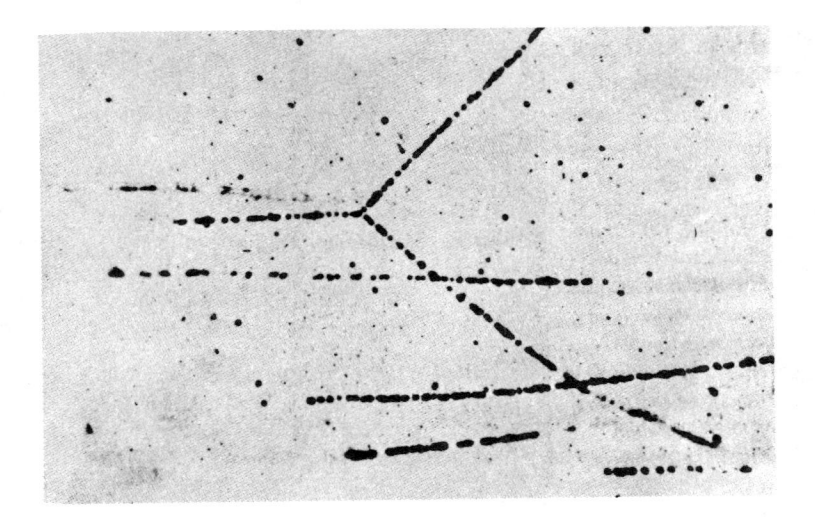

Fig. 5.6 Elastic collision between an incident proton and an initially stationary proton in a photographic emulsion. (From C.F. Powell and G.P.S. Occhialini, *Nuclear Physics in Photographs*, Oxford University Press, New York, 1947.)

θ', so that the other is at $\pi - \theta'$ on the other side of the line of approach. To get back to the laboratory frame we add the velocity $v/2$ to each proton, parallel to the original line of motion, as shown in Fig. 5.7(b). But the triangles ABC and AEF are both isosceles, so the directions θ_1 and θ_2 of the protons as observed in the laboratory are given by

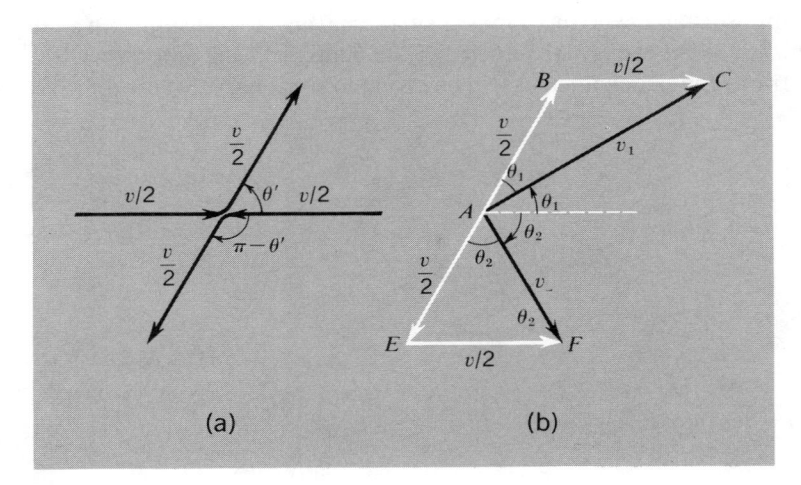

Fig. 5.7 (a) Elastic collision between two equal masses, as seen in the CM frame. (b) Transformation to the laboratory frame, showing a 90° angle between the final velocities.

$\theta_1 = \theta'/2$

$\theta_2 = (\pi - \theta')/2$

Therefore,

$\theta_1 + \theta_2 = \pi/2$

Moreover, we can easily find the laboratory velocities of the two protons after the collision, for we have

$v_1 = 2(v/2) \cos \theta_1 = v \cos \theta_1$

$v_2 = 2(v/2) \cos \theta_2 = v \sin \theta_1$

We see that in any such collision the total kinetic energy is conserved, because

Initial KE $= \frac{1}{2} m v^2$

Final KE $= \frac{1}{2} m (v \cos \theta_1)^2 + \frac{1}{2} m (v \cos \theta_2)^2$
$= \frac{1}{2} m v^2 (\cos^2 \theta_1 + \sin^2 \theta_1)$
$= \frac{1}{2} m v^2$

Example 2. Neutron-nucleus elastic collisions. In nuclear fission processes, neutrons are ejected with a variety of energies, but the average energy is of the order of 1 MeV (i.e. 1.6×10^{-13} J). These neutrons are, however, most effective in causing further fissions if they are reduced to energies of the order of 10^{-2} or 10^{-1} eV (thermal energies). Thus an essential feature of every slow-neutron reactor is a means of slowing down the neutrons. And elastic collisions of neutrons with other nuclei (those composing the moderator material of the reactor) do most of the job.

Suppose that a neutron of mass m makes an elastic collision with a nucleus of mass M. Let the initial velocity of the neutron in the laboratory be v_0; in this frame the struck nucleus will be assumed stationary. Figure 5.8(a) shows

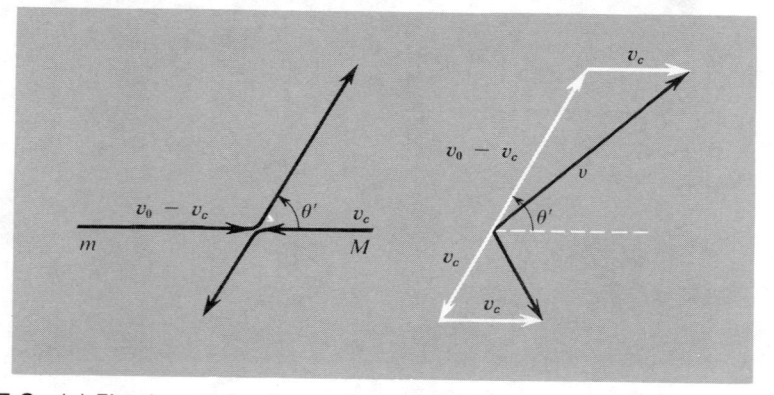

Fig. 5.8 (a) Elastic collision between two unequal masses, as seen in the CM frame. (b) Transformation to the laboratory frame; the angle between the velocities is different from 90°.

KINETIC ENERGY CHANGES IN COLLISIONS

the collision as seen in the centre-of-mass frame, which has a velocity v_c relative to the laboratory frame given by eqn (5.17):

$$v_c = \frac{m}{M+m} v_0 \tag{5.22}$$

If the collision turns the neutron through an angle θ' in the zero-momentum frame, its final velocity in the laboratory frame is the vector v shown in Fig. 5.8(b). The magnitudes of the final velocities, as measured in the laboratory, for any given value of θ', are readily calculated.

The biggest energy loss for a neutron as seen in the laboratory frame occurs if it is scattered straight backward ($\theta' = \pi$). In this case we have

$$v(\pi) = v_c - (v_0 - v_c) = -(v_0 - 2v_c)$$

$$v(\pi) = -v_0\left(1 - \frac{2m}{M+m}\right)$$

$$= -\frac{M-m}{M+m} v_0$$

For $\theta' = 0$ the neutron loses no energy at all. (What sort of a collision is this?) Thus the kinetic energy of the neutron after the collision lies between the following limits:

$$K_{max} = \tfrac{1}{2}mv_0^2$$
$$K_{min} = \tfrac{1}{2}m\left(\frac{M-m}{M+m}\right)^2 v_0^2 \tag{5.23}$$

Since K_{max} is independent of the mass M of a moderator nucleus, it is the expression for K_{min} that tells us what value of M is most likely to lead to the greatest reduction of the average neutron energy. And we see that $M = m$ makes K_{min} equal to zero; we cannot do as well as this for any other value of M, whether it be bigger or smaller than m. Thus if no other considerations were involved, ordinary hydrogen would make the best moderator, since in this case M (the proton mass) is equal to m within about 1 part in 10^3. Protons, however, also capture slow neutrons rather effectively, thereby making them unavailable for causing further fissions, and it turns out that certain other light nuclei (e.g. deuterium, beryllium and carbon) offer a better compromise between moderating and trapping of the fission neutrons.

Example 3.   The D—D reaction.   One of the most famous nuclear reactions (and an important one for the process of energy generation by nuclear fusion) is the reaction of two nuclei of deuterium (hydrogen 2) to form a helium 3 nucleus and a neutron:

$$^2_1H + ^2_1H \to ^3_2He + ^1_0n + 3.27 \text{ MeV}$$

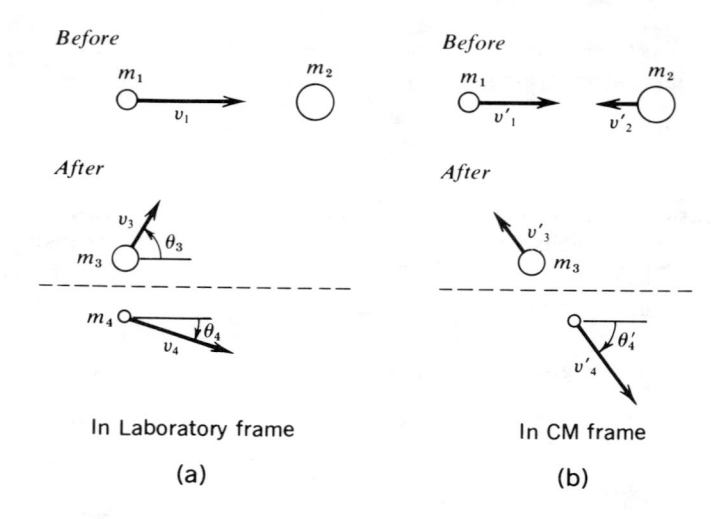

Fig. 5.9 (a) Reaction process, in which the collision of the particles m_1 and m_2 leads to the formation of two different particles, m_3 and m_4. (b) Same process as seen in the CM frame.

The 3.27 MeV represents the extra amount of kinetic energy, Q, made available because the masses of the product particles (their *rest masses*, to be precise) add up to a little less than the masses of the initial particles; the total energy, including mass equivalents, remains constant, of course.

Suppose now that a deuteron with a kinetic energy of 1 MeV strikes a stationary deuteron. What is the final state of affairs as viewed in the CM frame? (See Fig. 5.9.) First, let us calculate the velocity v_c. The mass of a deuteron is about 2 amu, or about 3.34×10^{-27} kg. Now, 1 MeV = 1.6×10^{-13} J. Therefore,

$$v_1 = \left(\frac{2K_1}{m_1}\right)^{1/2} \approx 1.0 \times 10^7 \text{ m s}^{-1}$$

Since $m_2 = m_1$ and m_2 is initially at rest, we have

$$v_c = v_1/2 = 0.5 \times 10^7 \text{ m s}^{-1}$$

In the CM frame the deuterons have equal and opposite velocities of magnitude equal to v_c. Hence we have

$$K'_{\text{initial}} = 2 \times \tfrac{1}{2}m_1 v_c^2 = \tfrac{1}{4}m_1 v_1^2$$

i.e.

$$K'_{\text{initial}} = \tfrac{1}{2}K_1 = 0.5 \text{ MeV}$$

(We see here a particular application of eqn (5.21). If a moving object collides with a stationary one of equal mass, only half the initial kinetic energy is available for their relative motion in the CM frame.)

Now consider the result of the nuclear reaction. The final total kinetic energy in the CM frame is given by

$$K'_{final} = K'_{initial} + 3.27 \text{ MeV}$$
$$= 3.77 \text{ MeV} = 6.03 \times 10^{-13} \text{ J}$$

This is partitioned between the ^{3}He and the neutron in such a way that the momenta are numerically equal. Denoting the masses and velocities of the ^{3}He and the neutron as m_3 and v'_3 and m_4 and v'_4, respectively, we have

$$m_3 v'_3 + m_4 v'_4 = 0$$
$$v'_4 = \frac{m_3}{m_4} v'_3$$

Then

$$K'_{final} = \tfrac{1}{2} m_3 (v'_3)^2 + \tfrac{1}{2} m_4 (v'_4)^2$$
$$= \tfrac{1}{2} m_3 (v'_3)^2 + \tfrac{1}{2} m_4 \left(\frac{m_3}{m_4}\right)^2 (v'_3)^2$$
$$= \frac{1}{2} \frac{m_3(m_3 + m_4)}{m_4} (v'_3)^2$$

Putting $m_3 \approx 3$ amu and $m_4 \approx 1$ amu, this gives us

$$v'_3 = \left(\frac{2 \times 6.03 \times 10^{-13}}{12 \times 1.67 \times 10^{-27}}\right)^{1/2} \approx 0.77 \times 10^7 \text{ m s}^{-1}$$

and so $v'_4 (= 3v'_3) \approx 2.3 \times 10^7$ m s^{-1}.

We thus have a full picture of the final situation as viewed in the CM frame for any specified direction θ'_4 of the outgoing neutron. To go back to the laboratory frame we have simply to add the CM velocity v_c to each of the vectors v'_3 and v'_4.

The great advantage of using the CM frame in this way is that, regardless of the final directions as specified by θ'_4, the *magnitudes* of v'_3 and v'_4 always have the same values, whereas in the laboratory frame v_3 (and also v_4) has a different magnitude for each direction. This does not mean, however, that it is *always* desirable or necessary to go into the CM frame. For example, one may wish to answer the question: what is the speed of a neutron emitted at some given direction θ_4 in the laboratory with respect to the initial direction of a deuteron beam? In such a case it is easiest to work directly from the equations for energies and momenta as measured in the laboratory:

$$K_1 + Q = K_3 + K_4$$
$$p_1 = p_3 + p_4 \tag{5.24}$$

(In the first equation Q represents the amount by which K_{final} differs from

K_{initial}, so that in the example we have just considered we have $Q = +3.27$ MeV.)

Note that eqn (5.24) represents three independent equations (one for kinetic energy and two for momentum—treating this as a two-dimensional problem). In the final state there are four unknowns: a magnitude and a direction for each of the vectors v_3 and v_4. The situation is indeterminate unless we put in one more piece of information, as for example the direction of one of the particles. Q is here taken to be a known quantity (as are K_1 and p_1), but one could *deduce* it from a complete set of measurements of v_3 and v_4.

Interacting particles subject to external forces

Having discussed the conservation of linear momentum, and the general relation of force to rate of change of momentum, we can now consider the motion of a system of interacting particles that are *not* free of external influence. This is, of course, a very important extension of our ideas, because in practice a system is never completely isolated from its surroundings.

It will suffice to look at a two-particle system. The extension to any number of particles is quite simple but will be deferred to Chapter 10. Let m_1 and m_2 be the masses of the two particles, F_1 and F_2 the *external* forces acting on them, and f_{21} and f_{12} the internal interaction forces—f_{21} the force exerted on particle 1 by particle 2, and f_{12} the force exerted on particle 2 by particle 1.

Newton's second law of motion applied to the particles individually states that

$$F_1 + f_{21} = m_1 \frac{dv_1}{dt}$$
$$F_2 + f_{12} = m_2 \frac{dv_2}{dt} \tag{5.25}$$

If the interactions are Newtonian, and we shall consider this to be the case, then the third law of Newton requires that

$$f_{12} + f_{21} = 0$$

Adding the two equations (vectorially), we then get

$$F_1 + F_2 = m_1 \frac{dv_1}{dt} + m_2 \frac{dv_2}{dt}$$

or

$$F_1 + F_2 = \frac{d}{dt}(m_1 v_1 + m_2 v_2) \tag{5.26}$$

in which the internal forces f_{12} and f_{21} have disappeared. This equation states that the resultant of all the external forces acting on the system equals the rate of change of the *total* vector momentum of the system.

We can express this result in another, very compact way by introducing the concept of the centre of mass of the system. If the individual positions of the two particles are given by the vectors r_1 and r_2, drawn from some origin, the position of the centre of mass is given by

$$r_c = \frac{m_1 r_1 + m_2 r_2}{m_1 + m_2} \tag{5.27}$$

This is simply a restatement of eqn (5.18) and corresponds to the vector velocity of the centre of mass as already defined by eqn (5.19), so that

$$m_1 v_1 + m_2 v_2 = M v_c$$

where $M (= m_1 + m_2)$ is the total mass of the system. Accordingly, we have

$$F_1 + F_2 = F = M \frac{dv_c}{dt} \tag{5.28}$$

and this proves the result we are after:

The motion of the centre of mass of a system of two particles is the same as the motion of a single particle of mass equal to the total mass of the system acted on by the resultant of all the external forces which act on the individual particles.

The implications of this result are significant. First, it suggests that a fundamental method for treating the motion of a system of particles is to analyse its motion as the combination of (1) the motion of its centre of mass and (2) the motions of the particles relative to their centre of mass. The latter motion, the internal motion of the system, is one of zero momentum, as we saw earlier. Furthermore, eqn (5.28) allows us to treat some aspects of the motions of extended objects by the laws of dynamics for a simple particle. In particular, when an extended object moves in translation, i.e. when there is no motion of any particle in the object relative to the centre of mass, eqn (5.28) tells the whole story. We shall return to such questions in Chapter 10.

Incidentally, eqns (5.26) and (5.28) also provide a basis for a criterion as to whether or not a system of colliding particles is effectively isolated. The conservation of momentum in a collision process holds good only to the extent that the effect of any external forces can be ignored. If external forces are indeed present, the duration Δt of the collision must be so short that the product $F \Delta t$ is negligible. A different way of stating this same condition is that the forces of interaction between the colliding particles must be much greater than any external forces which may be acting.

The neutrino

Any conservation law or conservation principle in physics is provisional, but if, in the face of apparent failure, it is finally vindicated, then its status may be greatly strengthened. The most dramatic success of the conservation laws of dynamics took place in connection with the neutrino—that elusive, neutral particle emitted in the process of radioactive beta decay. The prediction of its existence stemmed from an apparent non-conservation of energy and angular momentum, but perhaps the most beautiful and direct dynamical evidence for it is furnished by the apparent non-conservation of linear momentum.

The situation can be simply stated as follows: it is known that the process of beta decay involves the ejection of an electron from a nucleus, as a result of which the nuclear charge goes up by one unit (if the electron is an ordinary negative electron). If no other particles were involved, the process could be written

$$A \rightarrow B + e^-$$

where A is the initial nucleus and B the final nucleus. If A were effectively isolated, and initially stationary, our belief in linear momentum conservation would lead us to predict that, whatever the direction (or energy) of the ejected electron, the nucleus B would inevitably recoil in the opposite direction. Any departure from this, regardless of all other details, would demand the involvement of another particle.

Figure 5.10 shows a cloud-chamber photograph of the beta decay of helium 6. The decay takes place at the position of the sharp knee near the top of the picture. The short stubby track pointing in a 'northwesterly' direction is the recoiling nucleus of lithium 6; the other track is the electron. There *must*

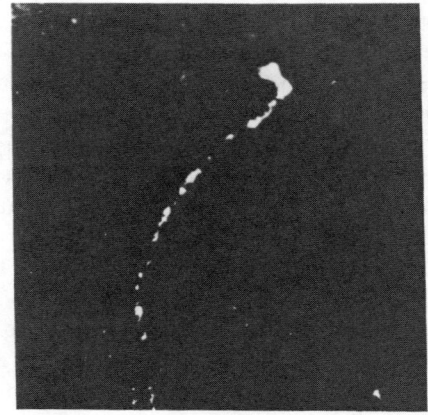

Fig. 5.10 Evidence for the neutrino. The visible tracks of the electron and the recoiling lithium 6 nucleus in the beta decay of helium 6 in a cloud-chamber are not collinear. [From J. Csikai and A. Szalay, *Soviet Physics JETP*, **8**, 749 (1959).]

be another particle—the neutrino—if the final momentum vectors are to add up to have the same resultant—i.e. zero—as the initially stationary ^{6}He nucleus. It fails to reveal itself because its lack of charge, or of almost any other interaction, allows it to escape unnoticed—so readily, in fact, that the chance would be only about 1 in 10^{12} of its interacting with any matter in passing right through the earth. (Nevertheless, neutrinos *have* been directly detected through their interactions within large masses of material.)

Problems

5.1 A particle of mass m, travelling with velocity v_0, makes a completely inelastic collision with an initially stationary particle of mass M. Make a graph of the final velocity v as a function of the ratio m/M from $m/M = 0$ to $m/M = 10$.

5.2 A particle of mass m_0, travelling at speed v_0, strikes a stationary particle of mass $2m_0$. As a result, the particle of mass m_0 is deflected through 45° and has a final speed of $v_0/2$. Find the speed and direction of the particle of mass $2m_0$ after this collision. Was kinetic energy conserved?

5.3 Find the average recoil force on a machine gun firing 240 rounds (shots) per minute, if the mass of each bullet is 10 g and the muzzle velocity is 900 m/s.

5.4 A 'standard fire stream' employed by a city fire brigade delivers 250 gallons of water per minute and can attain a height of 70 ft on a building whose base is 63 ft from the nozzle. Neglecting air resistance:
 (a) What is the nozzle velocity of the stream?
 (b) If directed horizontally against a vertical wall, what force would the stream exert? (Assume that the water spreads out over the surface of the wall without any 'rebound, so that the collision is effectively inelastic.)

5.5 A helicopter has a total mass M. Its main rotor blade sweeps out a circle of radius R, and air over this whole circular area is pulled in from above the rotor and driven vertically downward with a speed v_0. The density of air is ρ.
 (a) If the helicopter hovers at some fixed height, what must be the value of v_0?
 (b) One of the largest helicopters of the type described above weighs about 10 tons and has $R \approx 10$ m. What is v_0 for hovering in this case? Take $\rho = 1.3$ kg/m^3.

5.6 A rocket of initial mass M_0 ejects its burnt fuel at a constant rate $|dM/dt| = \mu$ and at a speed v_0 relative to the rocket.
 (a) Calculate the initial acceleration of the rocket if it starts vertically upward from its launch pad.
 (b) If $v_0 = 2000$ m/s, how many kilograms of fuel must be ejected per second to give such a rocket, of mass 1000 tons, an initial upward acceleration equal to 0.5 g?

5.7 This rather complicated problem is designed to illustrate the advantage that can be obtained by the use of multiple-stage instead of single-stage rockets as launching vehicles. Suppose that the payload (e.g. a space capsule) has mass m and is mounted on a two-stage rocket (see the figure). The *total* mass—both rockets fully fuelled, plus the payload—is Nm. The mass of the second-stage rocket plus the payload, after first-stage burnout and separation, is nm. In each stage the ratio of burnout mass (casing) to initial mass (casing plus fuel) is r, and the exhaust speed is v_0.

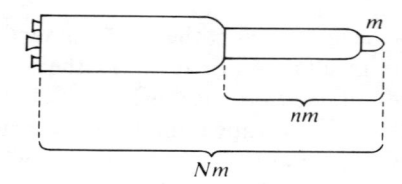

(a) Show that the velocity v_1 gained from first-stage burn, starting from rest (and ignoring gravity), is given by

$$v_1 = v_0 \ln \left[\frac{N}{rN + n(1-r)} \right]$$

(b) Obtain a corresponding expression for the additional velocity, v_2, gained from the second-stage burn.

(c) Adding v_1 and v_2, you have the payload velocity v in terms of N, n and r. Taking N and r as constants, find the value of n for which v is a maximum.

(d) Show that the condition for v to be a maximum corresponds to having equal gains of velocity in the two stages. Find the maximum value of v, and verify that it makes sense for the limiting cases described by $r = 0$ and $r = 1$.

(e) Find an expression for the payload velocity of a single-stage rocket with the same values of N, r and v_0.

(f) Suppose that it is desired to obtain a payload velocity of 10 km/s, using rockets for which $v_0 = 2.5$ km/s and $r = 0.1$. Show that the job can be done with a two-stage rocket but is impossible, however large the value of N, with a single-stage rocket.

(g) If you are ambitious, try extending the analysis to an arbitrary number of stages. It is possible to show that once again the greatest payload velocity for a given total initial mass is obtained if the stages are so designed that the velocity increment contributed by each stage is the same.

5.8 A block of mass m, initially at rest on a frictionless surface, is bombarded by a succession of particles each of mass δm ($\ll m$) and of initial speed v_0 in the positive x direction. The collisions are perfectly elastic and each particle bounces back in the negative x direction. Show that the speed acquired by the block after the nth particle has struck it is given very nearly by $v = v_0(1 - e^{-\alpha n})$, where $\alpha = 2\delta m/m$. Consider the validity of this result for $\alpha n \ll 1$ as well as for $\alpha n \to \infty$.

5.9 Newton calculated the resistive force for an object travelling through a fluid by supposing that the particles of the fluid (supposedly initially stationary) rebounded elastically when struck by the object.

(a) On this model, the resistive force would vary as some power, n, of the speed v of the object. What is the value of n?

(b) Suppose that a flat-ended object of cross-sectional area A is moving at speed v through a fluid of density ρ. By picturing the fluid as composed of n particles, each of mass m, per unit volume (such that $nm = \rho$), obtain an explicit expression for the resistive force if each particle that is struck by the object recoils elastically from it.

(c) If the object, instead of being flat-ended, were a massive sphere of radius r, travelling at speed v through a medium of density ρ, what would the magnitude of the resistive force be? The whole calculation can be carried out from the standpoint of a frame attached to the sphere, so that the fluid particles approach it with the velocity $-v$. Assume that in this frame the fluid particles are reflected as by a mirror—angle of reflection equals angle of incidence (see the figure). You must consider the surface of the sphere as divided up into circular zones corresponding to small angular increments $d\theta$ at the various possible values of θ.

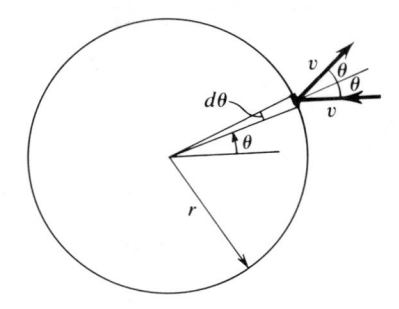

5.10   A particle of mass m_1 and initial velocity u_1 strikes a stationary particle of mass m_2. The collision is perfectly elastic. It is observed that after the collision the particles have equal and opposite velocities. Find
 (a) The ratio m_2/m_1.
 (b) The velocity of the centre of mass.
 (c) The total kinetic energy of the two particles in the centre of mass frame expressed as a fraction of $\frac{1}{2}m_1u_1^2$.
 (d) The final kinetic energy of m_1 in the centre of mass frame.

5.11   A mass m_1 collides with a mass m_2. Define relative velocity as the velocity of m_1 observed in the rest frame of m_2. Show the equivalence of the following two statements:
(1) Total kinetic energy is conserved.
(2) The magnitude of the relative velocity is unchanged.
(It is suggested that you solve the problem for a one-dimensional collision, at least in the first instance.)

5.12   A collision apparatus is made of a set of n graded masses suspended so that they are in a horizontal line and not quite in contact with one another (see the figure). The first mass is fm_0, the second is f^2m_0, the third f^3m_0, and so on, so that the last mass is f^nm_0. The first mass is struck by a particle of mass m_0 travelling at a speed v_0. This produces a succession of collisions along the line of masses.

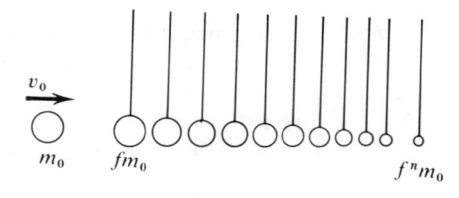

(a) Assuming that all the collisions are perfectly elastic, show that the last mass flies off with a speed v_n given by

$$v_n = \left(\frac{2}{1+f}\right)^n v_0$$

(b) Hence show that, if f is close to unity, so that it can be written as $1 \pm \epsilon$ (with $\epsilon \ll 1$), this system can be used to transfer virtually all the kinetic energy of the incident mass to the last one, even for large n.

(c) For $f = 0.9$, $n = 20$, calculate the mass, speed and kinetic energy of the last mass in the line in terms of the mass, speed and kinetic energy of the incident particle. Compare this with the result of a direct collision between the incident mass and the last mass in the line.

5.13   In a certain road accident (this is based on an actual case) a car of mass 2000 kg, travelling south, collided in the middle of an intersection with a truck of mass 6000 kg, travelling west. The vehicles locked and skidded off the road along a line pointing almost exactly southwest. A witness claimed that the truck had entered the intersection at 50 mph.

(a) Do you believe the witness?

(b) Whether or not you believe him, what fraction of the total initial kinetic energy was converted into other forms of energy by the collision?

5.14   A nucleus A of mass $2m$, travelling with a velocity u, collides with a stationary nucleus of mass $10m$. The collision results in a change of the total kinetic energy. After collision the nucleus A is observed to be travelling with speed v_1 at 90° to its original direction of motion, and B is travelling with speed v_2 at angle θ (sin $\theta = 3/5$) to the original direction of motion of A.

(a) What are the magnitudes of v_1 and v_2?

(b) What fraction of the initial kinetic energy is gained or lost as a result of the interaction?

5.15   In a historic piece of research, James Chadwick in 1932 obtained a value for the mass of the neutron by studying elastic collisions of fast neutrons with nuclei of hydrogen and nitrogen. He found that the maximum recoil velocity of hydrogen nuclei (initially stationary) was 3.3×10^7 m/s, and that the maximum recoil velocity of nitrogen 14 nuclei was 4.7×10^6 m/s with an uncertainty of $\pm 10\%$. What does this tell you about

(a) The mass of a neutron?

(b) The initial velocity of the neutrons used?

(Take the uncertainty of the nitrogen measurement into account. Take the mass of an H nucleus as 1 amu and the mass of a nitrogen 14 nucleus as 14 amu.)

5.16   A nuclear reactor has a moderator of graphite. The carbon nuclei in the atoms of this crystal lattice can be regarded as effectively free to recoil if struck by fast neutrons, although they cannot be knocked out of place by thermal neutrons. A fast neutron, of kinetic energy 1 MeV, collides elastically with a stationary carbon 12 nucleus.

(a) What is the initial speed of each particle in the centre of mass frame?

(b) As measured in the centre-of-mass frame, the velocity of the carbon nucleus is turned through 135° by the collision. What are the final speed and direction of the neutron as measured in the laboratory frame?

(c) About how many elastic collisions, involving random changes of direction, must a neutron make with carbon nuclei if its kinetic energy is to be reduced from 1 MeV to 1 keV? Assume that the mean energy loss is midway between maximum and minimum values.

5.17 (a) A moving particle of mass M collides perfectly elastically with a stationary particle of mass $m < M$. Show that the maximum possible angle through which the incident particle can be deflected is $\sin^{-1}(m/M)$. (Use of the vector diagrams of the collision in laboratory and CM systems will be found helpful.)

(b) A particle of mass m collides perfectly elastically with a stationary particle of mass $M > m$. The incident particle is deflected through 90°. At what angle θ with the original direction of m does the more massive particle recoil?

6 Energy conservation in dynamics; vibrational motions

Introduction

Of all the physical concepts, that of energy is perhaps the most far-reaching. Everyone, whether a scientist or not, has an awareness of energy and what it means. Energy is what we have to pay for in order to get things done. The word itself may remain in the background, but we recognize that each gallon of gasoline, each Btu of heating gas, each kilowatt-hour of electricity, each car battery, each calorie of food value, represents, in one way or another, the wherewithal for doing what we call *work*. We do not think in terms of paying for force, or acceleration, or momentum. *Energy* is the universal currency that exists in apparently countless denominations; and physical processes represent a conversion from one denomination to another.

The above remarks do not really *define* energy. No matter. It is worth recalling the opinion that the distinguished Dutch physicist H. A. Kramers once expressed: 'The most important and most fruitful concepts are those to which it is impossible to attach a well-defined meaning.' The clue to the immense value of energy as a concept lies in its transformation. It is *conserved*—that is the point. Although we may not be able to define energy in general, that does not mean that it is only a vague, qualitative idea. We have set up quantitative measures of various specific *kinds* of energy: gravitational, electrical, magnetic, elastic, kinetic, and so on. And whenever a situation has arisen in which it seemed that energy had disappeared, it has always been possible to recognize and define a new form of energy that permits us to save the conservation law. And conservation laws represent one of the physicist's most powerful tools for organizing his description of nature.

In this book we shall be dealing only with the two main categories of energy that are relevant to classical mechanics—the kinetic energy associated with the bodily motion of objects, and the potential energy associated with elastic deformations, gravitational attractions, electrical interactions, and the like. If energy should be transferred from one or another of these forms into chemical energy, radiation, or the random molecular and atomic motion we call heat, then from the standpoint of mechanics it is lost. This is a very important feature, because it means that, if we restrict our attention to the purely mechanical aspects, the conservation of energy is *not* binding; it must not be blindly assumed. Nevertheless, as we shall see, there are many physical situations for which the conservation of the total mechanical energy holds good, and in such contexts it is of enormous value in the analysis of physical problems.

It is interesting historically to note that in pursuing the subject of energy we are temporarily parting company with Newton, although not with what we may properly call Newtonian mechanics. In the whole of the *Principia*, with its awe-inspiring elucidation of the dynamics of the universe, the concept of energy is never once used or even referred to! For Newton, $F = ma$ was enough. But we shall see how the energy concept, although rooted in $F = ma$, has its own special contributions to make.

Work, energy, and power

This chapter is chiefly concerned with developing some general dynamical methods based upon the concepts of work and mechanical energy.

We must first emphasize a most important property of work and energy: they are *scalar* quantities. An object moving vertically with a speed v has exactly the same kinetic energy as if it were travelling horizontally at this same speed, although its vector momentum would be quite different. This scalar property of energy will be exploited repeatedly in our future work.

The practical application of methods using the concepts of work and energy will involve numerical measures of these quantities. The SI unit of work or energy is the joule:

$$1 \text{ J} = 1 \text{ N m} = 1 \text{ kg m}^2 \text{ s}^{-2}$$

Before going any further, we shall introduce a seeming diversion—the concept of *power*, defined as the *rate* of doing work (W);

$$\text{power} = \frac{dW}{dt}$$

Power is essentially a practical engineering concept; we shall not be using it in our development of the principles of dynamics. But one of our accepted measures of work is often expressed in terms of a unit of power—the *watt*. In terms of mechanical quantities,

$$1 \text{ W} = 1 \text{ J s}^{-1}$$

so that $1 \text{ W s} = 1 \text{ J}$.

The most familiar use of the watt is, of course, electrical, through the relation watts = volts × amperes, but it is important to realize that it is not specifically an electrical quantity. A convenient energy unit for domestic purposes (especially one's electricity bill) is the kilowatt-hour (kWh):

1 kWh = 3.6×10^6 J

In chemical and thermal calculations the standard unit is the Calorie, defined as the amount of energy required to raise 1 kg of water from 15 to 16°C:

1 Cal = 4.2×10^3 J

In atomic and nuclear physics, energy measurements are usually expressed in terms of the electron volt (eV) or its related units keV (10^3), MeV (10^6) and GeV (10^9). The electron volt is the amount of energy required to raise one elementary charge through 1 V of electric potential difference:

1 eV = 1.6×10^{-19} J

Finally, as Einstein first suggested and as innumerable observations have confirmed, there is an equivalence between what we customarily call mass and what we customarily call energy. In classical mechanics these are treated as entirely separate concepts, but it is perhaps worth quoting this equivalence here, so that we have our selection of energy measures all in one place:

1 kg of mass is equivalent to 9×10^{16} J

Energy conservation in one dimension

The equation $E = K + V(x)$, where E is the total mechanical energy, K is the kinetic energy and $V(x)$ is the potential energy, is a compact statement of the conservation of total mechanical energy in any one-dimensional problem for which the force acting on an object, due to its environment, depends only on the object's position. To derive this energy-conservation result, suppose that the environment supplies a force $F(x)$ that varies with position x in any arbitrary way—but has a unique value at any given value of x. We know that the change of kinetic energy of the object is directly related to the work done by this force, for we have

$$W = \int_{x_1}^{x_2} F\, dx = m \int_{x_1}^{x_2} \frac{dv}{dt}\, dx$$

ENERGY CONSERVATION IN ONE DIMENSION

But
$$\frac{dv}{dt} dx = \frac{dx}{dt} dv = v\, dv \tag{6.1}$$

Thus
$$\int_{x_1}^{x_2} F\, dx = m \int_{v_1}^{v_2} v\, dv = \tfrac{1}{2}m(v_2^2 - v_1^2) = K_2 - K_1$$

In order to cast this into the form of a conservation statement, we introduce an arbitary reference point x_0 and express the work integral as follows:

$$\int_{x_1}^{x_2} F(x)\, dx = \int_{x_0}^{x_2} F(x)\, dx - \int_{x_0}^{x_1} F(x)\, dx \tag{6.2}$$

Using this result, we can rewrite eqn (6.1) in the form

$$K_2 + \left[-\int_{x_0}^{x_2} F(x)\, dx\right] = K_1 + \left[-\int_{x_0}^{x_1} F(x)\, dx\right]$$

The left-hand side involves only the reference point and the position x_2, the right-hand side only the reference point and the position x_1. By *defining* the potential energy $V(x)$ at any point x by the equation

$$V(x) - V(x_0) = -\int_{x_0}^{x} F(x)\, dx \tag{6.3}$$

we then have the energy-conservation statement $K_2 + V_2 = K_1 + V_1$. Notice especially the minus sign on the right of eqn (6.3). The potential energy at a point, relative to the reference point, is always defined as the *negative* of the work done by the force as the object moves from the reference point to the point considered. The value of $V(x_0)$, the potential energy at the reference point itself, can be set equal to zero if we please, because in any actual problem we are concerned only with *differences* of potential energy between one point and another, and the associated changes of kinetic energy.

A simple and familiar example of the use of these ideas is, of course, the analysis of vertical motion under gravity (ignoring air resistance). If we take a y axis, positive upwards, the force F_g acting on a body is $-mg$, and eqn (6.3) gives us

$$V(y) - V(y_0) = -\int_{y_0}^{y} (-mg)\, dy = mg(y - y_0)$$

Thus, for conservation of mechanical energy between points at heights y_1 and y_2, measured from the arbitrary reference level y_0, we have

$$\tfrac{1}{2}mv_2^2 + mg(y_2 - y_0) = \tfrac{1}{2}mv_1^2 + mg(y_1 - y_0)$$

In obtaining eqn (6.3) we took the force as the primary quantity and the potential energy as the secondary one. Increasingly, however, as one goes deeper into mechanics, potential energy takes over the primary role, and force becomes the derived quantity—literally so, indeed, because by differentiation of both sides of eqn (6.3) with respect to x, we obtain

$$F(x) = - \frac{dV}{dx} \qquad (6.4)$$

This inversion of the roles is not just a formal one (although it does prove to be valuable theoretically) for there are many physical situations in which one's only measurements are of energy differences between two very distinct states, and in which one has no direct knowledge of the forces acting. The electronic work function of a metal, for example, and the dissociation energy of a molecule, represent the only directly observable quantities in these processes of removing a particle to infinity from some initial location. How the force varies from point to point may not be known well—perhaps not at all.

We shall often be spelling out the kinetic energy K in terms of m and v, so that the equation of energy conservation in one dimension is written as follows;

$$\tfrac{1}{2}mv^2 + V(x) = E \qquad (6.5)$$

Suppose we choose any particular value of x. Then eqn (6.5) becomes a quadratic equation for v with two equal and opposite roots:

$$v(x) = \pm \left(\frac{2[E - V(x)]}{m} \right)^{1/2} \qquad (6.6)$$

This is the expression of a familiar result, which we can discuss in terms of motion under gravity. If an object were observed to pass a certain point, travelling vertically upward with speed v, then (to the approximation that air resistance could be ignored) it would be observed, a little later, to pass the same point, travelling *downward* at the same speed v. The direction of the velocity has been reversed, but there has been no loss or gain of kinetic energy. In such a case the force is said to be *conservative*. We can see from eqn (6.6) that this result will hold as long as $V(x)$ is a unique function of x. It means that a particle, after passing through any given point at any speed, will be found to have the same kinetic energy every time it passes through that point again.

Under what conditions does the force have this conservative property? It will certainly *not* be conservative if $F(x)$ depends on the direction of motion of the object to which it is applied. Consider, for example, the addition of a resistive force to the gravitational force in the vertical motion of an object. As the object goes upward through a certain point, the net force on it (downward) is greater than F_g. After it reaches its highest point and begins moving down, the net force on it (again downward) is less than F_g. Hence the net negative work done on it as it rises is numerically greater than the net *positive* work as it descends. Thus on balance negative work has been done and the kinetic energy as the object passes back through the designated point is less than initially. The crucial feature is, indeed, that the net work done by $F(x)$ should be *zero* over any journey beginning and ending at any given value of x; only if this condition is satisfied can one define a potential-energy function. It might seem that an equivalent condition is that F be a unique function of position. In one-dimensional situations this is correct. In two- and three-dimensional situations, however, as we shall see later (Chapter 7), the condition that F be a single-valued function of r is necessary but not sufficient. The condition of zero net work over any *closed* path defines a conservative force in *all* circumstances and should be remembered as a basic definition.

The energy method for one-dimensional motions

The use of energy diagrams provides an excellent way of obtaining a complete, although perhaps qualitative, picture of possible motions in a one-dimensional system.

The general scheme is as follows: we plot $V(x)$ as a function of x, and on the same plot draw horizontal lines corresponding to different total energies. Figure 6.1 shows such a potential-energy curve and several values of the total energy E. Frequently the information in such a diagram suffices for obtaining physical insight into situations for which analytic solutions are complicated or even unobtainable. In fact, even when analytic solutions *can* be obtained in terms of unfamiliar functions, they often are of little help in revealing the essential physical characteristics of the motion.

The kinetic energy K of a particle is equal to $(E - V)$, i.e. to the vertical distance from one of the lines of constant energy to the curve $V(x)$ at any point x. For a low energy E_1, $V(x)$ is greater than E for all values of x; this would simply imply a negative value of K and hence an imaginary value of v. Such a situation has no place in classical mechanics—although it must not be discarded so lightly when one comes to atomic and nuclear systems requiring the use of quantum mechanics. For a higher total energy E_2, the motion can occur in two regions, between x_3 and x_4 or between x_7 and x_8. These represent two quite separate situations, because a particle cannot escape from one region to the other as long as its energyy is held at the value E_2. One way of seeing this is, of course, in terms of the impossible negative value of K between x_4 and x_7. But there is another way which is valuable as an example of how one 'reads' such an energy diagram.

Suppose that our particle, with total energy E_2, is at the point $x = x_3$ at some instant. Its potential energy $V(x_3)$ at this point is equal to the total energy, for this is where the curve of $V(x)$ and the line $E = E_2$ intersect. Thus

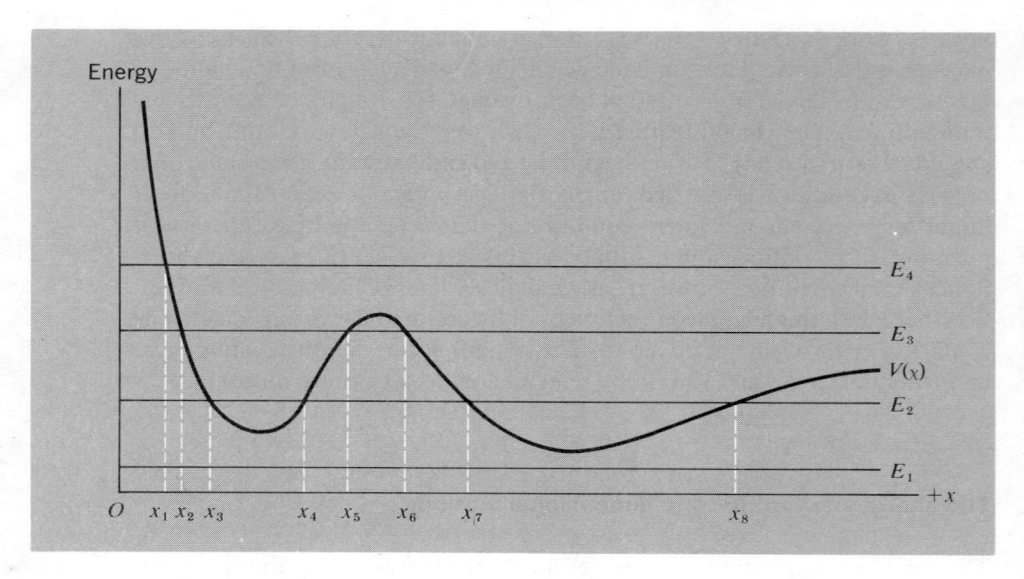

Fig. 6.1 Hypothetical energy diagram for a one-dimensional system.

the particle has zero kinetic energy and hence is instantaneously at rest. However, there *is* a force on it:

$$F(x) = -\frac{dV}{dx}$$

At $x = x_3$, dV/dx is negative and hence $F(x_3)$ is positive, i.e. in the $+x$ direction. Thus the particle accelerates to the right. The force on it, and hence its acceleration, decreases as the slope of the $V(x)$ curve decreases, falling to zero at the value of x at which $V(x)$ is minimum. At this point the speed of the particle is a maximum, and as it moves further in the $+x$ direction (with $dV/dx > 0$) it now experiences a force in the $-x$ direction. The diagram displays all this information before us, and shows the kinetic energy $E_2 - V$ continuing to decrease as the particle approaches x_4. Finally, at x_4 itself, the velocity has fallen to zero—but there is still a force acting in the $-x$ direction. What happens? The particle picks up speed again, travelling to the left, until it reaches x_3 with its velocity reduced to zero. This whole cycle of motion will continue to repeat itself indefinitely as long as the total energy does not decrease. We have, in short, a *periodic* motion, of which we can discern many of the principal features without solving a single equation—just by seeing what the energy diagram has to tell us. The motion between x_7 and x_8 is likewise periodic.

We can dispose of the other possibilities more briefly, having indicated the method. For a still higher energy E_3, two kinds of motion are possible; either a periodic motion between x_2 and x_5, or the unbounded motion of a particle coming in from large values of x, speeding up as it passes x_8, then slowing down and reversing its direction of motion at x_6, moving off to the right and duplicating all the changes of speed on the way in. Finally, for the still larger energy E_4, the only possible motion is unbounded; a particle coming in from large values of x, speeding up, slowing down, speeding up, slowing down again, and reversing its direction of motion at x_1, after which it proceeds inexorably in the direction of ever-increasing x. For each of these motions, the speed at any point can be obtained graphically by measuring the vertical distance from the appropriate line of constant energy to the corresponding point on the potential-energy curve.

Caution: The curve of $V(x)$ in Fig. 6.1 is almost *too* graphic. It tends to conjure up a picture of a particle sliding down the slopes and up the peak like a roller coaster. Do not forget that it is a *one*-dimensional motion that is the subject of the analysis. The vertical scale is energy, and has nothing necessarily to do with altitude.

After this general introduction, let us consider some specific examples of one-dimensional motions as analysed by the energy method.

Some examples of the energy method

Bouncing ball

Suppose that a ball, moving along a vertical line, bounces repeatedly on a horizontal floor. Let us first imagine that there is no dissipation (loss) of mechanical energy, so that this energy remains constant at some value E.

We shall use y to denote the position of the centre of gravity (CG) of the ball, and take $y = 0$ to be defined by the first contact of the ball with the floor. We shall take this configuration to correspond to $V = 0$. For $y > 0$ the potential energy of the ball is given by

$$V(y) = mgy \quad (y > 0)$$

Now $y = 0$ does not, in any real physical situation, represent the lowest point reached by the CG of the ball. The floor does not exert any force on the ball until it (the floor) has been compressed slightly. An equivalent remark can be made about the ball. Thus the ball certainly moves into the region $y < 0$. As it does this, however, it experiences a positive (upward) force that increases extremely rapidly as y becomes more negative, and completely overwhelms the (negative) gravitational force that exists, of course, at all values of y. This large positive force gives rise to a very steep increase of $V(y)$ for $y < 0$ (see Fig. 6.2a).

For any given value of the total energy, therefore, the ball oscillates between positions y_1 and y_2 as shown in the figure. The motion is

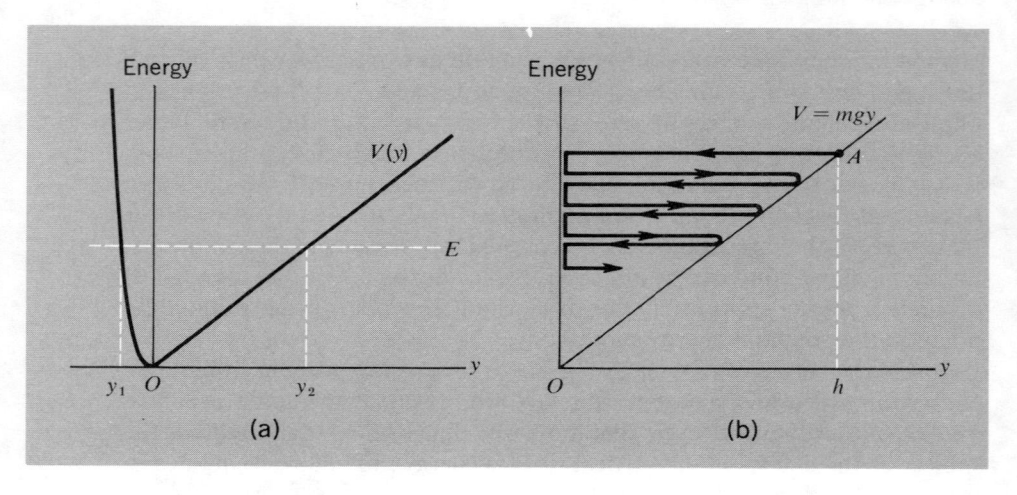

Fig. 6.2 (a) Energy diagram for a ball bouncing vertically. (b) Idealization of (a) to represent a situation in which the impact at $y = 0$ is completely rigid but in which there is some dissipation of energy at each bounce.

periodic—that is, there is some well-defined time T between successive passages *in the same direction* of the ball through any given point.

Now in practice y_1 may be numerically very small compared to y_2. For instance, if a steel ball bounces on a glass plate, we might easily have y_1 of the order of 0.01 cm and y_2 of the order of 10 cm. Thus for many purposes we can approximate the plot of $V(y)$ against y for $y < 0$ by a vertical line, coinciding with the energy axis of Fig. 6.2(a). This represents the physically unreal property of perfect rigidity—an arbitrarily large force is called into play for zero deformation. However, if we can justifiably use this approximation, then we have a simple quantitative description of the situation. The motion is confined to $y \geq 0$ and is defined by

$$\tfrac{1}{2}mv_y^2 + mgy = E \quad (y > 0) \tag{6.7}$$

The maximum height h is, of course, defined by putting $v_y = 0$:

$$h = \frac{E}{mg} \tag{6.8}$$

To find the period T of the motion we can calculate the time for the ball to travel from $y = 0$ to $y = h$ and then double it. That is, we have the following relation:

$$\frac{T}{2} = \int_{y=0}^{h} \frac{dy}{v_y} \tag{6.9}$$

because the elementary contribution dt to the time of flight is equal to dy

SOME EXAMPLES OF THE ENERGY METHOD

divided by the speed v_y at any given point.

Now from eqn (6.7) we have

$$v_y = \pm \left[\frac{2(E - mgy)}{m}\right]^{1/2}$$

Taking the positive root, to correspond to upward motion, we have, from eqn (6.9)

$$T = 2 \int_0^h \left[\frac{m}{2(E - mgy)}\right]^{1/2} dy$$

$$= \sqrt{\frac{2}{g}} \int_0^h \frac{dy}{[(E/mg) - y]^{1/2}}$$

We can simplify this by noting that E/mg is just the maximum height h (eqn 6.8). Thus we have

$$T = \sqrt{\frac{2}{g}} \int_0^h \frac{dy}{(h - y)^{1/2}}$$

This is an elementary integral (change the variable to $w = h - y$) yielding the result

$$T = 2\sqrt{\frac{2h}{g}}$$

You will, of course, recognize the correctness of this result from the simple kinematic problem of an object falling with constant acceleration, and could reasonably object that this is a case in which we have used a sledgehammer to kill a fly, as the saying goes. But it is the *method* that you should focus attention on, and perhaps the use of a familiar example will facilitate this. It should not be forgotten that most motions involve *varying* accelerations, so that the standard kinematic formulas for motion with constant acceleration do not apply. But eqn (6.9), in which v is defined at any point by the energy equation, can be used for any one-dimensional motion and can be integrated numerically if necessary.

Before leaving this example, let us use it to illustrate one other instructive feature of the energy diagram. We know that the total mechanical energy of a bouncing ball does not in fact stay constant but decreases quite rapidly. Although there is little loss of energy while the ball is in flight, there is a substantial loss at each bounce. Figure 6.2(b) shows how this behaviour can be displayed on the energy diagram. Starting at the point A, the history of the whole motion is obtained by following the arrows. The successive decreases in the maximum height of bounce, and the inevitable death of the motion at $y = 0$, are quite apparent by inspection of the figure.

Mass on a spring

There are very many physical systems—not just ordinary mechanical systems, but also atomic systems, and even electrical ones—that can be analysed by analogy with a mass on a spring. The reason for this lies in two features:

1. A mass typifies the property of inertia, which has its analogues in diverse systems and which acts as a repository of kinetic energy.
2. A spring represents a means of storing potential energy according to a particular law of force that has its counterparts in all kinds of physical interactions.

This basic system, the *harmonic oscillator*, is very usefully analysed from the standpoint of energy conservation because its description in terms of energy opens the way to a wide range of situations. Not only does it provide a pattern for the handling of more complex oscillatory problems in classical mechanics; it also supplies the foundation for formulating equivalent problems in quantum theory.

Our starting point will again be the restoring force of an ideal spring as described by Hooke's law:

$$F = -kx \tag{6.10}$$

where x is the position of the free end of the spring relative to its relaxed position, k the 'force constant' of the spring, measurable in N m^{-1}, and the negative sign gives the direction of the force, opposite to the displacement of the free end. No real spring obeys this law over more than a limited range. The properties of a real spring can be expressed by a graph such as Fig. 6.3(a), which represents the force $F(x)$ exerted *by the spring* as a function of its extension x. Within the linear range the potential energy stored in the spring is

$$V(x) = -\int_0^x F(x)\,dx = +\int_0^x kx\,dx = \frac{kx^2}{2} \tag{6.11}$$

where we have chosen $V = 0$ for $x = 0$, i.e. when the spring is relaxed. Figure 6.3(b) shows this potential energy plotted against x. Since the potential-energy change can be calculated as the work done by a force F_{ext} just sufficient to overcome the spring force itself, the increase of potential energy in the spring for any given increase of extension can be obtained as the area, between given limits, under a graph of F_{ext} against x (see Fig. 6.3c). F_{ext} can be measured as the force needed to maintain the spring at constant extension for various values of x. Outside the linear region (whose boundaries are indicated by dashed lines) $F(x)$ can be integrated graphically so as to obtain numerically the potential energy for an arbitrary displacement.

SOME EXAMPLES OF THE ENERGY METHOD 135

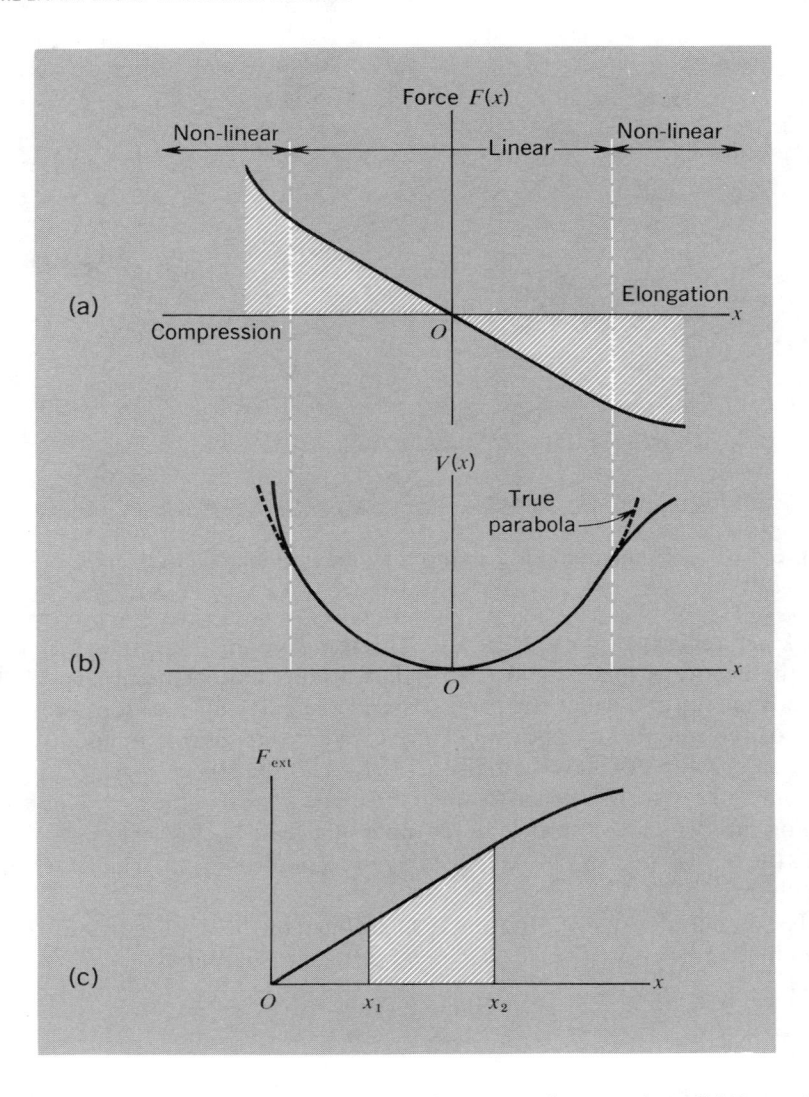

Fig. 6.3 (a) Restoring force versus displacement for a spring. (b) Potential-energy diagram associated with (a). (c) Graph of applied force versus extension in a static deformation of a spring.

Let us now consider in more detail the way in which the use of the energy diagram helps us to analyse the straight-line motion of a harmonic oscillator. In Fig. 6.4 are shown the potential energy $V = \frac{1}{2}kx^2$ plotted against x, and two different total energies, E_1 and E_2.

For a given energy E_1, as we have already discussed, the vertical distance from the horizontal line E_1 to the curve $V = \frac{1}{2}kx^2$ for any value of x is equal to

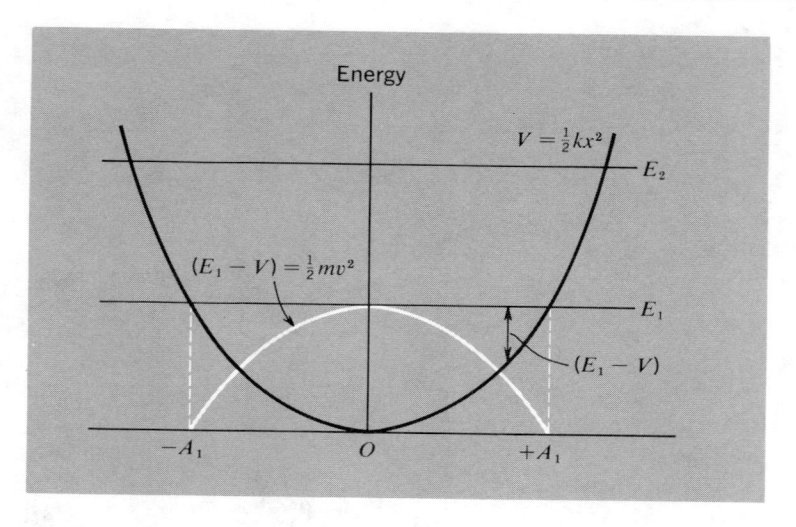

Fig. 6.4 Energy diagram for a spring that obeys Hooke's law.

the kinetic energy of the particle at x. This is maximum at $x = 0$; at this point all the energy is kinetic and the particle attains its maximum speed. The kinetic energy and hence the particle speed decreases for positions on either side of the equilibrium position O and is reduced to zero at the points $x = \pm A_1$. For values of x to the right of $+A_1$ or to the left of $-A_1$, $(E_1 - V)$ becomes negative; v^2 is negative and there exists no real value of v. This is the region into which the particle never moves (at least in classical mechanics); thus the positions $x = \pm A_1$ are *turning points* of the motion, which is clearly oscillatory.

The amplitude A_1 of the motion is determined by the total energy E_1. Since the kinetic energy $K_1 (= E_1 - V)$ is zero at $x = \pm A_1$, we have

$$\tfrac{1}{2}kA_1^2 = E_1$$

or

$$A_1 = \left(\frac{2E_1}{k}\right)^{1/2} \tag{6.12}$$

For a larger energy, E_2, the amplitude is larger in the ratio $(E_2/E_1)^{1/2}$, but the qualitative features of the motion are the same.

It is interesting to note that the general character of the motion as inferred from the energy diagram would be the same for any potential-energy curve that has a minimum at $x = 0$ and is symmetrical about the vertical axis through this point. All motions of this sort are periodic but differ one from the other in detail, e.g., the dependence of speed on position and the dependence of the period on amplitude. Suppose that the period of the motion is T.

The harmonic oscillator by the energy method

Then for any *symmetrical* potential-energy diagram, we can imagine this time divided up into four equal portions, any one of which contains the essential information about the motion. For suppose that, at $t = 0$, the particle is travelling through the point $x = 0$ in the positive x direction. Let its velocity at this instant be called v_m—it is the biggest velocity the particle will have during its motion. At $t = T/4$ the particle is at its maximum positive displacement ($x = +A$ in Fig. 6.4), and $v = 0$. It then retraces its steps, reaching $x = 0$ after a further time $T/4$ and passing through this point with $v = -v_m$. In two further intervals $T/4$ it goes to its extreme negative displacement ($x = -A$) and at $t = T$ is once again passing through the point $x = 0$ with $v = v_m$. This sequence will repeat itself indefinitely. Furthermore, knowing the symmetry of the problem, we could construct the complete graph of v against t from a detailed graph for the first quarter-period alone (see Fig. 6.5, in which the basic quarter-period is drawn with a heavier line than the rest).

In the next section we shall go beyond this rather general examination of the mass–spring system, including non-linear restoring forces, and shall redevelop the rest of the detailed results that apply to the ideal harmonic oscillator.

The harmonic oscillator by the energy method

We shall now return to the analysis of the oscillatory motion of an object attached to a spring that obeys Hooke's law. The basic energy equation for a mass on a spring with a restoring force proportional to displacement is

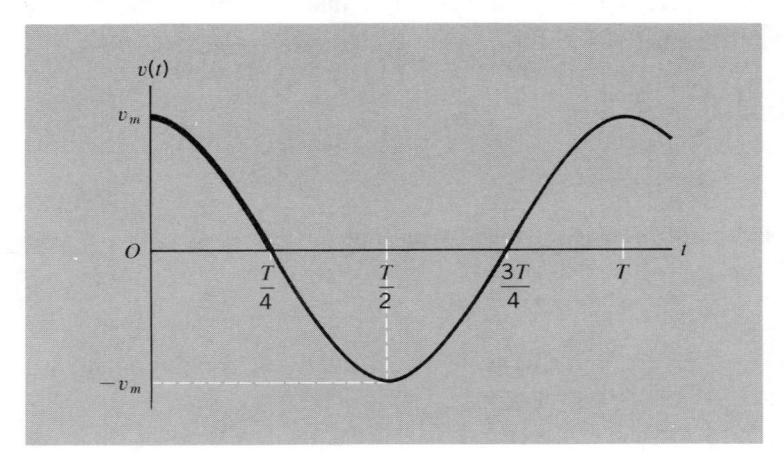

Fig. 6.5 Sinusoidal variation of velocity with time for a particle subjected to a restoring force proportional to displacement. The motion during the first quarter-period suffices to define the rest of the curve.

$$\tfrac{1}{2}mv^2 + \tfrac{1}{2}kx^2 = E \tag{6.13}$$

where E is some constant value of the total energy. Since $v = dx/dt$, this can be rewritten as

$$\tfrac{1}{2}m\left(\frac{dx}{dt}\right)^2 + \tfrac{1}{2}kx^2 = E \tag{6.14}$$

Equation (6.13) already gives us v as a function of x, but to have a full description of the motion we must solve eqn (6.14) so as to obtain x (and hence v) as functions of t.

We start out by dividing the equation throughout by E. Then we get

$$\frac{m}{2E}\left(\frac{dx}{dt}\right)^2 + \frac{k}{2E}x^2 = 1 \tag{6.15}$$

We notice that this is a sum of two terms involving the square of a variable (x) and the square of its derivative (with respect to t). The sum is equal to 1. Now we can relate this to a very familiar relationship involving trigonometric functions: If $s = \sin\varphi$, then

$$\frac{ds}{d\varphi} = \cos\varphi$$

and

$$\left(\frac{ds}{d\varphi}\right)^2 + s^2 = \cos^2\varphi + \sin^2\varphi = 1 \tag{6.16}$$

Equations (6.15) and (6.16) are exactly similar in form! We must be able to match them, term by term:

$$\frac{m}{2E}\left(\frac{dx}{dt}\right)^2 \equiv \left(\frac{ds}{d\varphi}\right)^2$$
$$\frac{k}{2E}x^2 \equiv s^2 \tag{6.17}$$

The second of these is satisfied by putting

$$x = \left(\frac{2E}{k}\right)^{1/2} s = \left(\frac{2E}{k}\right)^{1/2}\sin\varphi \tag{6.18}$$

What is φ? We can find it by evaluating dx/dt by differentiation of both sides of eqn (6.18) with respect to t:

$$\frac{dx}{dt} = \left(\frac{2E}{k}\right)^{1/2}\cos\varphi\,\frac{d\varphi}{dt}$$

But the first equation of (6.17) is satisfied by putting

THE HARMONIC OSCILLATOR BY THE ENERGY METHOD

$$\frac{dx}{dt} = \left(\frac{2E}{m}\right)^{1/2} \frac{ds}{d\varphi} = \left(\frac{2E}{m}\right)^{1/2} \cos\varphi$$

Comparing these two expressions for dx/dt we find the following condition on φ:

$$\frac{d\varphi}{dt} = \left(\frac{k}{m}\right)^{1/2} = \omega \tag{6.19}$$

where ω is an angular frequency (also called the circular frequency).

Integrating the last equation with respect to t we thus get

$$\varphi = \omega t + \varphi_0$$

where φ_0 is the initial phase. Substituting this expression for φ back into eqn (6.18) then gives us

$$x = \left(\frac{2E}{k}\right)^{1/2} \sin(\omega t + \varphi_0) \tag{6.20}$$

You will notice that since $(2E/k)^{1/2}$ is equal to the amplitude A of the motion, eqn (6.20) is the same as eqn (3.33a) which was developed from the different starting point of $F = ma$ (cf. pp. 62–65).

[If you have some prior knowledge of differential equations, you may regard our method of solution above as being rather cumbersome. You may prefer to proceed at once to the recognition that eqn (6.14) leads to the relationship

$$\left(\frac{dx}{dt}\right)^2 = \omega^2(A^2 - x^2)$$

and hence to the following solution by direct integration:

$$\frac{dx}{dt} = \omega(A^2 - x^2)^{1/2}$$

$$\omega \, dt = \frac{dx}{(A^2 - x^2)^{1/2}}$$

$$\omega t + \varphi_0 = \sin^{-1}\left(\frac{x}{A}\right)$$

and so $x = A \sin(\omega t + \varphi_0)$ as before.]

Equations (6.19) and (6.20) tell us something very remarkable indeed: *the period of a harmonic oscillator*, as typified by a mass on a spring, *is completely independent of the energy or amplitude of the motion*—a result that is not true of periodic oscillations under any other force law. The physical consequences of this are tremendously important. We depend heavily on the use of vibrating systems. If the frequency ν (defined as the number of complete oscillations per second, i.e. $1/T$ or $\omega/2\pi$) varied significantly with the amplitude for a given system, the situation would become vastly more complicated. Yet most vibrating systems behave, to some approximation, as harmonic oscillators with properties as described above. Let us see why.

Small oscillations in general

There are many situations in which an object is in what we call *stable equilibrium*. It is at rest at some point—under no net force—but if displaced in any direction it experiences a force tending to return it to its original position. Such a force, unless it has pathological properties (such as a discontinuous jump in value for some negligible displacement) will have the kind of variation with position shown in Fig. 6.6(a). The normal resting position is marked as x_0. This force function can be integrated to give the potential-energy graph of Fig. 6.6(b). One can then form a mental picture of the object sitting at the bottom, as it were, of the potential-energy hollow, the minimum of which is at $x = x_0$.

Now we can fit any curve with a polynomial expansion. Let us do this with the potential-energy function—but let us do it with reference to a new origin chosen at the point x_0, by putting

$$x = x_0 + s$$

where s is the displacement from equilibrium. The potential-energy curve, now appearing as in Fig. 6.7, can be fitted by the following expansion:

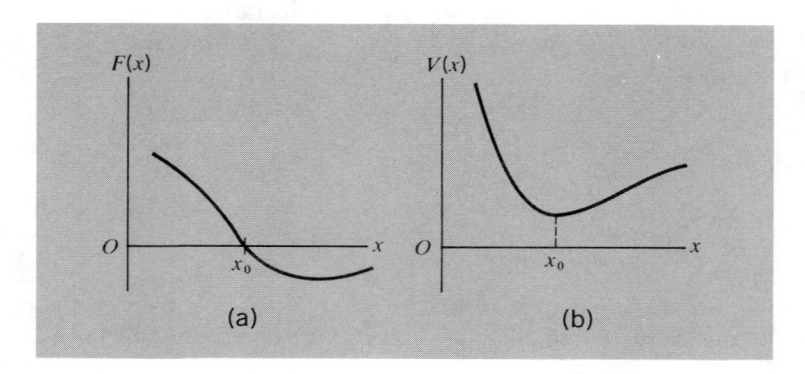

$$V(s) = V_0 + c_1 s + \tfrac{1}{2}c_2 s^2 + \tfrac{1}{3}c_3 s^3 + \ldots \tag{6.21}$$

Fig. 6.6 (a) Variation of force with displacement on either side of the equilibrium position in a one-dimensional system. (b) Potential-energy curve associated with (a).

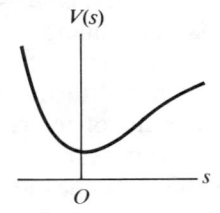

Fig. 6.7 Potential-energy curve of Fig. 6.6(b) referred to an origin located at the equilibrium position.

(The numerical factors are inserted for a reason that will appear almost immediately.)

The force as a function of s is obtained from the general relation

$$F(s) = -\frac{dV}{ds}$$

so that we have

$$F(s) = -c_1 - c_2 s - c_3 s^2 \ldots$$

However, by definition, $F(s) = 0$ at $s = 0$; this is the equilibrium position. Hence $c_1 = 0$, and so our equation for F becomes

$$F(s) = -c_2 s - c_3 s^2 \ldots \quad (6.22)$$

Now, whatever the relative values of the constants c_2 and c_3, there will always be a range of values of s for which the term in s^2 is much less than the term in s, for the ratio of the two is equal to $c_3 s/c_2$, which can be made as small as we please by choosing s small enough. A similar argument applies, even more strongly, to all the higher terms in the expansion. Hence, unless our potential-energy function has some very special properties (such as having $c_2 = 0$) we can be sure that for sufficiently small oscillations it will be just like the potential-energy function of a spring that obeys Hooke's law. We can write

$$V(s) \approx \tfrac{1}{2} c_2 s^2 \quad (6.23)$$

which means that the effective spring constant k for the motion is equal to the constant c_2, and is equal to d^2V/ds^2 evaluated at $s=0$.

The linear oscillator as a two-body problem

A statement of the form: '*the potential energy of an object* of mass m raised to a height h above the earth's surface is mgh' is perfectly legitimate for situations in which the mass of a particle is very small compared to the mass of the object (or objects) with which it interacts. In such a case, the centre of mass of the system is effectively determined by the position of the larger mass. A frame of reference anchored to this larger mass is both a zero-momentum frame and a fixed frame of reference. This is the case for the earth and an ordinary object moving near its surface. It is also the case for interactions between any two objects if one of them is rigidly attached to the earth. One must remember, however, that, strictly speaking, one is analysing a *two-body system* (the earth and the object which is raised): mgh is the increase of potential energy *of the system* when the *separation* between the earth and the object of mass m is increased by an amount h. In other words, the potential energy is a property of the two objects jointly; it cannot be associated with one or the other individually. If one has two interacting particles of comparable mass, both will accelerate and gain or lose kinetic energy as a result of the interaction between them. It is to this basic two-body aspect of the potential-energy problem that we now turn.

Suppose that we have two particles, of masses m_1 and m_2, connected by a spring of negligible mass aligned parallel to the x axis (Fig. 6.8). Let the particles be at positions x_1 and x_2, as shown, referred to some origin O. If the spring is effectively massless, the forces on it at its two ends must be equal and opposite (otherwise it would have infinite acceleration) and hence, accepting the equality of action and reaction in the contacts between the masses and the spring, the forces exerted on the masses are also equal and opposite. Thus, denoting the force exerted on mass 2 *by the spring* as F_{12}, the force F_{21} exerted on mass 1 by the spring is equal to $-F_{12}$.

We shall relate the changes in kinetic energy of the masses to the changes in stored potential energy in the spring.

The potential energy of the spring

First, suppose that m_1 moves a distance dx_1 while m_2 moves dx_2. The work done by the spring is given by

$$dW = F_{12}\, dx_2 + F_{21}\, dx_1$$
$$= F_{12}(dx_2 - dx_1) \quad \text{(since } F_{21} = -F_{12}\text{)}$$
$$= F_{12}\, d(x_2 - x_1)$$

Clearly the difference $x_2 - x_1$, rather than x_1 and x_2 separately, defines the elongation of the spring (and hence the energy stored in it). Let us introduce a special coordinate, r, to denote this:

$$r = x_2 - x_1$$

Then

THE LINEAR OSCILLATOR AS A TWO-BODY PROBLEM

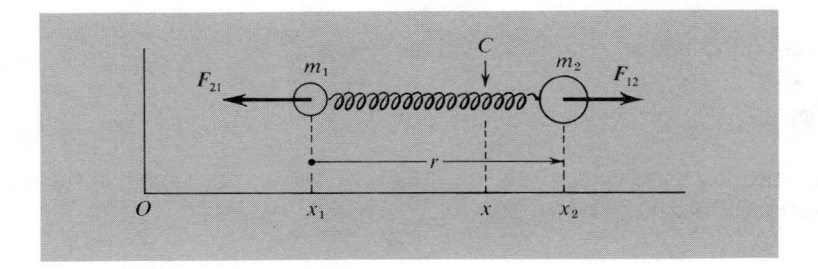

Fig. 6.8 System of two masses connected by a spring, showing the separate coordinates and forces.

$$dW = F_{12}\,dr \qquad \text{(work done } by \text{ spring)} \tag{6.24}$$

The change of potential energy of the spring is equal to $-dW$. Introducing the potential energy function $V(r)$ we have

$$dV = -F_{12}\,dr$$

$$V(r) = -\int F_{12}\,dr \tag{6.25}$$

The kinetic energy of the masses

Our discussion of two-body systems in Chapter 5 suggests clearly that we should introduce the centre of mass of the system and refer the motions of the individual masses to it. This allows us, as we have seen, to consider the dynamics in the CM frame without reference to the motion of the system as a whole. By eqn (5.20) we have

$$K = K' + \tfrac{1}{2}Mv_c^2$$

where K' is the total kinetic energy of the two masses as measured in the CM frame. Denoting the velocities relative to the CM by v_1' and v_2' as usual, we have

$$K' = \tfrac{1}{2}m_1 v_1'^2 + \tfrac{1}{2}m_2 v_2'^2 \tag{6.26}$$

We have seen previously that it is very convenient to express K' in terms of the *relative velocity*, v_r, and the *reduced mass*, μ, of the two particles:

$$v_r = v_2' - v_1'$$

$$\mu = \frac{m_1 m_2}{m_1 + m_2}$$

From the definition of the CM (zero-momentum) frame, we have

$$m_1v_1' + m_2v_2' = 0$$

Using this, together with the equation for v_r, we find

$$v_1' = -\frac{m_2}{m_1 + m_2}v_r\,; \qquad v_2' = \frac{m_1}{m_1 + m_2}v_r$$

Substituting these values into eqn (6.26) one arrives once again at the result expressed by eqn (5.21a):

$$K' = \frac{1}{2}\frac{m_1 m_2}{m_1 + m_2}v_r^2 = \tfrac{1}{2}\mu v_r^2 \tag{6.27}$$

We shall be considering the changes of kinetic energy of the masses as related to the work done on them by the spring. On the assumption that no external forces are acting, we have $v_c = $ const., in which case

$$dK = dK'$$

The motions

At this point we can assemble the foregoing results and equate the change of kinetic energy to the work done by the spring. We evaluated dW (in eqn 6.24) in terms of laboratory coordinates, although, as we saw (and could have predicted), the result depends only on $x_2 - x_1$, which is equal to $x_2' - x_1'$, both being equal to the relative coordinate r. Likewise, as we have just seen, $dK = dK'$. We can, in fact, put

$$dK' = F_{12}\,dr \qquad \text{(work done } by \text{ spring)}$$

Integrating,

$$K' = \int F_{12}\,dr + \text{const.}$$

And now, with the help of eqn (6.25), we can write this as a statement of the total mechanical energy E' in the CM frame:

$$K' + V(r) = E' \tag{6.28}$$

For the specific case of a spring of spring constant k and natural length r_0, we can put

$$r = r_0 + s$$
$$V(s) = \tfrac{1}{2}ks^2$$

Also

$$v_r = \frac{dr}{dt} = \frac{ds}{dt}$$

Thus the equation of conservation of energy (eqn 6.28) becomes

$$\tfrac{1}{2}\mu \left(\frac{ds}{dt}\right)^2 + \tfrac{1}{2}ks^2 = E' \qquad (6.29)$$

where

$$\mu = \frac{m_1 m_2}{m_1 + m_2}$$

This is exactly of the form of the linear oscillator equation; its angular frequency ω and its period T are given by

$$\omega = \left(\frac{k}{\mu}\right)^{1/2}; \qquad T = 2\pi \left(\frac{\mu}{k}\right)^{1/2} \qquad (6.30)$$

It is to be noted that the reduced mass μ is less than either of the individual masses, so that for a given spring the period is shorter in free oscillation than if one of the masses is clamped tight.

Problems

6.1 A railway waggon is loaded with 20 tons of coal in a time of 2 s while it travels through a distance of 10 m beneath a hopper from which the coal is discharged.
 (a) What average extra force must be applied to the car during this loading process to keep it moving at constant speed?
 (b) How much work does this force perform?
 (c) What is the increase in kinetic energy of coal?
 (d) Explain the discrepancy between (b) and (c).

6.2 A car is being driven along a straight road at constant speed v. A passenger in the car hurls a ball straight ahead so that it leaves his hand with a speed u *relative to him*.
 (a) What is the gain of energy of the ball as measured in the reference frame of the car? Of the road?
 (b) Relate the answers in (a) to the work done by the passenger and by the car. Satisfy yourself that you understand exactly what forces are acting on what objects, over what distances.

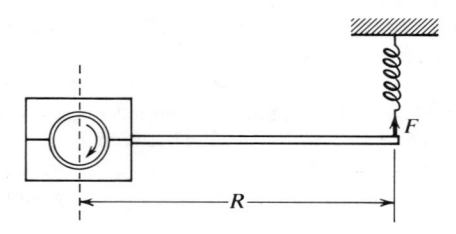

6.3 A common device for measuring the power output of an engine at a given rate of revolution is known as an absorption dynamometer. A small friction brake, called a Prony brake, is clamped to the output shaft of the engine (as shown), allowing the shaft to rotate, and is held in position by a spring scale a known distance R away.

(a) Derive an expression for the horsepower of the engine in terms of the quantities R, F (the force recorded at the spring scale), and ω (the angular velocity of the shaft).

(b) You will recognize the product of the force F times its lever arm R as the torque exerted by the engine. Does your relation between torque and horsepower agree with the data published for automobile engines? Explain why there may be discrepancies.

6.4 An electric pump is used to empty a flooded basement which measures 30 by 20 ft and is 15 ft high. In a rainstorm, the basement collected water to a depth of 4 ft.

(a) Find the work necessary to pump the water out to ground level.

(b) Supposing that the pump was driven by a 1-hp motor with a 50% efficiency, how long did the operation take?

(c) If the depth of basement below ground were 50 ft instead of 15 ft, how much work would be required of the pump, again for a 4-ft flood?

6.5 (a) Look up the current world records for shot put, discus and javelin. The masses of these objects are 7.15, 2.0 and 0.8 kg respectively. Ignoring air resistance, calculate the minimum possible kinetic energy imparted to each of these objects to achieve these record throws.

(b) What force, exerted over a distance of 2 m, would be required to impart these energies?

(c) Do you think that the answers imply that air resistance imposes a serious limitation in any of these events?

6.6 A perverse traveller walks *down* an ascending escalator, so as to remain always at the same vertical level. Does the motor driving the escalator have to do more work than if the man were not there? Analyse the dynamics of this situation as fully as you can.

6.7 The Great Pyramid of Gizeh when first erected (it has since lost a certain amount of its outermost layer) was about 150 m high and had a square base of edge length 230 m. It is effectively a solid block of stone of density about 2500 kg/m³.

(a) What is the total gravitational potential energy of the pyramid, taking as zero the potential energy of the stone at ground level?

(b) Assume that a slave employed in the construction of the pyramid had a food intake of about 1500 Cal/day and that about 10% of this energy was available as useful work. How many man-days would have been required, at a minimum, to construct the pyramid? (The Greek historian Herodotus reported that the job involved 100 000 men and took 20 years. If so, it was not very efficient.)

6.8 It is claimed that a rocket would rise to a greater height if, instead of being ignited at ground level (A), it were ignited at a lower level (B) after it had been allowed to slide from rest down a frictionless chute—see the figure. To analyse this claim, consider a simplified model in which the body of the rocket is represented by a mass M, the fuel is represented by a mass m, and the chemical energy released in the burning of the fuel is represented by a compressed spring between M and m which stores a definite amount of potential energy, V, sufficient to eject m suddenly with a velocity v relative to M. (This corresponds to instantaneous burning and ejection of all the fuel—i.e. an explosion.) Then proceed as follows:

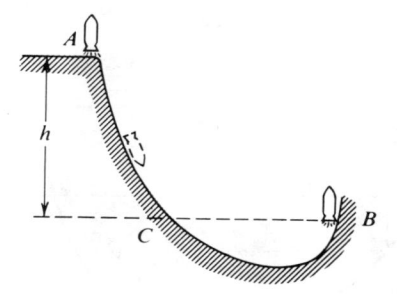

(a) Assuming a value of g independent of height, calculate how high the rocket would rise if fired directly upward from rest at A.

(b) Let B be at a distance h vertically lower than A, and suppose that the rocket is fired at B after sliding down the frictionless chute. What is the velocity of the rocket at B just before the spring is released? just after the spring is released?

(c) To what height *above* A will the rocket rise now? Is this higher than the earlier case? By how much?

(d) Remembering energy conservation, can you answer a sceptic who claimed that someone had been cheated of some energy?

(e) If you are ambitious, consider a more realistic case in which the ejection of the fuel is spread out over some appreciable time. Assume a constant rate of ejection during this time.

6.9 A neutral hydrogen atom falls from rest through 100 m in vacuum. What is the order of magnitude of its kinetic energy *in electron volts* at the bottom? (1 eV = 1.6 × 10^{-19} J. Avogadro's number = 6 × 10^{23}.)

6.10 A spring exerts a restoring force given by

$$F(x) = -k_1 x - k_2 x^3$$

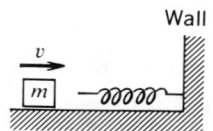

where x is the deviation from its unstretched length. The spring rests on a frictionless surface, and a frictionless block of mass m and initial velocity v hits a spike on the end of the spring and sticks (see the figure). How far does the mass travel, after being impaled, before it comes to rest? (Assume that the mass of the spring is negligible.)

6.11 A particle moves along the x axis. Its potential energy as a function of position is as shown in the figure on the next page. Make a careful freehand sketch of the force $F(x)$ as a function of x for this potential-energy curve. Indicate on your graph significant features and relationships.

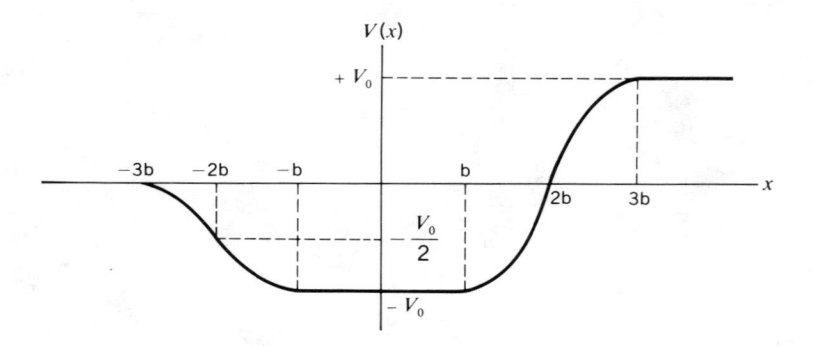

6.12 Two masses are connected by a massless spring as shown.

(a) Find the minimum downward force that must be exerted on m_1 such that the entire assembly will barely leave the table when this force is suddenly removed.

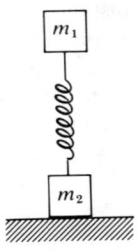

(b) Consider this problem in the time-reversed situation. Let the assembly be supported above the table by supports attached to m_1. Lower the system until m_2 barely touches the table and then release the supports. How far will m_1 drop before coming to a stop? Does knowledge of this distance help you solve the original problem?

(c) Now that you have the answer, check it against your intuition by (1) letting m_2 be zero and (2) letting m_1 be zero. Especially in the second case, does the theoretical answer agree with your common sense? If not, discuss possible sources of error.

6.13 A particle moves in a region where the potential energy is given by

$$V(x) = 3x^2 - x^3 \quad (x \text{ in m}, V \text{ in J})$$

(a) Sketch a freehand graph of the potential for both positive and negative values of x.

(b) What is the maximum value of the total mechanical energy such that oscillatory motion is possible?

(c) In what range(s) of values of x is the force on the particle in the positive x direction?

6.14 A highly elastic ball (e.g. a 'Superball') is released from rest a distance h above the ground and bounces up and down. With each bounce a fraction f of its kinetic energy just before the bounce is lost. Estimate the length of time the ball will continue to bounce if $h = 5$ m and $f = \frac{1}{10}$.

6.15 A perfectly rigid ball of mass M and radius r is dropped upon a deformable floor which exerts a force proportional to the distance of deformation, $F = ky$.

(a) Make a graph of the potential energy of the ball as a function of height y. (Take $y = 0$ as the undisturbed floor level.)

(b) What is the equilibrium position of the centre of the ball when the ball is simply resting on the floor? (Note that this corresponds to the minimum of the potential-energy curve.)

(c) By how much is the period of M increased over its period of bouncing on a perfectly rigid floor?

6.16 A spring of negligible mass exerts a restoring force given by

$$F(x) = -k_1 x + k_2 x^2$$

(a) Calculate the potential energy stored in the spring for a displacement x. Take $V = 0$ at $x = 0$.

(b) It is found that the stored energy for $x = -b$ is twice the stored energy for $x = +b$. What is k_2 in terms of k_1 and b?

(c) Sketch the potential-energy diagram for the spring as defined in (b).

(d) The spring lies on a smooth horizontal surface, with one end fixed. A mass m is attached to the other end and sets out at $x = 0$ in the positive x direction with kinetic energy equal to $k_1 b^2 / 2$. How fast is it moving at $x = +b$?

(e) What are the values of x at the extreme ends of the range of oscillation? (Use your graph from part (c) for this.)

6.17 An object of mass m, moving from the region of negative x, arrives at the point $x = 0$ with speed v_0. For $x \geq 0$ it experiences a force given by

$$F(x) = -ax^2$$

How far along the $+x$ axis does it get?

6.18 The potential energy of a particle of mass m as a function of its position along the x axis is as shown. (The discontinuous jumps in the value of V are not physically realistic but may be assumed to approximate a real situation.) Calculate the period of one complete oscillation if the particle has a total mechanical energy E equal to $3V_0/2$.

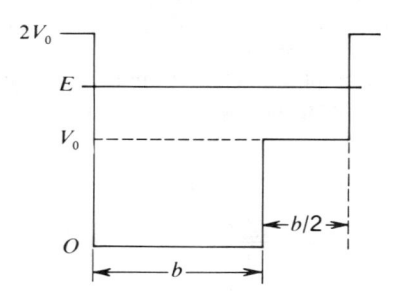

6.19 Consider an object of mass m constrained to travel on the x axis (perhaps by a frictionless guide wire or frictionless tracks), attached to a spring of relaxed length l_0 and spring constant k which has its other end fixed at $x = 0$, $y = l_0$ (see the figure).

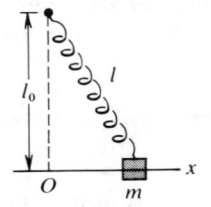

(a) Show that the force exerted on m in the x direction is

$$F_x = -kx\left[1 - \left(1 + \frac{x^2}{l_0^2}\right)^{-1/2}\right]$$

(b) For small displacements ($x \ll l_0$), show that the force is proportional to x^3 and hence

$$V(x) \approx Ax^4 \qquad (x \ll l_0)$$

What is A in terms of the above constants?

(c) The period of a simple harmonic oscillator is independent of its amplitude. How do you think the period of oscillation of the above motion will depend on the amplitude? (An energy diagram may be helpful.)

6.20 Consider a particle of mass m moving along the x axis in a force field for which the potential energy of the particle is given by $V = Ax^2 + Bx^4$ ($A > 0, B > 0$). Draw the potential-energy curve and, arguing from the graph, determine something about the dependence of the period of oscillation T upon the amplitude x_0. Show that, for amplitudes sufficiently small so that Bx^4 is always very small compared to Ax^2, the approximate dependence of period upon amplitude is given by

$$T = T_0\left(1 - \frac{3B}{4A}x_0^2\right)$$

6.21 A particle of mass m and energy E is bouncing back and forth between vertical walls as shown, i.e. over a region where $V = 0$. The potential energy is slightly changed by introducing a tiny rectangular hump of height $\Delta V (\ll E)$ and width Δx. Show that the period of oscillation is changed by approximately $(m/2E^3)^{1/2}(\Delta V \Delta x)$. (It is worth noting that the effect of the small irregularity in potential energy depends simply on the product of ΔV and Δx, not on the individual values of these quantities. This is typical of such small disturbing effects—known technically as perturbations.)

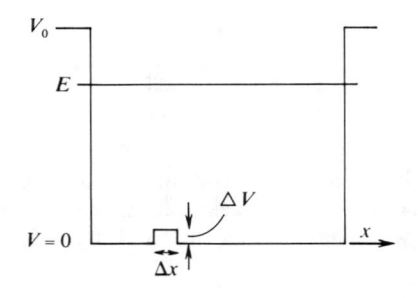

PROBLEMS

6.22 Two blocks of masses m and $2m$ rest on a frictionless horizontal table. They are connected by a spring of negligible mass, equilibrium length L, and spring constant k. By means of a massless thread connecting the blocks the spring is held compressed at a length $L/2$. The whole system is moving with speed v in a direction *perpendicular* to the length of the spring. The thread is then burned. In terms of m, L, k and v find
(a) The total mechanical energy of the system.
(b) The speed of the centre of mass.
(c) the maximum relative speed of the two blocks.
(d) The period of vibration of the system.
How do the quantities of parts (a) through (d) change if the initial velocity v is along, rather than perpendicular to, the length of the spring?

6.23 The mutual potential energy of a Li$^+$ ion and an I$^-$ ion as a function of their separation r is expressed fairly well by the equation

$$V(r) = \frac{-Ke^2}{r} + \frac{A}{r^{10}}$$

where the first term arises from the Coulomb interaction, and the values of its constants in MKS units are

$$K = 9 \times 10^9 \text{ N m}^2/\text{C}^2, e = 1.6 \times 10^{-19} \text{ C}$$

The equilibrium distance r_0 between the centres of these ions in the LiI molecule is about 2.4 Å. On the basis of this information,
(a) How much work (in eV) must be done to tear these ions completely away from each other?
(b) Taking the I$^-$ ion to be fixed (because it is so massive), what is the frequency ν (in Hz) of the Li$^+$ ion in vibrations of very small amplitude? (Calculate the effective spring constant k as the value of d^2V/dr^2 at $r = r_0$—see p. 141. Take the mass of the Li$^+$ as 10^{-26} kg.)

6.24 (a) If in addition to the van der Waals attractive force, which varies at r^{-7}, two identical atoms of mass M experience a repulsive force proportional to r^{-l} with $l > 7$, show that

$$V(r) = \frac{-A}{r^6} + \frac{B}{r^n} \quad (n > 6)$$

and plot your result versus r.
(b) Calculate the equilibrium separation r_0 in this molecule in terms of the constants by requiring

$$\left.\frac{dV(r)}{dr}\right|_{r=r_0} = 0$$

(c) The dissociation energy D of the molecule should be equal to $-V(r_0)$. What is its value in terms of A, n and r_0?
(d) Calculate the frequency of small vibrations of the molecule about the equilibrium separation r_0. Show that it is given by the mass M, the constant n, the equilibrium separation, and the dissociation energy of the molecule, as follows:

$$\omega^2 = \frac{12nD}{Mr_0^2}$$

7 Conservative forces and motion in space

Extending the concept of conservative forces

Throughout chapter 6 we consistently applied one important simplification or restriction, by confining our discussion to motion in one dimension only. This clearly prevented us from studying some of the most interesting and important problems in dynamics. In the course of the present chapter we shall free ourselves of this restriction and in the process show the energy method of analysis to still greater advantage.

To begin the discussion, let us consider a problem in motion under gravity near the earth's surface. Suppose we have two very smooth tubes connecting two points A and B at different levels in the same vertical plane (Fig. 7.1). A small particle, placed in either of these tubes and released from rest at A, slides down and emerges at B. If the tubes are effectively frictionless, the forces exerted by them on the particle are always at right angles to the particle's motion. Hence these forces do no work on the particle; whatever changes may occur in its kinetic energy cannot be ascribed to them. What these forces *do* achieve is to compel the particle to follow a particular path so that it emerges at B travelling in a designated direction. If it follows path 1, it emerges with a velocity v_1 as shown; if it follows path 2, it emerges with a velocity v_2. Of course the energy of the particle does change as it moves along

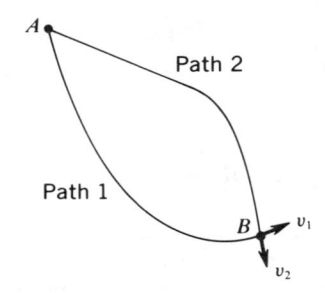

Fig. 7.1 Alternative paths between two given points for motion in a vertical plane.

EXTENDING THE CONCEPT OF CONSERVATIVE FORCES

either tube; the gravitational force is doing work on it. But we observe a very interesting fact; although the directions of the velocities v_1 and v_2 are quite different, their magnitudes are the same. *The kinetic energy given to the particle by the gravitational force is the same for all paths beginning at A and ending at B.*

How does this come about? It is not difficult to see. As the particle travels through some element of displacement ds the work dW done on it is given by

$$dW = |F_g|\,|ds|\cos\theta = F_g \cdot ds$$

where θ is the angle between the directions of F_g and ds. But the force F_g is a constant force (i.e. the same at any position) with the following components:

$$F_x = 0 \qquad F_y = -mg$$

(we take the y axis as positive upward). The element of displacement has components (dx, dy). Now from the basic definition of dW as $|F|\,|ds|\cos\theta$ we have

$$dW = F_x\,dx + F_y\,dy$$

To see this, consider any two vectors A and B in the xy plane, making angles α and β, respectively, with the x axis (Fig. 7.2). Then if the angle between them is θ we have, by a standard trigonometric theorem,

$$\cos\theta = \cos(\beta - \alpha) = \cos\beta\cos\alpha + \sin\beta\sin\alpha$$

The scalar product $S\,(= A\cdot B)$ is thus given by

$$S = |A|\,|B|\,(\cos\beta\cos\alpha + \sin\beta\sin\alpha)$$

But

$$|A|\cos\alpha = A_x \qquad |A|\sin\alpha = A_y$$

and similarly for the components of B. Thus we have

$$A\cdot B = A_x B_x + A_y B_y \qquad \text{(two dimensions only)}$$

More generally, if the vectors also have non-zero z components,

$$A\cdot B = A_x B_x + A_y B_y + A_z B_z$$

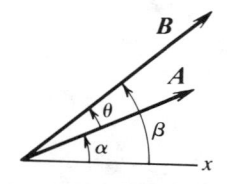

Fig. 7.2 Basis for obtaining the scalar product $A\cdot B$ in terms of individual components, as may be convenient in calculating the work $F\cdot ds$ in an arbitrary displacement.

Thus in general we shall have

$$dW = \mathbf{F} \cdot d\mathbf{s} = F_x\,dx + F_y\,dy + F_z\,dz \tag{7.1}$$

In the present two-dimensional problem, with $F_x = 0$ and $F_y = -mg$, this gives us

$$dW = -mg\,dy$$

Hence for a change of vertical coordinate from y_1 to y_2, regardless of the change of x coordinate or of the particular path taken, we have a change of kinetic energy given by

$$K_2 - K_1 = \int dW = -mg(y_2 - y_1) \tag{7.2}$$

which exactly reproduces the result that we derived in Chapter 6 for purely vertical motion and which permits us again to define a gravitational potential energy $V(y)$ equal to mgy.

This result makes for a great extension of our energy methods to situations where, in addition to gravity, we have so-called 'forces of constraint', which control the path of an object but, because they act always at right angles to its motion, do nothing to change its energy. Let us consider some specific examples.

Object moving in a vertical circle

Suppose that a particle P of mass m is attached to one end of a rod of negligible mass and length l, the other end of which is pivoted freely at a fixed centre C. Let us take an origin O at the normal resting position of the object (Fig. 7.3). Then the position of the object is conveniently described in terms of the single angular coordinate θ, or, if we prefer, by the displacement s

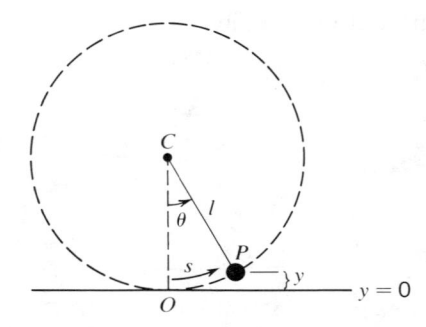

Fig. 7.3 Motion of a simple pendulum.

OBJECT MOVING IN A VERTICLE CIRCLE

along the circular arc of its path from O (assuming the rod to be of invariable length).

If the angular displacement is θ, the potential energy of the object is given by

$$V(\theta) = mgl(1 - \cos\theta) \tag{7.3}$$

Using our basic energy-conservation statement we have

$$K_1 + V_1 = K_2 + V_2$$

for any two points on the path. Substituting for K in terms of m and v and for V by eqn (7.3), we have

$$\tfrac{1}{2}mv_1^2 + mgl(1 - \cos\theta_1) = \tfrac{1}{2}mv_2^2 + mgl(1 - \cos\theta_2)$$

Therefore,

$$v_2^2 = v_1^2 + 2gl(\cos\theta_2 - \cos\theta_1) \tag{7.4}$$

Clearly, if $\theta_2 > \theta_1$, then $v_2 < v_1$.

If the object were started out at the lowest point with a velocity v_0, we should have

$$v_2^2 = v_0^2 - 2gl(1 - \cos\theta_2)$$

With the help of this result we can answer such questions as: what initial velocity is needed for the object to reach the top of the circle? and: what position does it reach if started out with less than this velocity?

Notice the great advantage that the energy method has over the direct use of $\mathbf{F} = m\mathbf{a}$ in this problem. The velocity of the object is changing in both magnitude and direction, it has radial and transverse components of acceleration, there is an unknown push or pull on the object from the rod—yet none of these things need be considered in calculating the speed v at any given θ (or y). Once v has been found by eqn (7.4) (or perhaps by going back to eqn (7.2) if that is more convenient), then one can proceed to deduce the acceleration components and other things.

There may be subtleties in such problems, however. Suppose, for example, that instead of a more or less rigid rod we had a string to constrain the object. This is now a constraint that works only one way; it can pull radially inward but it cannot push radially outward. One may have a situation in which the velocity is not great enough to take the object to the top of the circle (although it might be possible with a rigid rod); the object will fall away in a parabolic path (Fig. 7.4a). The breakaway point is reached when the tension in the string just falls to zero and the component of $\mathbf{F}_g$ along the radius of the circle is just equal to the mass m times the requisite centripetal acceleration v^2/l. An exactly similar situation can arise if an object moves along a circular track made of grooved metal, and Fig. 7.4(b) shows a stroboscopic photograph of an object falling away from such a track at the point where the normal

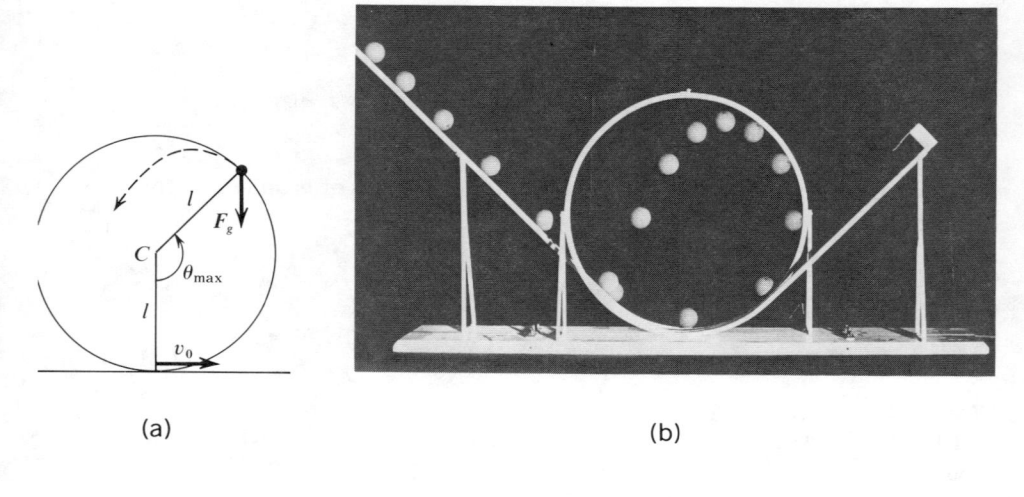

Fig. 7.4 (a) Path of the bob of a simple pendulum launched with insufficient velocity to maintain a tension in the string up to $\theta = \pi$. (b) Stroboscopic photograph of a ball falling away from a circular channel at the point where the contact force becomes zero. (Photograph by Jon Rosenfeld, Education Research Centre, MIT.)

reaction supplied by the track has fallen to zero. The angular position, θ_m, at which this occurs (cf. Fig. 7.4a) is defined by a statement of Newton's law:

$$mg \cos(\pi - \theta_m) = \frac{mv^2}{l} \quad \left(\text{assumes } \frac{\pi}{2} < \theta_m \leq \pi\right)$$

where

$$v^2 = v_0^2 - 2gl(1 - \cos \theta_m)$$

This leads to the result

$$\cos \theta_m = \frac{2}{3} - \frac{v_0^2}{3gl}$$

We can thus deduce that a particle that starts out from O with v_0 less than $\sqrt{5gl}$ will fail to reach the top of the circle, whereas with a rigid rod to support it an initial speed of $2\sqrt{gl}$ would suffice. Notice, therefore, that the energy-conservation principle should not be used blindly; one must always be on the alert as to whether Newton's law can be satisfied at every stage with the particular constraining forces available.

The simple pendulum

The simple pendulum is a special case of an object moving in a vertical circle. It is an important type of physical system, over and above its use in clocks, and it is not quite so simple as its traditional name implies. The simplicity is primarily in its structure—idealized as a point mass on a massless, rigid rod.

As we saw earlier (eqn 7.3), the potential energy, $V(\theta)$, expressed in terms of the angle that the supporting rod makes with the vertical, can be written in the form

$$V = mgl(1 - \cos\theta) = 2mgl \sin^2\left(\frac{\theta}{2}\right) \qquad (7.5)$$

If we plot this expression for V as a function of θ, we get Fig. 7.5.

In Fig. 7.5, θ can take all values from $-\infty$ to $+\infty$. However, all *positions* of the pendulum in space are described by values of θ between $-\pi$ and $+\pi$. Any value of θ outside this range corresponds to one and only one angle θ inside the range. The latter is obtained by adding to or subtracting from the former a whole-number multiple of 2π.

Two kinds of motion are possible, depending on whether E is less than or greater than $2mgl$ (referred to a zero of energy defined by an object at rest at the lowest position of the pendulum bob).

If the total energy is sufficiently great (e.g. as for E_2 in the figure), there are no turning points in the motion; θ increases (or decreases) without limit, corresponding to continued rotation of the pendulum rather than oscillation. The speed of the pendulum bob is maximum at $\theta = 0$ (or $\pm 2\pi$, $\pm 4\pi$, etc.) and minimum at $\theta = \pm\pi$ (or $\pm 3\pi$, $\pm 5\pi$, etc.) It is clear from Fig. 7.5 that to produce such rotational motion the pendulum must have a kinetic energy at least equal to $2mgl$ at the lowest point of the swing.

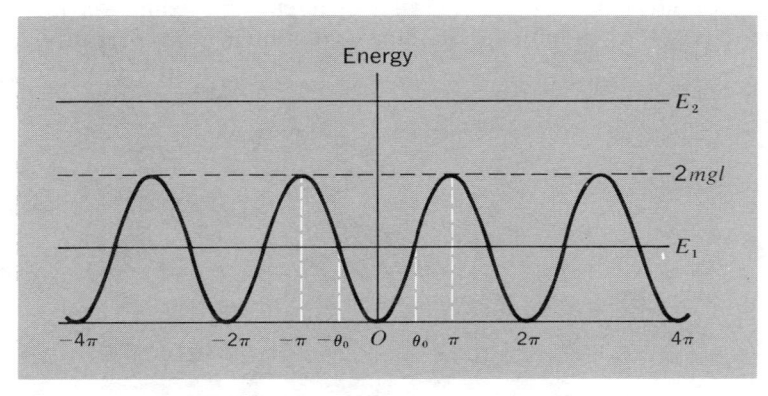

Fig. 7.5 Energy diagram for a rigid pendulum, using the angle θ as the coordinate. The pendulum is trapped in oscillatory motion about $\theta = 0$ if E is less than $2mgl$.

If $E < 2mgl$, say E_1, the motion is oscillatory and the angle θ changes from $+\theta_0$ to $-\theta_0$ and back again (Fig. 7.5). However, the motion is not harmonic except for sufficiently small amplitudes. In the neighbourhood of O, one can very nearly match the potential-energy curve of Fig. 7.5 by a parabola (Fig. 7.6), so that for these small amplitudes the oscillations are just those of the linear oscillator. One way of justifying this statement is to recall the general argument, made in Chapter 6, that almost any symmetrical potential-energy curve can be approximated by a parabola over some limited range of small displacements. Another way is to note that in eqn (7.5), if θ is small, one can set $\sin(\theta/2)$ approximately equal to $\theta/2$, so that we have

$$V(\theta) \approx \tfrac{1}{2}mgl\theta^2 \tag{7.6}$$

A third way depends upon using the same approximation that Newton used when he described the moon as a falling object. Applying it to the present problem, and taking the origin at O, the circular path is given by the equation

$$x^2 + (l - y)^2 = l^2$$

(see Fig. 7.7). Therefore,

$$y^2 - 2ly + x^2 = 0$$
$$y = l - (l^2 - x^2)^{1/2}$$
$$= l - l\left(1 - \frac{x^2}{l^2}\right)^{1/2}$$

If $x^2 \ll l^2$, we can use the binomial expansion to yield the approximate result

$$y \approx \frac{1}{2}\frac{x^2}{l} \tag{7.7}$$

Equation (7.7) suggests a somewhat different approximation for small oscillations of the pendulum, describing its motion in terms of its horizontal

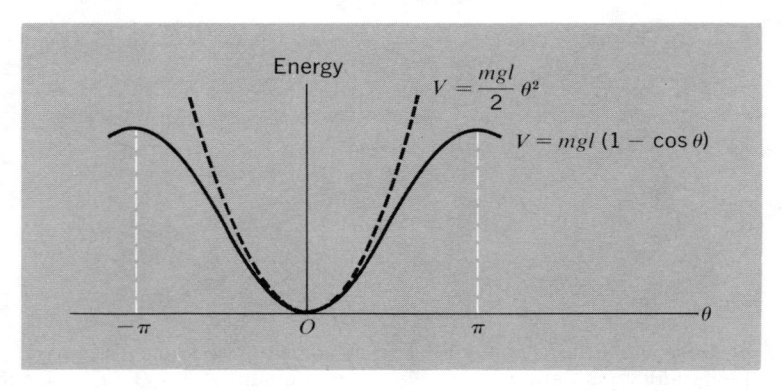

Fig. 7.6 Potential-energy diagram of a simple pendulum.

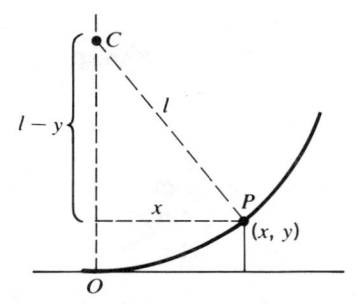

Fig. 7.7 Geometry of displacements of a simple pendulum.

displacement instead of its angular one. But whatever analysis we adopt, it is clear that the period must depend on the amplitude. We can also argue qualitatively which way it varies. Looking at Fig. 7.6, we can say that the curve of $V(\theta)$ for the pendulum describes a kind of spring that gets 'softer' at large extensions. Compared to the parabolic behaviour that would hold for an ideal spring, the restoring force is relatively less at larger displacements. Thus one might guess that the motion becomes more sluggish and hence that the period increases. Certainly there is one extreme case that leaves no room for doubt. If the energy is such that the pendulum just exactly arrives at $\theta_0 = \pi$, there is no restoring force at all; the pendulum would sit upside down indefinitely—although in practice this is of course an unstable equilibrium. Even at amplitudes short of this the increase of period is drastic. The section after next describes the analysis of larger-amplitude motion. But first we shall make a detailed study of the small-angle approximation.

The pendulum as a harmonic oscillator

The speed of the pendulum bob at any point is equal to $l\,d\theta/dt$, so that the energy-conservation equation is

$$\tfrac{1}{2}ml^2\left(\frac{d\theta}{dt}\right)^2 + V(\theta) = E \qquad \text{(exact)}$$

Using the approximate expression for $V(\theta)$ from eqn (7.6), we have

$$\tfrac{1}{2}ml^2\left(\frac{d\theta}{dt}\right)^2 + \tfrac{1}{2}mgl\theta^2 = E \qquad \text{(approx.)}$$

or

$$\left(\frac{d\theta}{dt}\right)^2 + \frac{g}{l}\theta^2 = \text{const.} \tag{7.8}$$

By now we have met this form of equation several times and can identify $(g/l)^{1/2}$ as being the quantity ω that defines the period of the oscillation:

$$T = \frac{2\pi}{\omega} = 2\pi \left(\frac{l}{g}\right)^{1/2} \tag{7.9}$$

[*Caution:* It is the angular displacement θ itself that undergoes a simple harmonic variation, described by the equation

$$\theta(t) = \theta_0 \sin(\omega t + \varphi_0) \tag{7.10}$$

This can be confusing; the actual angular displacement of the pendulum is described in terms of the sine of the purely mathematical phase angle $(\omega t + \varphi_0)$. The actual angular velocity of the pendulum is $d\theta/dt$—not the ω in $(\omega t + \varphi_0)$, which serves merely to define the periodicity.]

The isochronous behavior of the simple pendulum—the fact that its period is almost completely independent of amplitude over a wide range—provides a striking example of this remarkable and unique property of systems governed by restoring forces proportional to the displacement from equilibrium. Suppose we had two identical pendulums, each with a string of length 30 ft, suspended from a high support. Each would have a period of about 6 s. And if we set one swinging with an amplitude of only an inch or two, so that its motion was almost imperceptible, and set the other swinging with an amplitude of 5 ft, so that it swept through the central position at a speed of about 5 ft s^{-1}, the difference in their periods would be too small to put them significantly out of step in less than a hundred swings or so.

The simple pendulum with larger amplitude of swing

To find how the period of a simple pendulum departs from its ideal small-amplitude value, we write the equation for conservation of energy in the exact form

$$\left(\frac{d\theta}{dt}\right)^2 + 2\omega_0^2(1 - \cos\theta) = 2\omega_0^2(1 - \cos\theta_0) \tag{7.11}$$

where $\omega_0^2 = g/l$ and θ_0 is the maximum angle of deflection from the vertical.

The period of oscillation is then given by

$$T(\theta_0) = \frac{2}{\omega_0\sqrt{2}} \int_{-\theta_0}^{\theta_0} \frac{d\theta}{(\cos\theta - \cos\theta_0)^{1/2}} \tag{7.12}$$

For small amplitudes, we of course have $T_0 = 2\pi/\omega_0$. The integral of eqn (7.12) cannot be carried out exactly; one has to resort to numerical methods or to a series expansion of the integrand which gives, as a next approximation to the period, the following result:

$$T(\theta_0) \approx T_0 \left[1 + \frac{1}{4}\sin^2\left(\frac{\theta_0}{2}\right) \right] \tag{7.13}$$

If θ_0 is not too large, another acceptable form of this result, in terms of the horizontal amplitude $A\ (= l\sin\theta_0)$, is the formula

$$T(A) \approx T_0 \left(1 + \frac{A^2}{16l^2} \right) \tag{7.14}$$

Universal gravitation: a conservative central force

The problems of energy conservation that we have discussed so far have involved only the familiar force of gravity near the earth's surface—a force which, as experienced by any given object, has effectively the same magnitude and the same direction in a given locality, regardless of horizontal or vertical displacements (within wide limits). But, as we know, the basic gravitational interaction is a force varying as the inverse square of the distance between two particles and exerted along the line joining their centres. It is an example of the very important class of forces—*central* forces—which are purely radial with respect to a given point, the 'centre of force.' It has the further property of being spherically symmetric; that is, the magnitude of the force depends only on the radial distance from the centre of force, and not on the direction. We shall show that all such spherically symmetric cental forces are conservative, and shall then consider the special features of the $1/r^2$ force that holds for gravitation and electrostatics.

If a particle is exposed to a central force, the force vector F acting on it has only one component, F_r. If the force is also spherically symmetric, then F_r can be written as a function of r only:

$$F_r = f(r) \tag{7.15}$$

We shall prove that any such spherically symmetric central force is conservative by showing that the work done by the force on a test particle, as the latter changes its position from a point A to another point B (Fig. 7.8a), does not depend on the particular path connecting A and B, but only on these end points. If this result holds, then it will be true that the particle, if it went from A to B by this path, and returned from B to A by any other path, would have *no* net work done on it by the central force and would (if no work were done on it by other forces) return to A with the same kinetic energy that it had to start with. This then corresponds exactly to the conservative property as defined for one-dimensional motions.

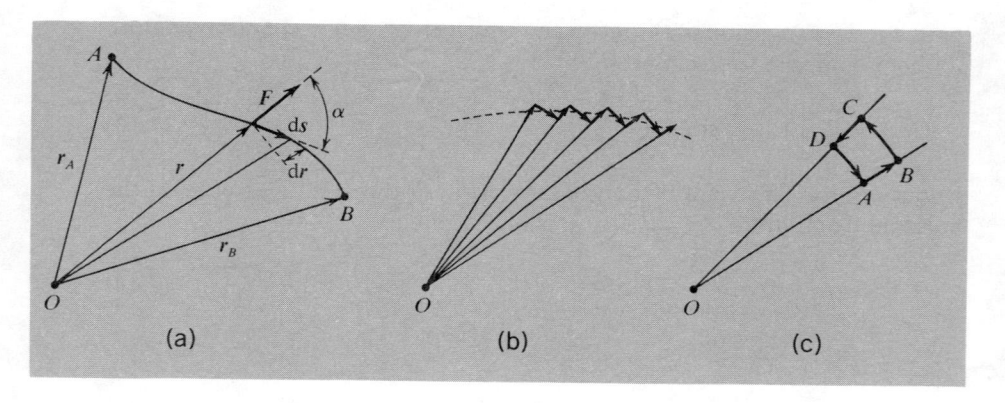

Fig. 7.8 (a) Diagram for consideration of potential-energy changes in a central force field. (b) Analysis of an arbitrary path into radial elements along which work is done, and transverse elements along which no work is done. (c) Closed path in a conservative central force field.

Let the centre of force be at O (Fig. 7.8a) and consider the work done by the central force F on the test particle as it undergoes a displacement ds along the path as shown. This work is given by

$$dW = \mathbf{F} \cdot d\mathbf{s} = F\,ds\cos\alpha \tag{7.16}$$

where α is the angle between the direction of $\mathbf{F}$, i.e. the direction of r, and the direction of $d\mathbf{s}$. From the figure, however, we see that

$$ds\cos\alpha = dr$$

where dr is the change of distance from O, resulting from the displacement ds. Inserting this value of $ds\cos\alpha$ into eqn (7.16), we have

$$dW = F\,dr$$

where the magnitude of the force ($F = f(r)$) depends only on r as indicated in eqn (7.15). We then have for the work done by $\mathbf{F}$ on the test object as it moves from A to B,

$$W = \int_{r_A}^{r_B} f(r)\,dr \tag{7.17}$$

Because this integral has a value that depends only on its limits and not on the path, we can conclude that the spherically symmetric central force is conservative. This result can be arrived at in slightly different terms by picturing the actual path as being built up from a succession of small steps, as shown in Fig. 7.8(b). One component of each step is motion along an arc at constant r, so that the force is perpendicular to the displacement and the contribution to W is zero, and the other component is a purely radial displacement so that the force and the displacement are in the same direction, resulting in the amount of work $F\,dr$ by the central force.

[A converse to the result we have just derived is the following important proposition:

A central force field that is also conservative must be spherically symmetric.

To show this suppose that a centre of force exists at the point O in Fig. 7.8(c). Imagine a closed path $ABCD$, formed by very short portions of two radial lines drawn from O, and by the two circular arcs BC and DA. Since, by definition, the force is purely central, it has no component perpendicular to the radial direction from O at any point. Thus, if we imagine a particle carried around $ABCDA$, it experiences no force along BC and DA. The condition that the force be conservative thus requires equal and opposite amounts of work along AB and CD. Since these lines are of equal length, the mean magnitude of the force must be the same on each. If we imagine the lengths of these elements of path to become arbitrarily small, we conclude that the value of the force at a particular scalar value of r is independent of the direction in which the vector r is drawn. The most important sources of such spherically symmetric force fields are spherically symmetric distributions of mass or electric charge.]

It is important to realize that a force *may* be conservative without being central or spherically symmetric. For example, the combined gravitational effect of a pair of concentrated unequal masses, separated by some distance, has a complicated dependence on position and direction, but we know that it is conservative beause it is the superposition of two individually conservative force fields of the separate masses.

Given the result expressed by eqn (7.17), we can proceed to define a potential energy $V(r)$ for any object exposed to a spherically symmetric central force:

$$V_B - V_A = -\int_{r_A}^{r_B} f(r)\,dr \tag{7.18}$$

If the kinetic energy of the object at points A and B has the values K_A and K_B, respectively, then (if no work is done by other forces) we have

$$K_B - K_A = W$$

and

$$K_A + V_A = K_B + V_B = E \tag{7.19}$$

Thus we have established an energy-conservation statement for an object moving under the action of any central force.

For an inverse-square force, we have

$$F(r) = \frac{C}{r^2} \tag{7.20}$$

In such a case, therefore, eqn (7.18) gives us

$$V_B - V_A = -C \int_{r_A}^{r_B} r^{-2}\, dr$$

from which we get

$$V_B - V_A = C \left(\frac{1}{r_B} - \frac{1}{r_A} \right) \tag{7.21}$$

There remains only the choice of the zero of potential energy. It is convenient to set $V = 0$ for $r = \infty$, i.e. at points infinitely far from the source at O, since the force vanishes at these points, and in colloquial terms one can say that the existence of either particle is of no consequence to the other one under these conditions. We now apply eqn (7.21) to the case where $r_A = \infty$; $V_A = 0$, and the potential energy of the test particle becomes $V_B = C/r_B$—or, if we drop the now redundant subscript B, there follows

$$V(r) = \frac{C}{r} \tag{7.22}$$

for the potential energy of a test particle as a function of its position.

Equation (7.22) is valid for either an attractive or a repulsive force, the constant C being negative for attractive forces and positive for repulsive forces. In particular, for a particle of mass m under the gravitational attraction of a point mass M (which we shall suppose to be fixed at the origin O) we have

$$F(r) = -\frac{GMm}{r^2} \tag{7.23}$$

$$V(r) = -\frac{GMm}{r} \tag{7.24}$$

Note that these last two equations refer only to the interaction between two objects that can be regarded as though they were points, i.e. their linear dimensions are small compared to their separation. As we saw in Chapter 4, however, some of the most interesting and important gravitational problems concern the gravitational forces exerted by large spherical objects such as the earth or the sun. In our earlier discussion of such problems, we saw that the basic problem is the interaction between a point particle and a thin spherical shell of material. In Chapter 4 we presented a frontal attack on this problem, going directly to an evaluation of an integral over all contributions to the net force. Now we shall approach the problem by way of a consideration of

A gravitating spherical shell

potential energies, and in the process we shall see the great value of the potential energy concept in such calculations.

A gravitating spherical shell

Suppose, as in our earlier treatment of the problem, that we have a thin, uniform shell of matter, of radius R and mass M (Fig. 7.9). Let a particle of mass m be placed at some point P a distance r from the centre of the shell. If we deal directly in terms of forces, then, as we saw, the force from material in the vicinity of a point such as A must be resolved along the line OP. In other words, a vector sum of all the force contributions is necessary. If we deal in terms of potential energy, however, we can exploit its most important property: *Potential energy is a scalar quantity*. We can just add up the contributions to V from all parts of the shell. The force on m is then obtained simply from the relation

$$F_r = -\frac{dV}{dr} \tag{7.25}$$

As in our direct calculation of the force, we take advantage of the symmetry of the system by considering the zone of material marked off on the sphere between the angles θ and $\theta + d\theta$ (Fig. 7.9). For this zone we have

$$\text{area} = 2\pi R^2 \sin\theta \, d\theta$$

$$\text{mass} = \frac{2\pi R^2 \sin\theta \, d\theta}{4\pi R^2} M$$

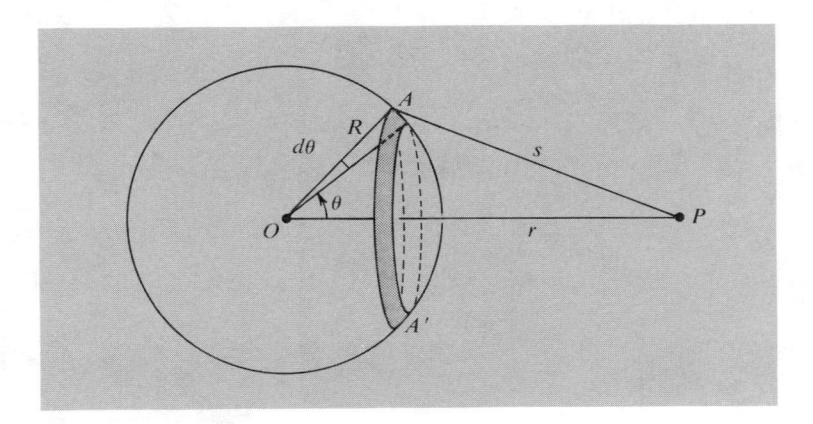

Fig. 7.9 Diagram for considering the gravitational potential energy due to a circular zone of a thin spherical shell of matter.

i.e.

$$dM = \tfrac{1}{2}M \sin\theta \, d\theta$$

All of this material in dM is at the same distance, s, from P. Thus the contribution that dM makes to V is given, according to eqn (7.24), by

$$dV = -\frac{Gm \, dM}{s} = -\frac{GMm}{2} \frac{\sin\theta \, d\theta}{s}$$

The total potential energy of m is thus

$$V(r) = -\frac{GMm}{2} \int \frac{\sin\theta \, d\theta}{s} \qquad (7.26)$$

To evaluate this integral we note that

$$s^2 = R^2 + r^2 - 2Rr\cos\theta$$

Differentiating both sides with respect to θ, remembering that R and r are constants for the purpose of the integration in eqn (7.26), we have

$$2s \frac{ds}{d\theta} = 2Rr \sin\theta$$

Therefore,

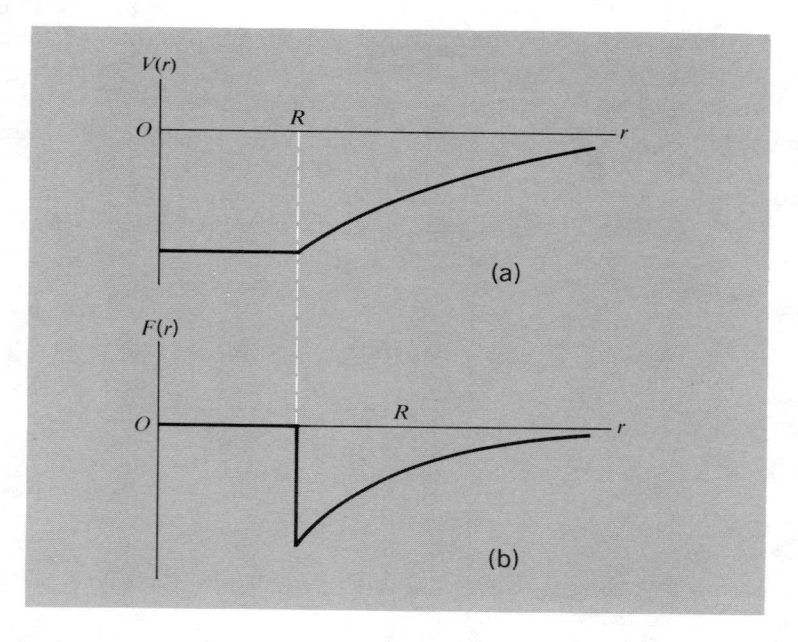

Fig. 7.10 (a) Gravitational potential energy of a point particle as a function of its distance from the centre of a thin spherical shell of radius R. (b) Variation of F with r, derived from (a).

$$\frac{\sin\theta\,d\theta}{s} = \frac{ds}{Rr}$$

But the left-hand side of this is just the integrand of eqn (7.26), which thus becomes

$$V(r) = -\frac{GMm}{2Rr}\int_{\theta=0}^{\pi} ds \tag{7.27}$$

Equation (7.27) is the key result in calculating gravitational potential energies and forces due to spherical objects. In evaluating the integral of ds, however, we must distinguish two cases:

Case 1. Point *P outside* the shell (i.e. $r > R$, as in Fig. 7.9). The limits on s are defined as follows:

$\theta = 0$ giving $s_{min} = r - R$
$\theta = \pi$ giving $s_{max} = r + R$

Hence

$s_{max} - s_{min} = 2R$

$$V(r) = -\frac{GMm}{r} \quad \text{(point } outside \text{ shell)} \tag{7.28}$$

Case 2. Point *P inside* the shell (i.e. $r < R$). The limits now become

$\theta = 0$ giving $s_{min} = R - r$
$\theta = \pi$ giving $s_{max} = R + r$

Hence

$s_{max} - s_{min} = 2r$

$$V(r) = -\frac{GMm}{R} \quad \text{(point } inside \text{ shell)} \tag{7.29}$$

These two results are disarmingly alike. The forces derived from them are, however, quite different. If we proceed now to calculate the forces from eqn (7.25), we must remember that R (the radius of the shell) is a *constant*. It is only r, the distance of the mass m from O, that can vary. Hence we have

Case 1.

$$F_r = -\frac{GMm}{r^2} \quad \text{(point } outside \text{ shell)} \tag{7.30}$$

Case 2.

$$F_r = 0 \quad \text{(point } inside \text{ shell)} \tag{7.31}$$

In Fig. 7.10 we show the graphs of potential energy and force as functions of the radial distance r of the test mass m from the centre of the shell. The discontinuities at $r = R$ are easily seen.

Once we have obtained these results, it is a simple matter to consider a solid sphere of material.

A gravitating sphere

Our primary assumption is that the sphere can be regarded as made up of a whole succession of *uniform* spherical shells, even though the density may vary with radial distance from the centre.

Granted this assumption of symmetry, we can at once draw some conclusions about the gravitational force exerted by such a sphere, of total mass M and radius R.

Case 1. Observation point *outside* the sphere ($r > R$). Since each component spherical shell acts as though its whole mass were at the centre, the same can be said for the sphere as a whole. Regardless of the way in which the density of the material varies with radius, we have

$$F(r) = -\frac{GMm}{r^2} \qquad \text{(point *outside* sphere)} \qquad (7.32)$$

This is the result that we have already used in Chapter 4 in discussing Newton's famous comparison of the gravitational accelerations of the moon and the apple.

Case 2. Observation point *inside* sphere ($r < R$). Here we must be more careful, but we can at once make two clear statements:

a. For all the spherical shells lying *outside* the observation point, the contribution to the force is *zero*, by eqn (7.31).

b. For all shells lying *inside* the radius defined by the observation point, the mass is effectively concentrated at the centre, by eqn (7.30).

To specify the force in this case, therefore, we must know how much mass is enclosed within the sphere of radius r drawn through the observation point.

Case 2'. (special) The same as case 2, but with the extra proviso that the density of the sphere is the same for all r—the sphere is *homogeneous*. In this case, we know that the fraction of the mass contained within radius r is equal to r^3/R^3, because this is the ratio of the partial volume to the whole volume. Hence the amount of mass to be considered is equal to Mr^3/R^3, effectively at the centre, i.e. a distance r from the position of the test mass. This then give us

$$F(r) = -\frac{GMm}{R^3}r \qquad \text{(point *inside* homogeneous sphere)} \qquad (7.33)$$

The combined results of eqns (7.32) and (7.33) for a homogeneous sphere are shown in Fig. 7.11(a). One can also construct the graph of potential energy

A GRAVITATING SPHERE

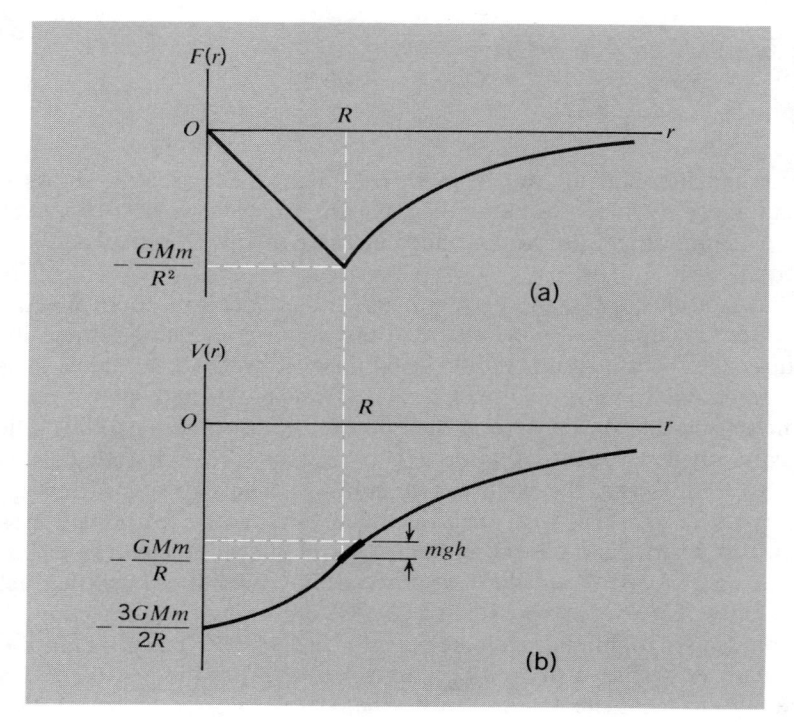

Fig. 7.11 (a) Force on a point particle as a function of its distance from the centre of a uniform solid sphere of radius R. (b) Variation of mutual potential energy with r, obtained from (a) by integration. The approximately linear increase of V with r just outside $r = R$ corresponds to gravity as observed near the earth's surface.

versus r for all values of the distance of a particle of mass m from the centre of the sphere. This is shown in Fig. 7.11(b). For interior points ($r < R$) we have

$$V(r) - V(R) = \frac{GMm}{R^3}\int_R^r r\,dr$$

$$= -\frac{GMm}{2R^3}(R^2 - r^2)$$

But we already know, by putting $r = R$ in eqn (7.28), that

$$V(R) = -\frac{GMm}{R}$$

Thus we get

$$(r < R) \qquad V(r) = -\frac{GMm}{2R^3}(3R^2 - r^2) \qquad (7.34)$$

In particular, at $r = 0$, we have

$$V(0) = -\frac{3GMm}{2R}$$

If one imagines starting out from $r = 0$, then, as Fig. 7.11(b) shows, the potential energy increases parabolically with distance up to $r = R$ and then goes over smoothly into the continued increase of V with r that is described by eqn (7.28).

The variations of $F(r)$ and $V(r)$ for $r > R$ in Fig. 7.11 hold good, as we have seen, for any spherically symmetrical distribution of matter; there is no requirement that the sphere should be homogeneous. On the other hand, one must be careful to remember that eqns (7.33) and (7.34) for $r < R$, refer only to the special case of a sphere of the *same* density throughout. Thus it does *not* correctly describe the variation of gravitational force with radial distance inside such objects as the earth and the sun, which have drastic variations of density with r (see Fig. 7.12). This invalidates a favourite textbook exercise: 'show that a particle would execute simple harmonic motion in a tunnel bored along a diameter of the earth.' The practical impossibility of making such a tunnel may, however, be considered as a far more powerful objection.

If one is used to thinking that the potential energy of an object of mass m, a distance h above the earth's surface, is given by the formula $V = mgh$, it may seem hard to reconcile this result with the result expressed by eqn (7.28):

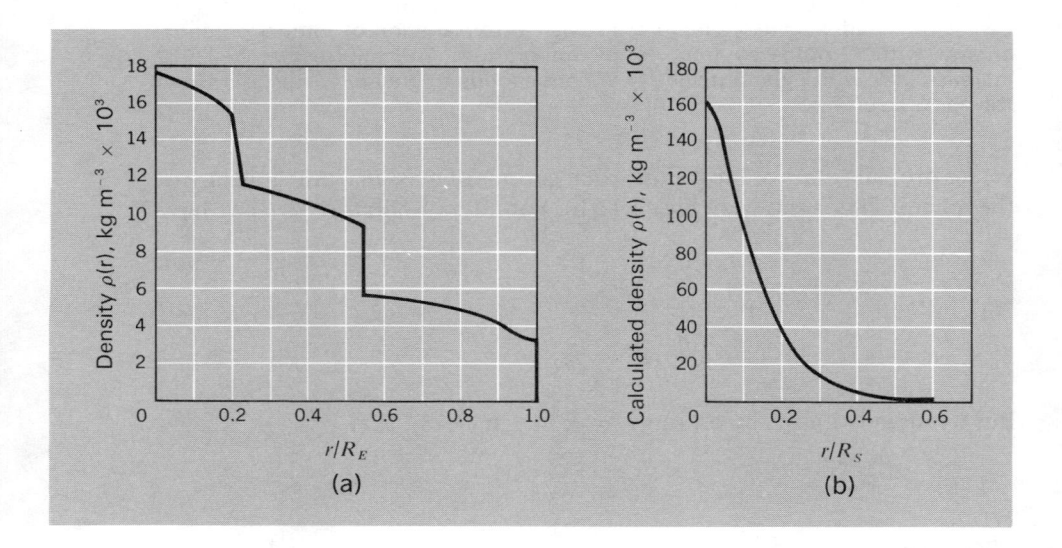

Fig. 7.12 (a) Radial variation of density inside the earth. (After E.C. Bullard.) (b) Calculated variation of density with radial distance inside the sun (After I. Iben and Z. Abraham.)

$$V(r) = -\frac{GMm}{r} \qquad (r > R)$$

There is really no difficulty, however, once one recognizes that the zero of potential energy is arbitrary and that the simple linear formula applies only to objects raised through distances that are exceedingly small compared to the earth's radius. We have, in fact, from the more general formula (i.e. eqn 7.28),

$$V(R + h) - V(R) = -\frac{GMm}{R + h} + \frac{GMm}{R}$$

$$\approx \frac{GMm}{R^2} h$$

Since, however, GMm/R^2 is the gravitational force mg exerted on m, this gives us

$$V(R + h) - V(R) \approx mgh$$

By putting $V(R) = 0$ as an arbitrary reference level of potential energy, we see that $V = mgh$ is an acceptable approximation that applies to small displacements near the earth's surface. This approximately linear increase of potential energy with distance just outside a sphere is indicated in Fig. 7.11(b).

Escape speeds

Suppose we have an object at the surface of a large gravitating sphere, such as a planet, and we want to shoot the object off into outer space so that it never returns. What speed must we give it? This problem is easily considered in terms of energy conservation. At the surface of the planet ($r = R$) the potential energy is given by

$$V(R) = -\frac{GMm}{R}$$

At $r = \infty$ we have $V = 0$. The particle, to reach $r = \infty$, must survive to this distance with some kinetic energy $K(\infty)$. Thus at launch it must have a kinetic energy $K(R)$ defined through the conservation equation

$$K(R) - \frac{GMm}{R} = K(\infty)$$

Therefore,

$$K(R) \geq \frac{GMm}{R}$$

The critical condition is reached if we take the equality in the above statement; the object would then just reach infinity with zero residual speed. The minimum escape speed at radius R is thus given by

$$\tfrac{1}{2}mv_0^2 = \frac{GMm}{R}$$

Therefore,

$$v_0^2 = \frac{2GM}{R} \tag{7.35}$$

As in the previous section, we can put

$$\frac{GMm}{R_E^2} = mg$$

So, from eqn (7.35), we have

$$v_0 = (2gR_E)^{1/2} \tag{7.36}$$

Putting in the familiar values $g \approx 9.8$ m s^{-2}, $R_E \approx 6.4 \times 10^6$ m, we have

$$v_0 \approx 11.2 \text{ km s}^{-1}$$

Notice once again the remarkable implications of energy conservation. For the purpose of calculating *complete escape*, we need specify nothing about the direction in which the escaping object is fired. It could be radially outward, or tangentially, or anything in between; the same value of v_0 applies to all cases (Fig. 7.13).

It is interesting that the magnitude of the escape speed v_0 is exactly $\sqrt{2}$ times

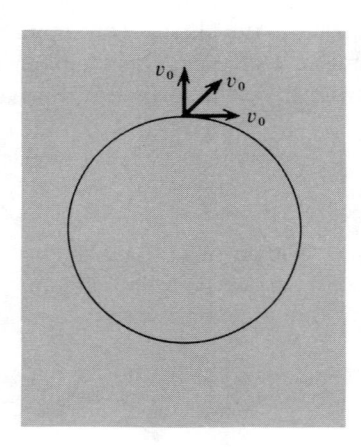

Fig. 7.13 The possibility of escape from a gravitating sphere depends on the magnitude of the velocity, not on its direction.

as great as the orbital speed that a particle would have if it could skim around the earth's surface in a circular orbit of radius R_E.

More about the criteria for conservative forces

In our discussions of forces and potential energy, we have pointed out that the fundamental criterion for a force to be conservative is that the net work done by the force be zero over any closed path. As an aside we pointed out that in one-dimensional problems, but only in one-dimensional problems, this condition is automatically met if the force is a unique function of position. We shall now give a simple example to illustrate how, in two dimensions, this latter condition is not sufficient. A force $F(x, y)$ that is a unique function of x and y may nevertheless be non-conservative.

Our example is this: suppose that a particle, at any point in the xy plane, finds itself exposed to a force F given by

$$F_x = -ky$$
$$F_y = +kx \qquad (7.37)$$

where k is a constant and x and y are the coordinates of the particle. Such a force evidently depends only on the position of the particle. Now let us calculate the work done by this force on the particle if the latter moves counterclockwise around the closed path shown in Fig. 7.14(a). It may well be objected that the particle would never follow this path under the action of the force F and nothing else, but we can imagine that F is exactly balanced by another force, supplied for example by a spring, so that the object can be

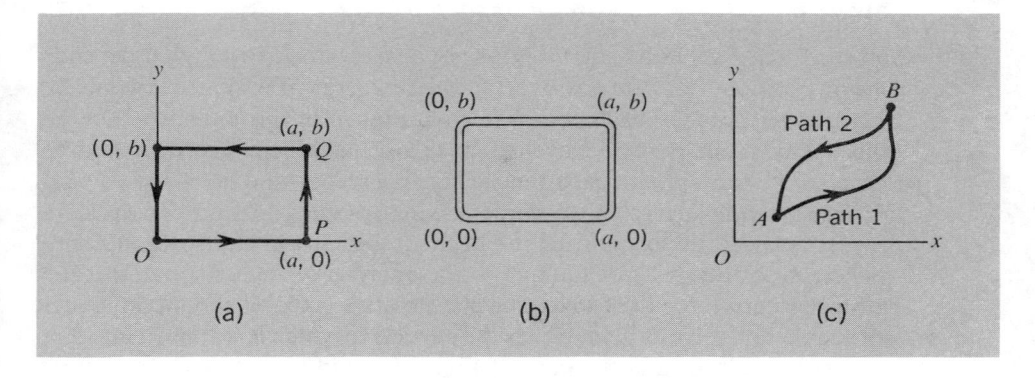

Fig. 7.14 (a) Rectangular path in a plane. (b) Smooth tube shaped to force a particle to follow the path shown in (a). (c) Closed path made up of two different paths between the given points A and B. If the force is conservative, the net work is zero.

freely moved around without any net work being done. Or, alternatively, we can imagine a very smooth pipe, as in Fig. 7.14(b), which compels the particle to travel along the sides of a rectangle and yet does no work on the particle because the force exerted by the pipe is always at right angles to the particle's motion. We should also add the proviso, in this case, that the particle has enough kinetic energy to carry it around the path even if F is taking energy away from it.

Starting at O, let the particle move along the x axis a distance a to the point P with coordinates $x = a$, $y = 0$. The work done by F during this motion is

$$W_1 = \int_{x=0}^{a} F_x \, dx = -ky \int_0^a dx = 0$$

since $y = 0$ everywhere on this portion of the closed path. Next, the particle moves from P to $Q(x = a, y = b)$. The work done by F along this path is

$$W_2 = \int_{y=0}^{b} F_y \, dy = +ka \int_0^b dy = kab$$

For the path from Q to $R(0, b)$, the work done is

$$W_3 = \int_{x=a}^{0} F_x \, dx = -kb \int_a^0 dx = +kab$$

and finally, the work W_4 done as the particle moves from R back to the origin O is zero, since x and hence F_y is zero everywhere on this path:

$$W_4 = 0$$

The total work done in the round trip is therefore

$$W = W_1 + W_2 + W_3 + W_4 = 2kab \neq 0$$

and the force F given by eqn (7.37) is *not* conservative, although it depends only on position. If k were positive, the kinetic energy of the particle would be increased by $2kab$ for each complete circuit taken in this direction. On the other hand, if the particle travelled clockwise, its kinetic energy would be *decreased* by this amount each time. This may seem—and indeed is—a very artificial example; nevertheless, forces having precisely this non-conservative property play an important role in physics, especially in electromagnetism.

There is another way of looking at the analysis of such a situation. The physical criterion for the force to be conservative is that it should do no net work, either positive or negative, as the particle to which it is applied makes a complete circuit around the path of Fig. 7.14(a) in either direction. Let us consider a more general situation (Fig. 7.14c) in which a particle travels from a point A to a point B along one path and returns from B to A along a different path (again imagine that constraining forces, doing no work, are applied

MORE ABOUT THE CRITERIA FOR CONSERVATIVE FORCES

as necessary). We shall assume that the force we are studying (i.e. not including any constraining forces) is a function of the position of the particle only. If this force is conservative, we then have

$$W = \int_{\substack{A \\ \text{Path 1}}}^{B} \mathbf{F} \cdot d\mathbf{s} + \int_{\substack{B \\ \text{Path 2}}}^{A} \mathbf{F} \cdot d\mathbf{s} = 0$$

It follows from this that

$$\int_{\substack{A \\ \text{Path 1}}}^{B} \mathbf{F} \cdot d\mathbf{s} = - \int_{\substack{B \\ \text{Path 2}}}^{A} \mathbf{F} \cdot d\mathbf{s}$$

But now, if the force is a function only of position, we can interchange the limits of the integral on the right and reverse its sign. Hence, if $\mathbf{F}$ is conservative, we must have

$$\int_{\substack{A \\ \text{Path 1}}}^{B} \mathbf{F} \cdot d\mathbf{s} = \int_{\substack{A \\ \text{Path 2}}}^{B} \mathbf{F} \cdot d\mathbf{s}$$

If this condition is satisfied we can put

$$W_{AB} = \int_{A}^{B} \mathbf{F} \cdot d\mathbf{s} \tag{7.38a}$$

without reference to the particular path. And if W_{AB} is the same for all paths from A to B, we can conclude that the force is conservative and that the potential-energy function can be defined through the equation

$$V_B - V_A = - \int_{A}^{B} \mathbf{F} \cdot d\mathbf{s} \tag{7.38b}$$

In evaluating work integrals such as that in eqn (7.38a), one may wish to resolve the force $\mathbf{F}$ into its components F_x and F_y, and resolve the element of path $d\mathbf{s}$ into its components dx and dy, thus getting (for a two-dimensional problem)

$$W = \int_{x_1}^{x_2} F_x \, dx + \int_{y_1}^{y_2} F_y \, dy$$

If one uses this equation, however, one should always remember that, basically, *the integrals are defined as being taken along an actual designated path of the particle.* This might seem to be obvious, because the force must necessarily be applied wherever the particle is. However, there are many

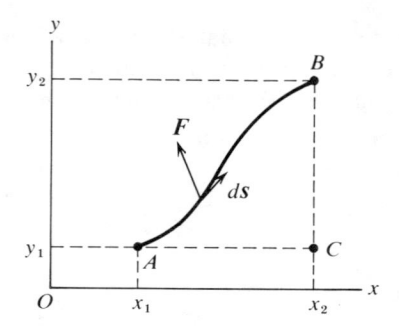

Fig. 7.15 Consideration of work done along a specified path between two points.

situations in which an object experiences a force wherever it happens to be (gravitational force, for example). One can then set up a purely *mathematical* statement that defines a value of F for any given x and y. In general each component of F is a function of both x and y. And unless one already knows that the force is conservative, it is essential to follow the given path, as in Fig. 7.15, and not take the simpler but perhaps unjustified course of (for example) finding the integral of F_x from x_1 to x_2 along the line AC ($y = y_1 =$ const.), followed by the integral of F_y from y_1 to y_2 along the line CB ($x = x_2$ = const.).

It is perhaps worth ending this discussion with the remark that, in many situations, one may use the concept of potential energy even when additional non-conservative forces act on the particle. For example, a satellite moving in the gravitational field of the earth may be subject to the frictional drag of the earth's upper atmosphere. This drag is a dissipative force, non-conservative, and the total mechanical energy of the motion will not be constant but will decrease as the motion proceeds. Nevertheless, one may still properly talk of the gravitational potential energy that the satellite possesses at any given point.

Fields

You will no doubt already have come across the concept of a *field of force* in describing gravitational, electric and magnetic interactions in which there is no apparent contact between interacting bodies.

In the case of a gravitational interaction, for example, we can assign to each point of space around a body a vector, of magnitude equal to the attraction of the body on a test particle of unit mass. The totality of such vectors is called a *field* and the vectors themselves are called the *field strengths* or *intensities* of the field.

As an aid to visualizing a field, use is often made of the concept of a *field line*. Starting from an arbitrary point, we draw an infinitesimal line element in the direction of the field at that point. We thus arrive at a neighbouring point in the field from which we draw another line element in the direction of the field at the new point, and so on. In the limit of making the line elements vanishingly short, we obtain a smooth curve, the tangent to which, at any point, is the direction of the field at that point.

Hand in hand with the concept of a line of force goes that of a line or surface of constant potential energy. Suppose we know the potential energy V of a test particle at each point of space, i.e. we have a relation of the form $V = f(x, y, z)$, where the single-valued function $f(x, y, z)$ depends on the particular field of force under consideration. If we wish to know at what points of space the test particle will have a given value of potential energy, say V_0, we set $V = V_0$ and obtain an equation of the form

$$f(x, y, z) = \text{const.} = V_0$$

This is the equation of a surface, and this surface is called an energy *equipotential* surface. There exists a whole family of these equipotential surfaces, one for each value of V_0. Since, by definition, it requires no work to move our test particle from one point to another on the same equipotential surface, it follows that *the lines of force are everywhere perpendicular to the equipotentials*.

We should draw attention to a distinction between two quantities here. Just as we define *field* as the *force per unit mass* (or charge, etc.) so we define *potential* (ϕ) as the *potential energy per unit mass* (or charge, etc.) To take the specific case of gravitation, we thus have the following paired quantities: gravitational force F (newtons) with gravitational potential energy V (joules), and gravitational field g (m s^{-2}) with gravitational potential ϕ (m^2 s^{-2}).

The complete array of field lines and equipotential surfaces (or, in two dimensions, equipotential lines) provides a graphic picture of a complete field pattern. In two dimensions we see two sets of curves which, however complicated their appearance may be, are everywhere orthogonal to one another. The gravitational field of a spherical object has a simple pattern, in which the field lines are radial lines and the equipotentials are a set of concentric spheres (Fig. 7.16a). The field pattern due to two spheres close together is far less simple (see Fig. 7.16b). In making drawings of the equipotentials, it is often convenient to draw them for equal successive increments of the potential (or potential energy); this makes the picture very informative, just like a contour map—which in effect it is (see Fig. 7.16).

To obtain a complete specification of the force field, we must be able to get the magnitude of the force vector, as well as its direction, at every point of space. Consider a test particle at a point P and let it be displaced an amount ds to a neighbouring point. Its potential energy will change by an amount

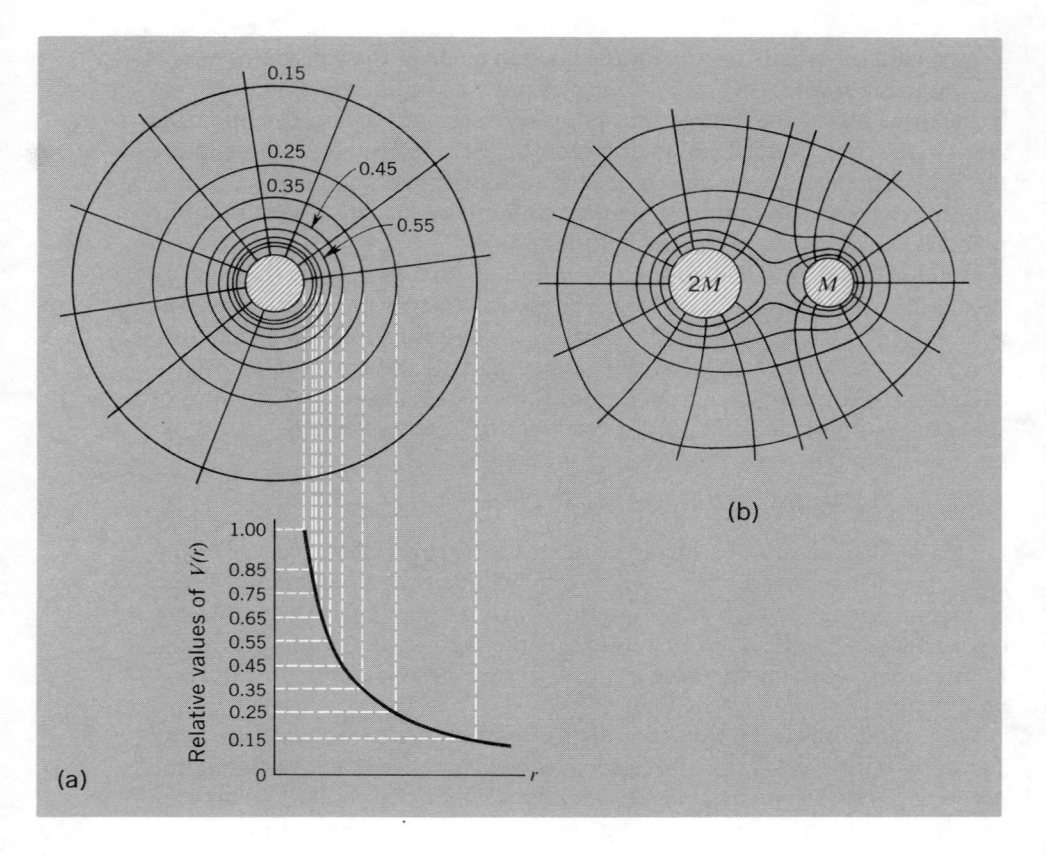

Fig. 7.16 (a) Equipotentials and field lines due to a sphere with a $1/r^2$ force law. (b) Equipotentials and field lines due to a system of two nearby spheres, of masses $2M$ and M.

$$dV = -\mathbf{F} \cdot d\mathbf{s} = -F_s \, ds$$

This may be written in the form

$$F_s = -\frac{dV}{ds} \tag{7.39}$$

In words, the component of the force F in any direction equals the negative rate of change of potential energy with position *in that direction*. The spatial derivative on the right-hand side of eqn (7.39) is called a *directional* derivative, because its value depends on the direction in which ds is chosen at the point P. (Strictly speaking, we should be using the notation of *partial* derivatives: $F_s = -\partial V/\partial s$.) If we move from P to a neighbouring point on the same equipotential surface as that on which P lies, then dV/ds is zero for this direction. If, however, we move to a neighbouring point not on the same equipotential surface as that containing P, dV/ds will be different from zero. That particular direction for which dV/ds has its *maximum* value at a given point defines the direction of the line of force at that point, and the magnitude of this maximum value of dV/ds is the magnitude of the vector force at the point in question.

This maximum value of dV/ds, together with its associated direction, is called the *gradient* of the potential energy; it is a vector directed at right angles to the equipotential surface. In symbols, we write

$$F = -\mathbf{grad}\ V \tag{7.40}$$

To help clarify the idea of the gradient, consider a conservative field in two dimensions. Here the equipotentials are lines, rather than surfaces as they would be in three dimensions. In Fig. 7.17 we show two equipotentials, one for $V = V_0$ and one for $V = V_0 + \Delta V$. Starting at P on V_0, we can move to the equipotential $V_0 + \Delta V$ by any of an infinite number of displacements. However, for a given change of potential energy ΔV one moves along a line of force to attain this change in the shortest possible displacement Δs. The rate of change of potential energy with position is maximum for this direction and this is the direction of the gradient of potential energy. This is indicated in Fig. 7.17, where three directions from P are shown. It is clear that Δs is shorter in length than either Δs_1 or Δs_2 and hence that $\Delta V/\Delta s$ is larger than $\Delta V/\Delta s_1$ or $\Delta V/\Delta s_2$.

Motion in conservative fields

We now turn our attention to the problem of the *motion* of particles in a conservative field of force. We have of course already discussed a number of problems involving motion under gravity, but here we shall try to indicate the value of the energy method as it applies to more complex situations.

If the only force acting on a particle is that due to the field, the law of

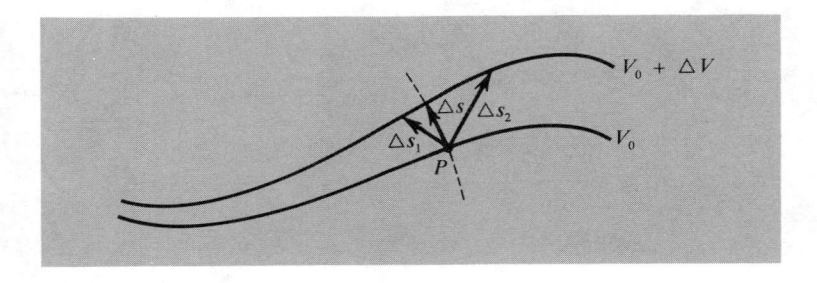

Fig. 7.17 Different paths between two neighbouring equipotentials.

conservation of mechanical energy provides a first step in solving for the motion. This law, expressed as

$$\tfrac{1}{2}mv^2 + V(x,y,z) = E \tag{7.41}$$

where $v^2 = v_x^2 + v_y^2 + v_z^2$, does not, however contain the whole story. It is the result of combining three statements of Newton's law as applied to the three independent coordinate directions, and the synthesis of these three vector equations into one scalar equation involves the discarding of information that in principle is still available to us. Given that V is a conservative potential, we can always find its gradient and hence the vector force (eqn. 7.40). With this, plus the initial conditions (values of r_0 and v_0 at $t = 0$) everything that one needs for a solution of the problem is provided.

Our interest here, however, lies in taking advantage of the energy-conservation equation as far as possible, and supplementing it with whatever other information may be necessary. This extra information may take the form of the explicit use of one or two of the Newton's law statements, as we shall see in the example about to be discussed. Or it may, in suitable circumstances, be contained in a conservation law for quantities besides energy—in particular, angular momentum. The exploitation of this latter property is one of the principal concerns of Chapter 9.

As a good example of the methods, we shall consider the motion of a charged particle in a combined electric and magnetic field. Suppose we have a pair of parallel metal plates mounted inside an evacuated tube and connected to a battery shown in Fig. 7.18, so that a uniform electric field $E\ (= \phi/d)$ exists between the plates. The plates are placed between the poles of a magnet that produces a uniform magnetic field B in a direction perpendicular to the page. We shall assume that electrons start out with negligible energy and velocity from the lower plate. This could, for example, be arranged by giving the

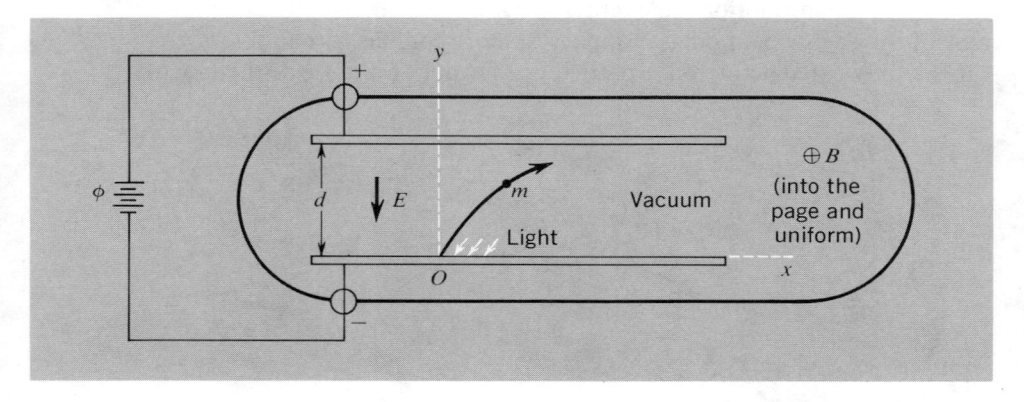

Fig. 7.18 Motion of an electron in vacuum under the combined action of electric and magnetic fields.

lower plate a photosensitive coating and shining light onto it, as in a commercial phototube.

Consider one electron of charge $Q = -e$ emitted at O. It will be accelerated vertically by a constant force equal to $e\phi/d$ directed along the positive y axis, and will be deflected into the path indicated in the figure. Since the magnetic deflecting force, always acting perpendicular to the particle's velocity, does no work and hence does not directly affect the particle's energy, the statement of energy conservation can be written simply as follows:

$$\tfrac{1}{2}m(v_x^2 + v_y^2) - \frac{e\phi}{d}y = 0 \tag{7.42}$$

where we have chosen $V\,[=\,-(e\phi/d)y]$ equal to zero for $y = 0$, i.e. at the bottom plate where the kinetic energy at $y = 0$ is negligible.

As stated at the beginning of this section, we need more information to determine the motion. Since the electric force is directed along the y axis, the only x component of force on the electron is the x component of the magnetic force. To evaluate this force, think of the velocity vector of the electron at some point of its path resolved into x and y components (Fig. 7.19). The component of magnetic force associated with v_x in this situation is parallel to the y axis and does not concern us here, but the velocity component v_y gives rise to a magnetic force component F_x given by

$$F_x = ev_y B$$

and Newton's law of motion for the x component of motion is accordingly

$$ev_y B = m\frac{dv_x}{dt} \tag{7.43}$$

Equations (7.42) and (7.43) can be solved for the motion of particle, as follows:

Since $v_y = dy/dt$, eqn (7.43) can be written as

$$eB\,dy = m\,dv_x$$

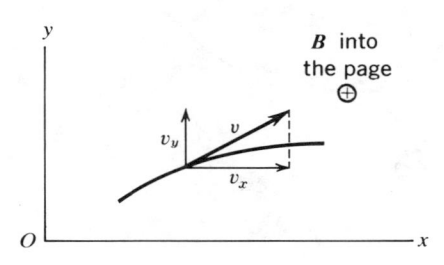

Fig. 7.19 Analysis of the magnetic force into components associated with the separate x and y components of v.

or, integrated,
$$mv_x = eBy \tag{7.44}$$
since $v_x = 0$ when $y = 0$. Now we use the value of v_x from eqn (7.44) in the energy equation (7.42) and get
$$\frac{mv_y^2}{2} + \frac{m}{2}\left(\frac{eB}{m}\right)^2 y^2 - \frac{e\phi}{d}y = 0 \tag{7.45}$$
This is of the form of the energy equation in one dimension; the total energy is zero and the effective potential energy $V'(y)$ is given by
$$V'(y) = -\frac{e\phi}{d}y + \tfrac{1}{2}m\left(\frac{eB}{m}\right)^2 y^2$$
This effective potential-energy curve is shown in Fig. 7.20. It is precisely of the form of a harmonic oscillator potential (centred on a point at a certain positive value of y) and its characteristic angular frequency ω is given by
$$\omega = \frac{eB}{m}$$
This implies a very interesting result. Equation (7.45) and Fig. 7.20 show that, as y increases from zero, the kinetic energy of the electron increases to some maximum value and then decreases again. Mathematically, eqn (7.45) defines a maximum value of y given by putting $v_y = 0$:
$$y_{max} = 2\,\frac{m}{e}\,\frac{\phi}{B^2 d}$$
If this is less than the separation d between the plates (a condition that can be obtained, as the above equation shows, by making B large enough), the y

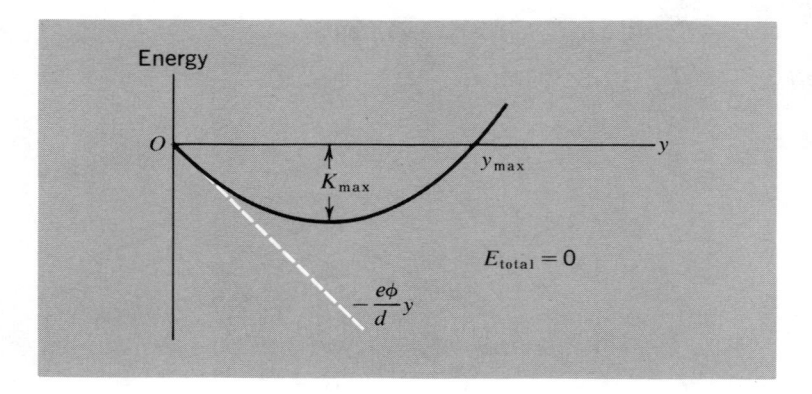

Fig. 7.20 Effective potential-energy curve for the y component of motion in the arrangement of Fig. 7.18.

displacement of the electron will perform simple harmonic motion with the angular frequency ω and with an amplitude A equal to $\frac{1}{2}y_{max}$. Taking $t = 0$ at the instant the electron leaves the lower plate, we can put

$$y = A(1 - \cos \omega t) \tag{7.46a}$$

What about the x component of the motion? If we look at eqn (7.44), we see that v_x is proportional to y. Since y is always positive, v_x is always in the positive x direction. When one couples this with the harmonic oscillation along y, one sees that the path of the electron is a succession of rabbit-like hops, as shown in Fig. 7.21. Specifically, we have, from eqn (7.44),

$$\frac{dx}{dt} = \frac{eB}{m} y = \omega y$$

where ω is the same angular frequency, eB/m, that we introduced earlier. Thus, substituting the explicit expression for y from eqn (7.46a), we have

$$\frac{dx}{dt} = \omega A(1 - \cos \omega t)$$

Integrating this and putting $x = 0$ at $t = 0$ we then find that

$$x = A(\omega t - \sin \omega t) \tag{7.46b}$$

Equations (7.46a and b) taken together show that the path of the electron is in fact a cycloid—just as if it were on the rim of a wheel of radius A rolling along the axis.

The effect of dissipative forces

We mentioned earlier how the conservative properties of a force field may be useful even in circumstances in which dissipative forces are also present. The

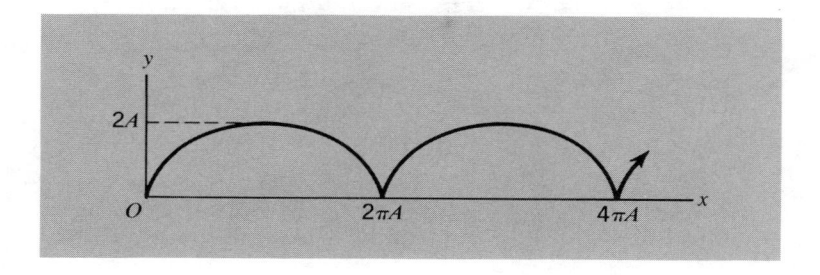

Fig. 7.21 Cycloidal path of electron between charged parallel plates in a magnetic field.

slow decay of the orbits of artificial earth satellites provides a particularly interesting example of this.

We all know that a satellite placed in orbit a few hundred miles above the earth's surface will eventually come down. The descent is, however, spread over many thousands of revolutions. Thus, although the actual path is a continuous inward spiral, as shown in Fig. 7.22(a), the motion during a single revolution is very nearly a closed orbit, which we will take to be circular for simplicity. The statement of Newton's law at any stage is thus given, with extremely little error, by the usual equation for uniform circular motion:

$$F = \frac{GMm}{r^2} = \frac{mv^2}{r}$$

Thus the kinetic energy at any particular value of r is given by

$$K = \tfrac{1}{2}mv^2 = \frac{GMm}{2r} \tag{7.47}$$

This exposes a very curious feature. As the orbital radius gets smaller, the kinetic energy increases—in other words, the satellite speeds up. This happens in spite of the fact that the motion of the satellite is continually opposed by a resistive force. If there were no such force the orbital radius and the speed would remain constant.

We know, however, that the satellite must have lost energy, and the amount of this loss is well defined. This becomes clear when we consider the potential energy:

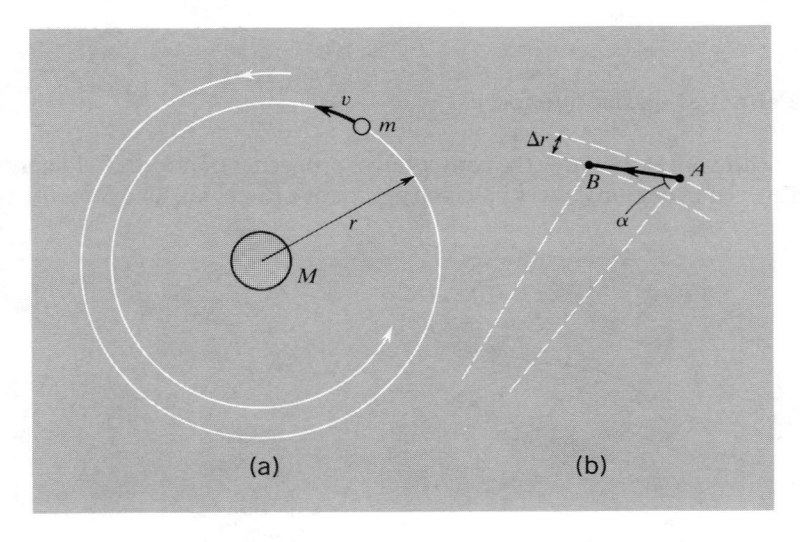

Fig. 7.22 (a) Earth satellite spiralling inward as it loses energy. (b) Small element of the path shown in (a).

THE EFFECT OF DISSIPATIVE FORCES

$$V = -\frac{GMm}{r}$$

The total energy, E, when the orbit radius is r, is given by

$$E = K + V = -\frac{GMm}{2r} \tag{7.48}$$

As r decreases, E becomes more strongly negative. We may note the following relationships in any such circular orbit:

$$E = -K = \tfrac{1}{2}V \tag{7.49}$$

But still one may wonder: why does the satellite *accelerate* in the face of a resistive force? Figure 7.22(b) suggests the answer. The line AB represents a small segment, of length Δs, of the path of the satellite. The starting point A lies on a circle of radius r; the end point B lies on a circle of radius $r + \Delta r$ (Δr being actually negative). We greatly exaggerate the magnitude of Δr, for the sake of clarity. We then see that as the satellite travels from A to B, it feels a component of the gravitational attraction along its path. If the resistive force is $R(v)$, the total force acting on the satellite in the direction of Δs is given by

$$F = \frac{GMm}{r^2} \cos\alpha - R(v)$$

Thus, in the distance Δs, the gain of kinetic energy, equal to the work $F\,\Delta s$, is given by

$$\Delta K = F\,\Delta s = \frac{GMm}{r^2}\Delta s \cos\alpha - R(v)\,\Delta s$$

In the first term on the right, we can substitute

$$\Delta s \cos\alpha = -\Delta r$$

In the second term on the right, we can substitute

$$\Delta s = v\,\Delta t$$

If we also put $\Delta K = mv\,\Delta v$, we arrive at the following equation:

$$mv\,\Delta v = -\frac{GMm}{r^2}\Delta r - R(v)v\,\Delta t$$

However, by differentiation of both sides of eqn (7.47) we have

$$mv\,\Delta v = -\frac{GMm}{2r^2}\Delta r$$

Using this result, we can substitute for the value of $-(GMm/r^2)\,\Delta r$ in the previous equation, and we get

$$mv \, \Delta v = 2mv \, \Delta v - R(v)v \, \Delta t$$

Hence

$$\frac{\Delta v}{\Delta t} = + \frac{R(v)}{m}$$

This is a most intriguing result. The positive sign is not an error. The rate of *increase* of speed of the satellite is directly proportional to the magnitude of the resistive force that opposes its motion! The seeming paradox is resolved when we recognize that the resistance, by changing the orbital path from a circle to an inward spiral, acts as an agent that allows the gravitational force to do *positive* work on the satellite, the amount of which is numerically twice as great as the amount of negative work done by the resistive force itself.

Problems

7.1 Show that if a mass on the end of a string is allowed to swing down in a circular arc from a position in which the string is initially horizontal (and the mass is at rest), then at the lowest point of the swing the tension in the string is three times as great as when the mass simply hangs there. (This result is quoted by Huygens at the end of his book on pendulum clocks, published in 1673.)

7.2 A pendulum bob of mass m, at the end of a string of length l, starts from rest at the position shown in the figure, with the string at 60° to the vertical. At the lowest point of the arc the bob strikes a previously stationary block, of mass nm, that is on a frictionless horizontal surface. The collision is perfectly elastic.
  (a) What is the speed of the bob just before the impact occurs?
  (b) What is the tension in the string at this instant?
  (c) What velocity is given to the block by the impact?
  (d) In the oscillations of the pendulum after the collision, what maximum angle θ to the vertical does the string make? (Obtain your answer in the form $\cos \theta = \cdots$.)

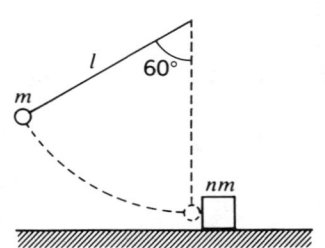

7.3 A small ball of putty, of mass m, is attached to a string of length l fastened to an upright on a wooden board resting on a horizontal table (see the figure). The combined mass of the board and upright is M. The friction coefficient between the board and the table is μ. The ball is released from rest with the string in a horizontal position. It hits the upright in a completely inelastic collision. While the ball swings down, the board does not move.

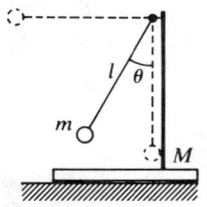

(a) How far does the board move after the collision?
(b) What is the minimum value that μ must have to prevent the board from moving to the left while the ball swings down? Assume that $m \ll M$, which gives the critical condition at $\theta \simeq 45°$.

7.4 An object of mass m slides on a frictionless loop-the-loop apparatus. The object is released from rest at a height h above the top of the loop (see the figure).

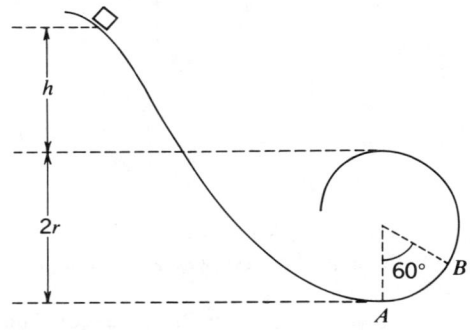

(a) What is the magnitude and direction of the force exerted on the object by the track as the object passes through the point A?
(b) Draw an isolation diagram showing the forces acting on the object at point B, and find the magnitude of the radial acceleration at that point.
(c) Show that the object must start from $h \geq r/2$ to successfully complete the loop.
(d) For $h < r/2$ the object will begin to fall away from the track before reaching the top of the circle. Show that this happens at a position such that $3 \cos \alpha = 2 + 2h/r$, where α is the angular distance from the (upward) vertical.

7.5 A mass slides down a very smooth curved chute as shown. At what horizontal distance from the end of the chute does it hit the ground?

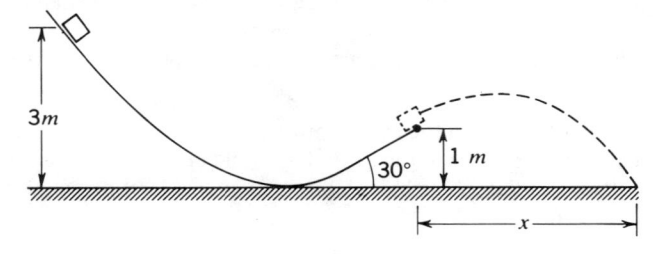

7.6 A daredevil astronomer stands at the top of his observatory dome (see figure) wearing roller skates, and starts with negligible velocity to coast down over the dome surface.

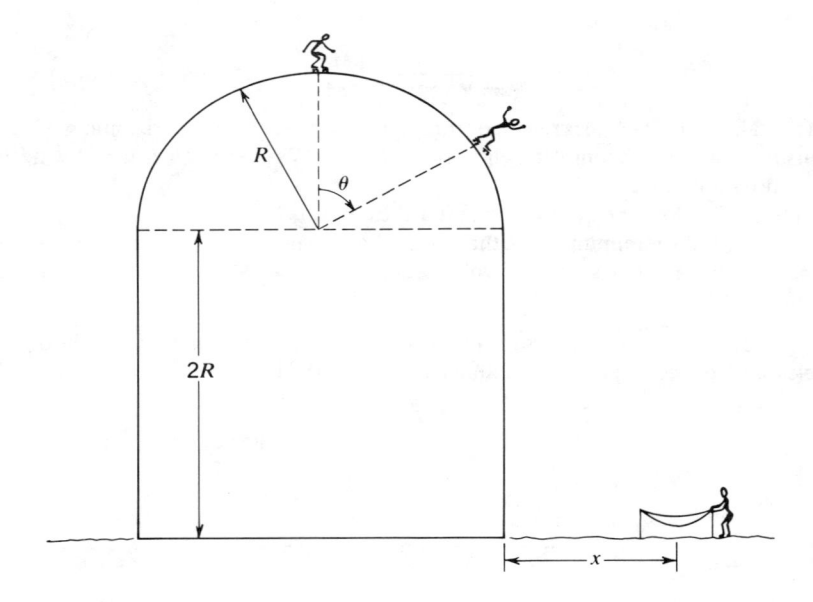

(a) Neglecting friction, at what angle θ does he leave the dome's surface?
(b) If he were to start with an initial velocity v_0, at what angle θ would he leave the dome?
(c) For the observatory shown, how far from the base should his assistant position a net to break his fall, in situation (a)? Evaluate your answer for $R = 8$ m, and use $g \approx 10$ m/s².

7.7 A frictionless airtrack is deformed into the parabolic shape $y = \frac{1}{2}Cx^2$. An object oscillates along the track in almost perfect simple harmonic motion. (Assume $y \ll x$.)
(a) If the period is 30 s, by what distance is the track at $x = \pm 2$ m higher than at $x = 0$?
(b) If the amplitude of oscillation is 2 m, how long does it take for the object to pass from $x = -0.5$ m to $x = +1.5$ m travelling in the positive x direction? (The description of SHM as a projection of uniform circular motion is useful for this.)

7.8 A pendulum clock keeps perfect time at ground level. Approximately how many seconds *per week* would it gain or lose at a height of 20 m above this level? (Assume a simple pendulum, with $T = 2\pi\sqrt{l/g}$. Earth's radius = 6.4×10^6 m. 1 week $\approx 6 \times 10^5$ s.)

7.9 (a) Show that free fall in the earth's gravitational field from infinity results in the same velocity at the surface of the earth that would be achieved by a free fall from a height $H = R_E$ (= radius of earth) under a constant acceleration equal to the value of g at the earth's surface.

(b) Show that the speed at the earth's surface of an object dropped from height h ($h \ll R_E$) is given approximately by

$$\frac{dr}{dt} = v = (2gh)^{1/2}\left(1 - \frac{1}{2}\frac{h}{R_E}\right)$$

(c) Verify the statement in the text (p. 172) that the speed needed for escape from the surface of a gravitating sphere is $v_0\sqrt{2}$, where v_0 is the speed of a particle skimming the surface of the sphere in circular orbit.

7.10 A physicist plans to determine the mass of the earth by observing the change in period of a pendulum as he descends a mine shaft. He knows the radius R of the earth, and he measures the density ρ_s of the crustal material that he penetrates, the distance h he descends, and the fractional change $(\Delta T/T)$ in the pendulum's period.

(a) What is the mass of the earth in terms of these measurements and the earth's radius?

(b) Suppose that the mean over-all density of the earth is twice the mean density of the portion above 3 km. (This supposition is actually rather accurate.) How many seconds per day would a pendulum clock at the bottom of a deep mine (3 km) gain, if it had kept time accurately at the surface?

7.11 The figure shows a system of two uniform, thin-walled, concentric spherical shells. The smaller shell has radius R and mass M, the larger has radius $2R$ and mass $2M$, and the point P is their common centre. A point mass m is situated at a distance r from P.

(a) What is the gravitational force $F(r)$ these shells exert on m in each of the three ranges of r: $0 < r < R$, $R < r < 2R$, $r > 2R$?

(b) What is the gravitational potential energy of the point mass m when it is at P? (Take the potential energy to be zero when m is infinitely far from P.)

(c) If the particle is released from rest very far away from the spheres, what is its speed when it reaches P? (Assume that the particle can pass freely through the walls of the shells.)

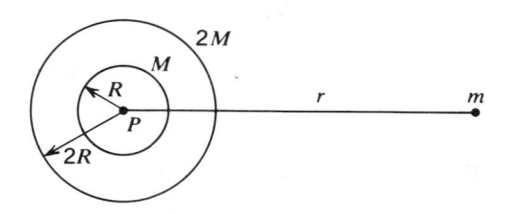

7.12 Assume the moon to be a sphere of uniform density with radius 1740 km and mass 7.3×10^{22} kg. Imagine that a straight smooth tunnel is bored through the moon so as to connect *any two* points on its surface.

(a) Show that the motion of objects along this tunnel under the action of gravity would be simple harmonic.

(b) Calculate the period of oscillation.

(c) Compare this with the period of a satellite travelling around the moon in a circular orbit at the moon's surface.

7.13 The escape speed from the surface of a sphere of mass M and radius R is $v_e = (2GM/R)^{1/2}$; the mean speed $\bar{v}$ of gas molecules of mass m at temperature T is about $(3kT/m)^{1/2}$, where k (= 1.38×10^{-23} J/K) is Boltzmann's constant. Detailed calculation shows that a planetary atmosphere can retain for astronomical times (10^9 years) only those gases for which $\bar{v} \leq 0.2v_e$. Using the data below, find which, if any, of the gases H_2, NH_3, and CO_2 could be retained for such periods by the earth; by the moon; by Mars. $M_E = 6.0 \times 10^{24}$ kg, $R_E = 6.4 \times 10^3$ km, $T_E = 250$ K. For the moon, $M = M_E/81$, $R = 0.27R_E$, and $T = T_E$. For Mars, $M = 0.11 M_E$, $R = 0.53R_E$, and $T = 0.8T_E$.

7.14 A double-star system consists of two stars, of masses M and $2M$, separated by a distance D centre to centre. Draw a graph showing how the gravitational potential energy of a particle would vary with position along a straight line that passes through the centres of both stars. What can you infer about the possible motions of a particle along this line?

7.15 A double-star system is composed of two identical stars, each of mass M and radius R, the centres of which are separated by a distance of $5R$. A particle leaves the surface of one star at the point nearest to the other star and escapes to an effectively infinite distance.

(a) Ignoring the orbital motion of the two stars about one another, calculate the escape speed of the particle.

(b) Assuming that the stars always present the same face to one another (like the moon to the earth), calculate the orbital speed of the point from which the particle is emitted. How would the existence of this orbital motion affect the escape problem?

7.16 Two stars, each of mass M, orbit about their centre of mass. The radius of their common orbit is r (their separation is $2r$). A planetoid of mass m ($\ll M$) happens to move along the axis of the system (the line perpendicular to the orbital plane which intersects the centre of mass), as shown in the figure.

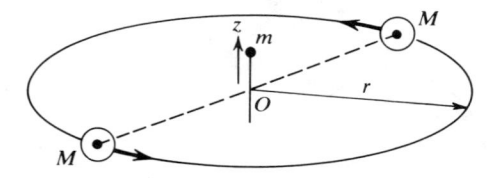

(a) Calculate directly the force exerted on the planetoid if it is displaced a distance z from the centre of mass.

(b) Calculate the gravitational potential energy as a function of this displacement z and use it to verify the result of part (a).

(c) Find approximate expressions for the potential energy and the force in the cases $z \gg r$ and $z \ll r$.

(d) Show that if the planetoid is displaced slightly from the centre of mass, simple harmonic motion occurs. Compare the period T_P of this oscillation with the orbital period T_0 of the binary system.

7.17 The nuclear part of the interaction of two nucleons (protons or neutrons) is described pretty well by a potential $V(r) = -\lambda e^{-r/r_0}/r$ when the separation r is greater than 1 F (1 F = 10^{-15} m). In this expression, r_0 = 1.4 F and λ = 70 MeV-F. (The potential was proposed by H. Yukawa and is named after him.)

(a) Find an expression for the nuclear force that acts on (each of) two nucleons separated by $r > 1$ F.

(b) Evaluate this force for two protons 1.4 F apart, and compare this with the repulsive Coulomb force at that separation.

(c) Estimate the separation at which the nuclear force has dropped to 1% of its value at $r = 1.4$ F. What is the Coulomb force at this separation?

(d) The last result indicates why the Yukawa potential is not part of our macroscopic experience. Note that r_0 characterizes the range of distance over which the interaction is important. In contrast to the nuclear force, gravitational and Coulomb forces have been called forces of infinite range. Indicate why such a description is appropriate.

7.18 A science fiction story concerned a space probe that was retrieved after passing near a neutron star. The unfortunate monkey that rode in the probe was found to be dismembered. The conclusion reached was that the strong gravity gradient (dg/dr) which the probe had experienced had pulled the monkey apart. Given that the probe had passed within 200 km of the surface of a neutron star of mass 10^{30} kg (about half the sun's mass) and radius 10 km, was this explanation reasonable? (The probe was in free fall at the time.)

7.19 A straight chute is to be used to transport articles a given horizontal distance l. The vertical drop of the chute can be freely chosen. The articles are to arrive at the top of the chute with negligible velocity and the chute is to be chosen such that the transit time is a minimum.

(a) If the surface is frictionless, what is the angle of the chute for which the time is minimized?

(b) What is the corresponding angle if the coefficient of friction is μ?

(c) If a playground slide were designed to give minimum duration of ride for given horizontal displacement, and if the coefficient of fricton of the child-slide surface is 0.2, what angle would the slide make with the horizontal? (Ignore the curved portion at the lower end of the slide.)

(d) If the optimization problem of (a) is encountered and if curved chutes are allowed, can you guess roughly what form the best design would have?

7.20 (a) Obtain an expression for the gravitational field due to a thin disc (thickness d, radius R and density ρ) at a distance h above the centre of mass of the disc. In calculating the field, assume the surface density $\sigma = \rho d$ to be concentrated in a disc of negligible thickness a distance h beneath the test point. This will give an accurate result whenever $h \gg d$, and reduces the complexity of the calculation.

(b) Express the field obtained in part (a) as a fraction of $2\pi G\sigma$ for the cases $R = 2h$, $R = 5h$ and $R = 25h$.

(c) How many seconds per year would a pendulum clock gain when suspended with the bob 5 cm above a lead floor 1 cm thick, if it keeps correct time in the absence of the floor?

7.21 Any mass of matter has a gravitational 'self-energy' arising from the gravitational attraction among its parts.

(a) Show that the gravitational self-energy of a uniform sphere of mass M and radius R is equal to $-3GM^2/5R$. You can do this either by calculating the potential energy of the sphere directly, integrating over the interactions between all possible pairs of thin spherical shells within the sphere (and remembering that each shell is counted twice in this calculation) or, perhaps more simply, by imagining the sphere to be built up from scratch by the addition of successive layers of matter brought in from infinity.

(b) Calculate the order of magnitude of the gravitational self-energy of the earth.

8 Inertial forces and non-inertial frames

Forces that can be held responsible for such phenomena as the motion of a Foucault pendulum, the effects in a high-speed centrifuge, the so-called *g* forces on an astronaut during launching, and the preferred direction of rotation of cyclones in the northern and southern hemispheres form an important class, called *inertial forces*, whose origins cannot be traced to some other physical system, as is possible for all the forces previously considered. Gravitational, electromagnetic and contact forces, for example, have their origins in other masses, other charges, or the 'contact' of another object. But the additional forces that make their appearance when an object is being accelerated have no such physical objects as sources. Are these inertial forces real or not? That question, and the answer to it, is bound up with the choice of reference frame with respect to which we are analysing the motion. Let us, therefore, begin this analysis with a reminder of dynamics from the standpoint of an unaccelerated frame.

Motion observed from unaccelerated frames

An unaccelerated reference frame belongs to the class of reference frames that we have called *inertial*. To a good first approximation, as we know, the surface of the earth defines an inertial frame. So also does any system moving at constant speed over the earth. If we considered an object that falls from rest relative to a moving ship, its path would be vertical in the ship's frame and parabolic in the earth's frame. More generally, if we considered an object projected with some arbitrary velocity relative to the earth, its subsequent path would have diverse shapes as viewed from different inertial frames (see Fig. 8.1) but all of them would be parabolic, and all of them, when analysed, would show that the falling object had the vertical acceleration, g, associated with the force F_g due to gravity. Let us now contrast this with what one finds if the reference frame itself has an acceleration.

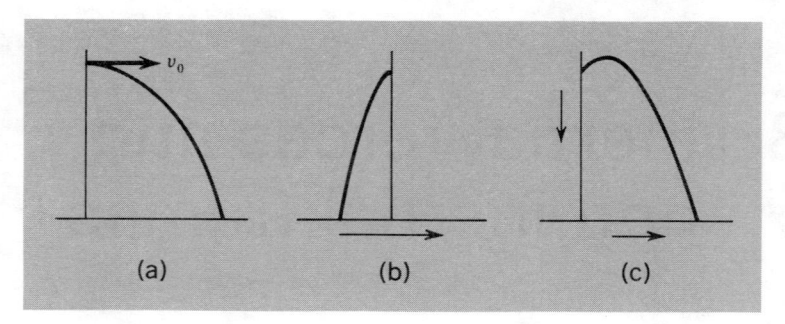

Fig. 8.1 (a) Parabolic trajectory under gravity, as observed in the earth's reference frame. The initial velocity v_0 is horizontal. (b) Same motion observed from a frame with a horizontal velocity greater than v_0. (c) Same motion observed from a frame having both horizontal and vertical velocity components.

Motion observed from an accelerated frame

Suppose that an object is released from rest in a reference frame that has a constant horizontal acceleration with respect to the earth's surface. Let us consider the subsequent motion as it appears with respect to the earth and with respect to the accelerating frame. We shall take the direction of the positive x axis in the direction of the acceleration and will set up two rectangular coordinate systems: system S, at rest relative to the earth, and S', fixed in the accelerating frame (Fig. 8.2). Take the origins of the frames to coincide at $t = 0$, and suppose that the velocity of S' with respect to S at this instant is equal to v_0. The vertical axes of the two systems are taken as positive upward, and the object is released at $t = 0$ from a point for which $x = x' = 0$, $y = y' = h$.

What will the trajectories in S and S' look like? For an observer in S, we already know the answer. To him, the object is undergoing free fall with initial horizontal velocity v_0 (Fig. 8.3a). Thus we have

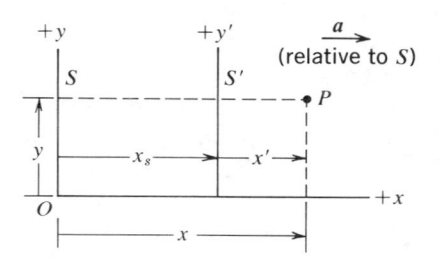

Fig. 8.2 Relationship of coordinates of a particle in two frames that are in accelerated relative motion.

MOTION OBSERVED FROM AN ACCELERATED FRAME

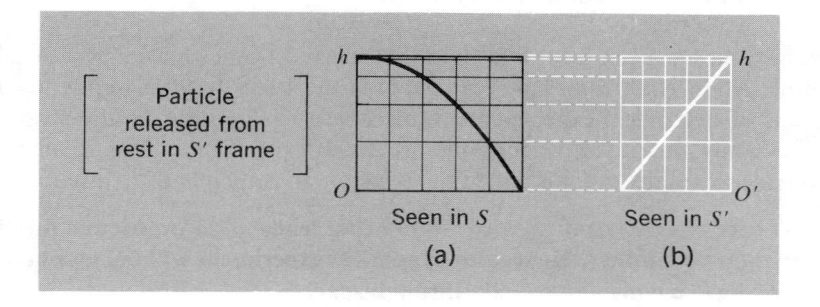

[Figure: Particle released from rest in S' frame. (a) Seen in S — parabolic curve from height h. (b) Seen in S' — straight line from height h.]

Fig. 8.3 (a) Parabolic trajectory of a particle under gravity, as observed in the earth's reference frame S. (b) Same motion observed in a frame S' that has a constant horizontal acceleration.

(as observed in S) $\begin{cases} x = v_0 t \\ y = h - \frac{1}{2}gt^2 \end{cases}$

These two equations uniquely define the position of the object at time t, but to describe the motion as observed in S' we must express the results in terms of the coordinates x' and y' as measured in S'. To transform to the S' frame, we substitute

$$x' = x - x_s$$
$$y' = y$$

where x_s is the separation along the x axis of the origins of S and S' (see Fig. 8.2). We know that

$$x_s = v_0 t + \tfrac{1}{2}at^2$$

Substituting these values we find

(as observed in S') $\begin{cases} x' = v_0 t - (v_0 t + \frac{1}{2}at^2) = -\frac{1}{2}at^2 \\ y' = h - \frac{1}{2}gt^2 \end{cases}$

Thus the path of the particle as observed in S' is a *straight line* given by the equation

$$x' = -\frac{a}{g}(h - y')$$

This is shown in Fig. 8.3(b). In the accelerated frame, the object appears to have not only a constant downward component of acceleration due to gravity, but also a constant horizontal component of acceleration in the $-x$ direction which causes the particle to follow a non-vertical straight-line path.

Accelerated frames and inertial forces

From what has been said, it is clear that inertial frames have a very special status. All inertial frames are *equivalent* in the sense that it is impossible by means of dynamical experiments to discover their motions in any absolute sense—only their relative motions are significant. Out of this dynamical equivalence comes what is called the Newtonian principle of relativity:

> **There is no dynamical observation that leads us to prefer one inertial frame to another. Hence, no dynamical experiment will tell us whether we have a constant velocity through space.**

As we have just seen, however, a relative acceleration between two frames *is* dynamically detectable. As observed in accelerating frames, objects have unexpected accelerations. It follows at once, since Newton's law establishes a link between force and acceleration, that we have a quantitative basis for calculating the magnitude of the inertial force associated with a measured acceleration. Conversely, and more importantly, we have a dynamical basis for inferring the magnitude of an acceleration from the inertial force associated with it. This is the underlying principle of all the instruments known as accelerometers. They function because of the inertial property of some physical mass.

To make the analysis explicit, consider the motion of a particle P with respect to two reference frames like those considered in the last section and shown in Fig. 8.2: an inertial frame S and an accelerated frame S'. The acceleration a of P, as measured in S, is related to its acceleration a' as measured in S' and the acceleration a_s of S' itself by the vector statement:

$$a = a' + a_s \tag{8.1}$$

Multiplying eqn (8.1) throughout by m, we recognize the left-hand side as giving the real (net) force F that is acting on the particle, since this defines the true cause of its acceleration as measured in an inertial frame. That is, in the S frame,

$$F = ma \tag{8.2}$$

but, using eqn (8.1), this gives us

$$F = ma' + ma_s \tag{8.3}$$

We now come to the crucial question: how do we interpret eqn (8.3) from the standpoint of observations made within the accelerated frame S' itself?

Newton's viewpoint—that the net force on an object is the cause of accelerated motion ($F_{net} = ma$)—is so deeply ingrained in our thinking that we are strongly motivated to preserve this relationship at all times. When we observe an object accelerating, we interpret this as due to the action of a net force on the object. Can we achieve a mathematical format that has the appearance of

$F_{net} = ma$ for the present case of an accelerated frame of reference? Yes. By transferring all terms but ma' to the left and treating these terms as forces that act on m, and have a resultant F', which is of the correct magnitude to produce just the observed acceleration a':

$$F' = F - ma_s = ma' \tag{8.4}$$

The net force in the S' frame is thus made up of two parts: a 'real' force, F, with components F_x and F_y, and a 'fictitious' force equal to $-ma_s$, which has its origin in the fact that the frame of reference itself has the acceleration $+a_s$. An important special case of eqn (8.4) is that in which the 'real' force F is zero, in which case the particle, as observed in S', moves under the action of the inertial force $-ma_s$ alone.

The result expressed by eqn (8.4) is not merely a mathematical trick. From the standpoint of an observer in the accelerating frame, the inertial force is actually present. If one took steps to keep an object 'at rest' in S', by tying it down with springs, these springs would be observed to elongate or contract in such a way as to provide a counteracting force to balance the inertial force. To describe such a force as 'fictitious' is therefore somewhat misleading. One would like to have some convenient label that distinguishes inertial forces from forces that arise from true physical interactions, and the term 'pseudo-force' is often used. Even this, however, does not do justice to such forces as experienced by someone who is actually in the accelerating frame. Probably the original, strictly technical name, 'inertial force', which is free of any questionable overtones, remains the best description.

Accelerating frames and gravity

In all our discussions of accelerated frames, we have assumed that the observers know 'which way is up', i.e. they know the direction and magnitude of the force of gravity and treat it (as we have done) as a real force, whose source is the gravitating mass of the earth. But suppose our frame of reference to be a completely enclosed room with no access to the external surroundings. What can one then deduce about gravity and inertial forces through dynamical experiments wholly within the room?

We shall suppose once again that there is an observer in a frame, S, attached to the earth. This observer is not isolated; he is able to verify that the downward aceleration of a particle dropped from rest is along a line perpendicular to the earth's surface and hence is directed toward the *centre* of the earth. He is able to draw the orthodox conclusion that this acceleration is due to the gravitational attraction from the large mass of the earth. Our second observer is shut up in a room that defines the frame S'. Initially it is known that the floor of his room is horizontal and that its walls are vertical. In subsequent measurements, however, the observer in S' finds that a plumbline hangs at an angle to what he had previously taken to be the vertical, and that objects dropped from rest travel parallel to his plumbline. The observers in S and S' report their findings to one another by radio. The observer in S' then concludes that he has two alternative ways of accounting for the component of force, parallel to the floor, that is now exerted on all particles as observed in *his* frame:

1. In addition to the gravitational force, there is an inertial force in the $-x$ direction due to the acceleration of his frame in the $+x$ direction.

2. His frame is not accelerating, but a large massive object has been set down in the $-x$ direction outside his closed room, thus exerting an additional gravitational force on all masses in his frame.

In supposing that both hypotheses work equally well to explain what happens in S', we must assume that the additional massive object, postulated in alternative 2, produces an effectively uniform gravitational field throughout the room.

From dynamical experiments made entirely within the closed room, there is no way to distinguish between these hypotheses. The acceleration of the frame of reference produces effects that are identical to those of gravitational attraction. Inertial and gravitational forces are both proportional to the mass of the object under examination. The procedures for detecting and measuring them are identical. Moreover, they are both describable in terms of the properties of a *field* (an acceleration field) that has a certain strength and direction at any given point. An object placed in this field experiences a certain force without benefit of any contact with its surroundings. Is all this just an interesting parallel, or does it have a deeper significance?

Einstein, after pondering these questions, concluded that there was indeed something fundamental here. In particular, the completely exact proportionality (as far as could be determined) between gravitational force and inertial mass suggested to him that no physical distinction could be drawn, at least within a limited region, between a gravitational field and a general acceleration of the reference frame (see Fig. 8.4). He announced this—his famous principle of equivalence—in 1911. The proportionality of gravitational force to inertial mass now becomes an exact necessity, not an empirical and inevitably approximate result. It is also implied that *anything* traversing a gravitational field must follow a curved path, because such a curvature would appear on purely kinematic and geometrical grounds if we replaced the gravitational field by the equivalent acceleration of our own reference frame. In particular, this should happen with rays of light (see Fig. 8.5). With the help of these ideas Einstein proceeded to construct his general theory of relativity.

ACCELERATING FRAMES AND GRAVITY

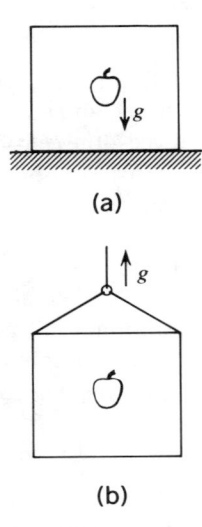

Fig. 8.4 (a) Apple falling inside a box that rests on the earth. (b) Indistinguishable motion when the apple is inside an accelerated box in outer space.

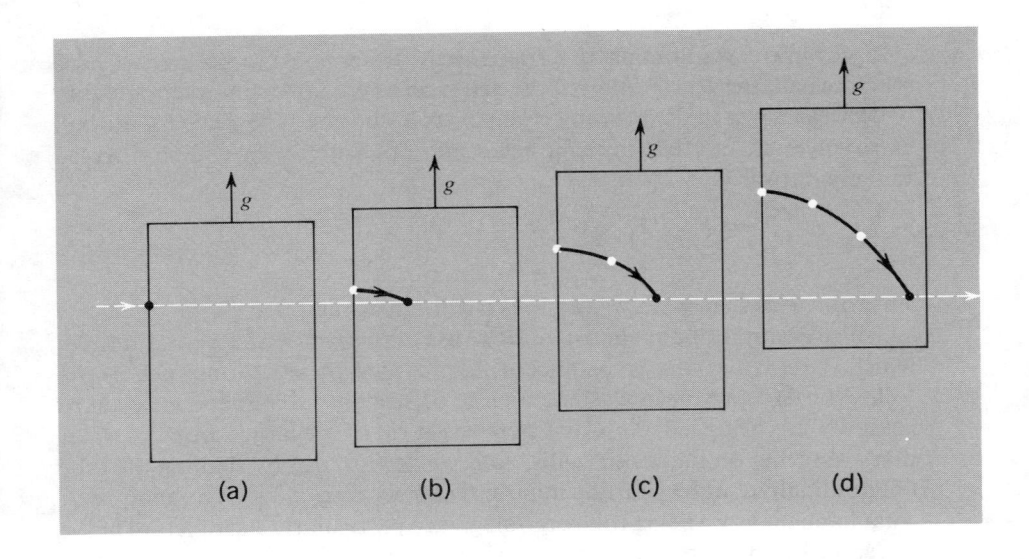

Fig. 8.5 Successive stages in the path of a horizontally travelling object as observed within an enclosure accelerating vertically upward. This illustrates the equivalence of gravity and a general acceleration of the reference frame.

Centrifugal force

We shall now consider a particular kind of inertial force that always appears if the motion of a particle is described and analysed from the standpoint of a *rotating* reference frame. This force—the centrifugal force—is familiar to us as the force with which, for example, an object appears to pull on us if we whirl it around at the end of a string. To introduce it, we shall consider a situation of just this kind.

Suppose that a 'tether ball' is being whirled around in horizontal circular motion with constant speed (Fig. 8.6). We shall analyse the motion of the ball as seen from two viewpoints: a stationary frame S, and a rotating frame S' that rotates with the same (constant) rotational speed as the ball. For convenience, we align the coordinate systems with their z and z' axes (as well as origins) coincident. The rotational speed of S' relative to S will be designated ω (in rad s^{-1}). Figure 8.6 shows the analysis with respect to these two frames. The essential conclusions are these;

1. From the standpoint of the stationary (inertial) frame, the ball has an acceleration $(-\omega^2 r)$ toward the axis of rotation. The force, F_r, to cause this acceleration is supplied by the tethering cord, and we must have

(in S) $\quad F_r = -m\omega^2 r$

2. From the standpoint of a frame that rotates so as to keep exact pace with the ball, the acceleration of the ball is zero. We can maintain the validity of Newton's law in the rotating frame if, in addition to the force F_r, the ball experiences an inertial force F_i, equal and opposite to F_r, and so directed radially outward:

(in S') $\quad \begin{cases} F'_r = F_r + F_i = 0 \\ F_i = m\omega^2 r \end{cases}$

The force F_i is then what we call the centrifugal force.

A nice example of our almost intuitive use of this force, under conditions in which there is nothing to balance it, is provided by situations such as the following: we have been washing a piece of straight tubing, and we want to get it dry on the inside. As a first step we get rid of the larger drops of water that are sitting on the inside walls. And we do this, not by shaking the tube longitudinally, but by whirling it in a circular arc (Fig. 8.7a). The analysis of what happens as we begin this rotation gives a particularly clear picture of the difference between the descriptions of the process in stationary and rotating frames. It also provides us with a different way of deriving the formula for the centrifugal force itself.

Suppose that a drop, of mass m, is sitting on the inner wall of the tube at a point A (Fig. 8.7b), a distance r from the axis of rotation. Assume that the tube is very smooth, so that the drop encounters no resistance if it moves along the tube. The drop must, however, be carried along in any transverse

CENTRIFUGAL FORCE

Procedure	Viewed from stationary frame S	Viewed from rotating frame S'
Pictorial sketch of problem	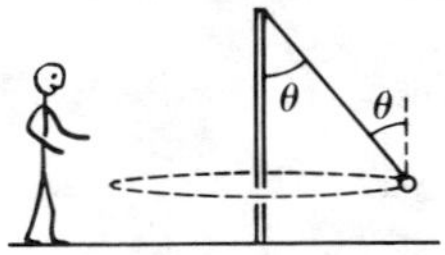 Ball is observed to move with speed v in a circle of radius r (angular speed ω)	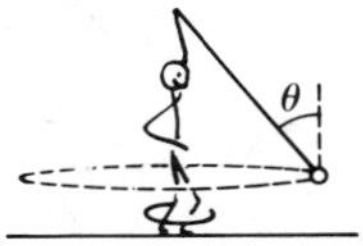 Boy turns around at the same angular speed ω as the ball; from his point of view, the ball is at rest
"Isolate the body." Draw all forces that act on the ball	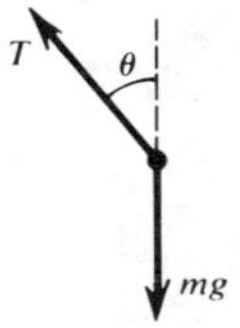 (forces: T, mg)	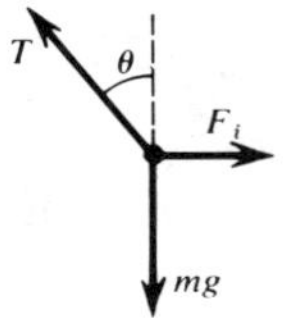 (forces: T, mg, F_i) T = tension in cord mg = force of gravity F_i = inertial force due to viewing the problem from a rotating frame
For ease of calculation we resolve forces into components in mutually perpendicular directions	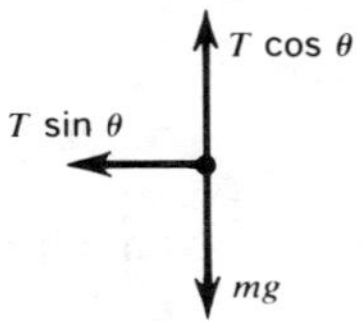 Components: $T\cos\theta$, $T\sin\theta$, mg	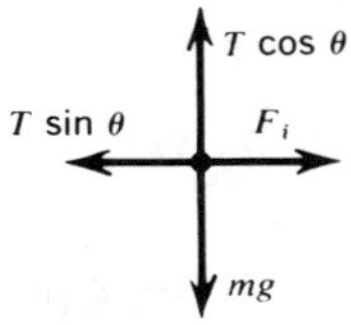 Components: $T\cos\theta$, $T\sin\theta$, F_i, mg
We now analyse the problem in terms of $F = ma$	**Vertical Direction** Because there is no vertical acceleration, we conclude that the net vertical force must be zero; hence $T\cos\theta = mg$ **Horizontal Direction** The object is travelling in a circle, therefore accelerating; the net force (i.e. the sum of all three forces) is horizontal towards the centre of the circle, and must be equal in magnitude to mv^2/r hence $$T\sin\theta = \frac{mv^2}{r} = m\omega^2 r$$ This force is directed radially inward	**Vertical Direction** Because there is no vertical acceleration, we conclude that the net vertical force must be zero; hence: $T\cos\theta = mg$ **Horizontal Direction** The object is "at rest," therefore the sum of all the forces on it must be zero; hence F_i is equal in magnitude to $T\sin\theta$. From the analysis in the left column, it is given by $$F_i = \frac{mv^2}{r} = m\omega^2 r$$ and is directed outward. We call F_i the centrifugal force

Fig. 8.6 Motion of a suspended ball, which is travelling in a horizontal circle, as analysed from the earth's reference frame and from a frame rotating with the ball.

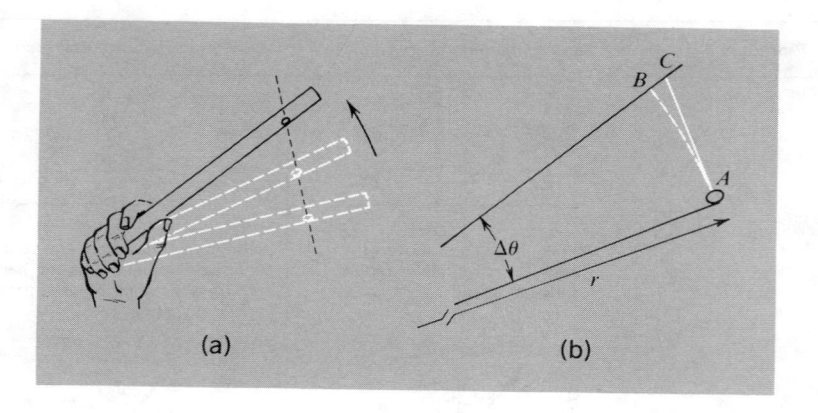

Fig. 8.7 (a) Shaking a drop of water out of a tube. (b) Analysis of initial motion in terms of centrifugal forces.

movement of the tube resulting from the rotation. Then if the tube is suddenly set into motion and rotated through a small angle $\Delta\theta$, the drop, receiving an impulse normal to the wall of the tube at A, moves along the straight line AC. This, however, means that it is now farther from the axis of rotation than if it had been fixed to the tube and had travelled along the circular arc AB. We have

$$BC \approx \tfrac{1}{2} r (\Delta\theta)^2$$

We can, however, express $\Delta\theta$ in terms of the angular velocity ω and the time Δt: $\Delta\theta = \omega \, \Delta t$. Thus we have

$$BC = \tfrac{1}{2} \omega^2 r (\Delta t)^2$$

This is then recognizable as the radial displacement that occurs in time Δt under an acceleration $\omega^2 r$. Hence we can put

$$a_{\text{centrifugal}} = \omega^2 r$$

and so

$$F_{\text{centrifugal}} = m\omega^2 r$$

Notice, then, that what is, in fact, a small transverse displacement in a straight line, with no real force in the radial direction, appears in the frame of the tube as a small, purely radial displacement under an unbalanced centrifugal force. The physical fact that the drop is moved outward along the tube is readily understood in terms of either description. (We should add, however, that our analysis as it stands does only apply to the initial step of the motion. Once the drop has acquired an appreciable radial velocity, things become more complicated.)

The term 'centrifugal force' is frequently used incorrectly. For example, one may read such statements as 'The satellite does not fall down as it moves around the earth because the centrifugal force just counteracts the force of gravity and hence there is no net force to make it fall.' Any such statement flouts Newton's first law—A body with no net force on it travels in a *straight line*. . . . For if the satellite is described as *moving* in a curved path around the earth, it must also have an unbalanced force on it. The only frame in which the centrifugal force does balance the gravitational force is the frame in which the satellite appears not to move at all. One can, of course, consider the description of such motions with respect to a reference frame rotating at some arbitrary rate different from that of the orbiting object itself. In this case, however, the centrifugal contribution to the inertial forces represents only a part of the story, and the simple balancing of 'real' and centrifugal forces does not apply. In particular, let us reemphasize *that in a non-rotating frame of reference there is no such thing as centrifugal force*.

General equation of motion in a rotating frame

We have seen how the centrifugal force, $m\omega^2 r$, exerted on a particle of mass m in a reference frame rotating at a given angular speed ω, depends only on the distance r of the particle from the axis of rotation. Under more general conditions, however, another inertial force appears in a rotating frame. This is the *Coriolis force*, whose magnitude depends *only* on the velocity of the particle (and not on its position). It is always a *deflecting* force exerted at right angles to the direction of the motion of the particle as observed in the rotating frame.

The behaviour of objects in motion within a rotating frame under the influence of both centrifugal and Coriolis effects can run very much counter to one's intuitions. We shall now seek, however, to analyse the general situation mathematically. By introducing *vector* expressions for rotational motion a succinct notation can be developed that gives both centrifugal and Coriolis forces in a form valid in three dimensions using any type of coordinate system.

The aim of the following discussion is to relate the time derivatives of the displacement of a moving object as observed in a stationary frame S and in a rotating frame S'. To set the stage, we shall introduce the idea that angular velocity may be represented as a vector.

Consider first a point P on a rotating disc (Fig. 8.8a). It has a purely tangential velocity, v_θ, in a direction at right angles to the radius OP. We can describe this velocity, in both magnitude and direction, if we define a *vector* according to the same convention that we introduced for torque in Chapter 2. That is, if the fingers of the *right* hand are curled around in the sense of

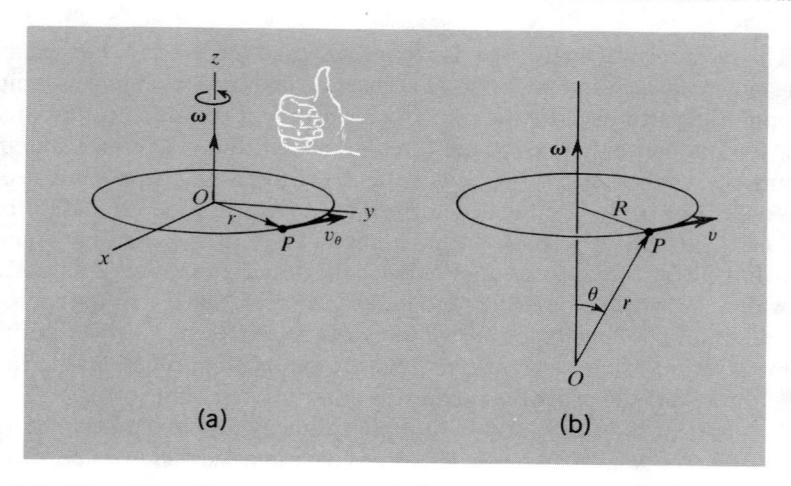

Fig. 8.8 Use of angular velocity as a vector to define the linear velocity of a particle on a rotating table: $v = \omega \times r$.

rotation, keeping the thumb extended as shown in the figure, then ω is represented as a vector, of length proportional to the angular speed, in the direction in which the thumb points. Thus with ω pointing along the positive z direction, one is defining a rotation that carries each point such as P from the positive x direction towards the positive y direction. The rotation of the disc is in this case counterclockwise as seen from above.

The velocity of P is now given by the vector (cross) product of ω with the radius vector r:

$$v = \omega \times r \qquad (8.5)$$

This vector-product expression is valid in three dimensions also, if the position vector r of P is measured from any point on the axis of rotation, as shown in Fig. 8.8(b). The radius of the circle in which P moves is $R = r \sin \theta$. Thus we have $v = v_\theta = \omega r \sin \theta$, in a direction perpendicular to the plane defined by ω and r. That is precisely what eqn (8.5) gives us.

Next, we consider how the change of *any* vector during a small time interval Δt can be expressed as the vector sum of two contributions:

1. The change that would occur if it were simply a vector of constant length embedded in the rotating frame S'.
2. The further change described by its change of length and direction as observed in S'.

In Fig. 8.9(a) we show this analysis for motion confined to a plane. The vector A at time t is represented by the line CD. If it remains fixed with respect to a rotating table, its direction at time $t + \Delta t$ is given by the line CE, where $\Delta \theta = \omega \Delta t$. Thus its change due to the rotation alone would be represented by

GENERAL EQUATION OF MOTION IN A ROTATING FRAME

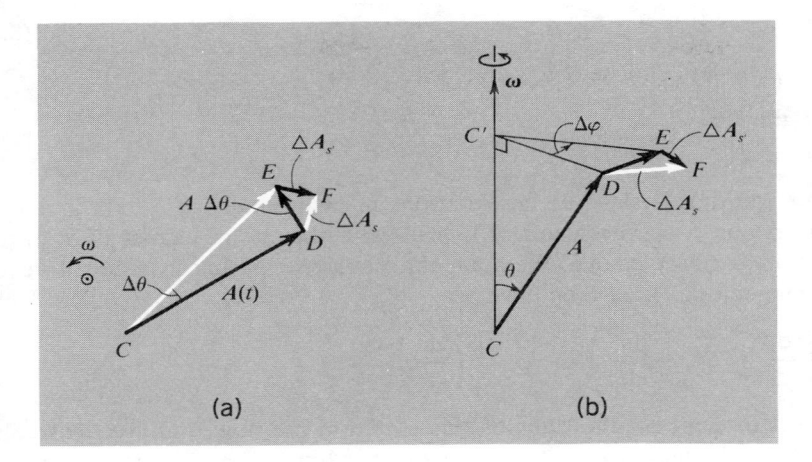

Fig. 8.9 (a) Change of a vector, analysed in terms of its change as measured on a rotating table, together with the change due to rotation of the table itself. (b) Similar analysis for an arbitrary vector referred to any origin on the axis of rotation.

DE, where $DE = A\,\Delta\theta = A\omega\,\Delta t$. From the standpoint of frame S' this change would not be observed. There might, however, be a change represented by the line EF; we shall denote this as $\Delta A_{S'}$—the change of A as observed in S'. The vector sum of DE and EF, i.e. the line DF, then represents the true change of A as observed in S. We therefore denote this as ΔA_S.

In Fig. 8.9(b) we show the corresponding analysis for three dimensions. The length of DE is now equal to $A\sin\theta\,\Delta\varphi$; its direction is perpendicular to the plane defined by ω and A. Since $\Delta\varphi = \omega\,\Delta t$, we can put

vector displacement $DE = (\omega \times A)\Delta t$

The displacement $\Delta A_{S'}$ may be in any direction with respect to DE, but the two again combine to give a net displacement DF which is to be identified with ΔA_S. Thus we have

$$\Delta A_S = \Delta A_{S'} + (\omega \times A)\Delta t$$

We can at once proceed from this to a relation between the rates of change of A as observed in S and S', respectively.

$$\left(\frac{dA}{dt}\right)_S = \left(\frac{dA}{dt}\right)_{S'} + \omega \times A \tag{8.6}$$

This is a very powerful relation because A can be any vector we please.

First, we shall choose A to be the position vector r. Then $(dA/dt)_S$ is the true velocity, v, as observed in S, and $(dA/dt)_{S'}$ is the apparent velocity, v', as observed in S'. Thus we immediately have

$$v = v' + \omega \times r \tag{8.7}$$

Next, we shall choose A to be the velocity v:

$$\left(\frac{dv}{dt}\right)_S = \left(\frac{dv}{dt}\right)_{S'} + \omega \times v \tag{8.8}$$

Now $(dv/dt)_S$ is the true acceleration, a, as observed in S. The quantity $(dv/dt)_{S'}$ is, however, a sort of hybrid—it is the rate of change in S' of the velocity as observed in S. We can make more sense of this if we substitute for v from eqn (8.7); we then have

$$\left(\frac{dv}{dt}\right)_{S'} = \left(\frac{dv'}{dt}\right)_{S'} + \omega \times \left(\frac{dr}{dt}\right)_{S'}$$

The two terms on the right of this equation are now quite recognizable; $(dv'/dt)_{S'}$ is the acceleration, a', as observed in S', and $(dr/dt)_{S'}$ is just v'. Thus we have

$$\left(\frac{dv}{dt}\right)_{S'} = a' + \omega \times v'$$

Substituting this in eqn (8.8) we thus get

$$a = a' + \omega \times v' + \omega \times v$$

We do not need to have both v and v' on the right-hand side, and we shall again substitute for v from eqn (8.7). This gives us finally

$$a = a' + 2\omega \times v' + \omega \times (\omega \times r) \tag{8.9}$$

A remark is in order regarding the last term, which involves the cross product of three vectors. According to the rules of vector algebra, the cross product inside the parentheses is to be taken first, then the other cross product performed. A non-zero answer will result for all cases where the angle formed by ω and r is other than 0° or 180°. Performing the cross products in the reverse (incorrect) order, however, would result in zero for all cases, regardless of the angle between these vectors.

Multiplying eqn (8.9) throughout by the mass m of the object, we recognize the left side as the net external force on the mass as seen in the stationary system.

$$ma = F_{net} = ma' + 2m(\omega \times v') + m[\omega \times (\omega \times r)]$$

In the rotating frame of reference, the object m has the acceleration a'. We may preserve the format of Newton's second law in this accelerated frame of reference by rearranging the above equation, so as to be able to write

$$F'_{net} = ma' \tag{8.10a}$$

where

$$F'_{net} = F_{net} - 2m(\omega \times v') - m[\omega \times (\omega \times r)] \qquad (8.10b)$$

The above result shows that the dynamics of motion as observed in a uniformly rotating frame of reference may be analysed in terms of the following three categories of forces:

'Real': F_{net}
: This is the sum of all the 'real' forces on the object such as forces of contact, tensions in strings, the force of gravity, electrical forces, magnetic forces, and so on. Only these forces are seen in a stationary frame of reference.

Coriolis: $-2m(\omega \times v')$
: The Coriolis force is a *deflecting* force always at right angles to the velocity v' of the mass m. If the object has no velocity in the rotating frame of reference, there is no Coriolis force. It is an inertial force *not* seen in a stationary frame of reference. Note minus sign.

Centrifugal: $-m[\omega \times (\omega \times r)]$
: The centrifugal force depends on position only and is always radially outward. It is an inertial force *not* seen in a stationary frame of reference. We could equally well write it as $-m[\omega \times (\omega \times r')]$. Note the minus sign.

The specification of F' in eqn (8.10) can be made entirely on the basis of measurements of position, velocity, and acceleration as observed within the rotating frame itself. The centrifugal term, involving the vector r, might seem to contradict this, but we could just as well put r' instead of r, because observers in the two frames do agree on the vector position of a moving object at a given instant, granted that they use the same choice of origin.

The earth as a rotating reference frame

The local value of g

If a particle P is at rest at latitude λ near the earth's surface, then as judged in the earth's frame it is subjected to the gravitational force F_g and the centrifugal force F_{cent} shown in Fig. 8.10(a). The magnitude of the latter is given, according to eqn (8.10b), by the equation

$$F_{cent} = m\omega^2 R \sin\theta = m\omega^2 R \cos\lambda$$

where R is the earth's radius. We have already discussed in Chapter 4 the way in which this centrifugal term reduces the local magnitude of g and also

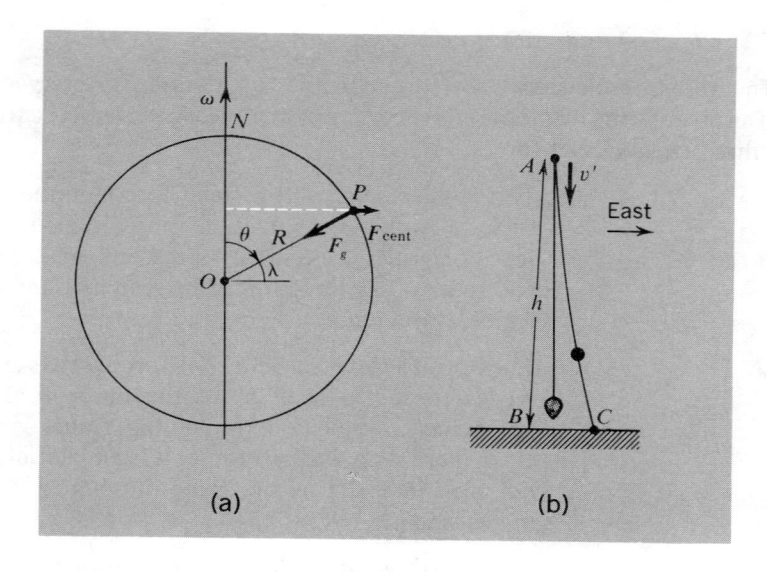

Fig. 8.10 (a) Forces on an object at rest on the earth, as interpreted in a reference frame that rotates with the earth. (b) An object falling from rest relative to the earth undergoes an eastward displacement.

modifies the local direction of the vertical as defined by a plumbline. The analysis is in fact much simpler and clearer from the standpoint of our natural reference frame as defined by the earth itself. We have, as Fig. 8.10(a) shows, the following relations:

$$F'_r = F_g - F_{\text{cent}} \cos \lambda = F_g - m\omega^2 R \cos^2 \lambda$$
$$F'_\theta = F_{\text{cent}} \sin \lambda = m\omega^2 R \sin \lambda \cos \lambda \tag{8.11}$$

Deviation of freely falling objects

If a particle is released from rest at a point such as P in Fig. 8.10(a), it begins to accelerate downward under the action of a net force F' whose components are given by eqn (8.11). As soon as it has any appreciable velocity, however, it also experiences a Coriolis force given by the equation

$$F_{\text{Coriolis}} = -2m\omega \times v' \tag{8.12}$$

Now the velocity v' is in the plane PON containing the earth's axis. The Coriolis force must be perpendicular to this plane, and a consideration of the actual directions of ω and v' shows that it is eastward. Thus if we set up a local coordinate system defined by the local plumbline vertical and the local easterly direction, as in Fig. 8.10(b), the falling object deviates eastward

THE EARTH AS A ROTATING REFERENCE FRAME

from a plumbline AB and hits the ground at a point C. The effect is very small but has been detected and measured in careful experiments.

To calculate what the deflection should be for an object falling from a given height h, we use the fact that the value of v' to be inserted in eqn (8.12) is extremely well approximated by the simple equation of free vertical fall:

$$v' = gt$$

where v' is measured as positive downward. Thus if we label the eastward direction as x', we have

$$m\frac{d^2x'}{dt^2} = (2m\omega \cos \lambda)gt$$

Integrating this twice with respect to t, we have

$$x' = \tfrac{1}{3}g\omega t^3 \cos \lambda \tag{8.13}$$

For a total distance of vertical fall equal to h, we have $t = (2h/g)^{1/2}$, which thus gives us

$$x' = \frac{2\sqrt{2}}{3} \frac{\omega \cos \lambda}{g^{1/2}} h^{3/2} \tag{8.14}$$

Inserting approximate numerical values ($\omega = 2\pi$ day$^{-1} \approx 7 \times 10^{-5}$ s^{-1}), one finds

$$x' \approx 2 \times 10^{-5} h^{3/2} \cos \lambda \quad (x' \text{ and } h \text{ in m})$$

Thus, for example, with $h = 50$ m at latitude $45°$, one has $x' \approx 5$ mm, or about $\tfrac{1}{4}$ in.

Patterns of atmospheric circulation

Because of the Coriolis effect, air masses being driven radially inward towards a low-pressure region, or outwards away from a high-pressure region, are also subject to deflecting forces. This causes most cyclones to be in a counterclockwise direction in the northern hemisphere and clockwise in the southern hemisphere. In the northern hemisphere the horizontal components of the Coriolis force deflect the moving air toward the right. Thus, as the air masses converge on the centre of the low-pressure region, they produce a net counterclockwise rotation. For air moving north or south over the earth's surface the Coriolis force is due east or due west, parallel to the earth's surface. If we consider a 1-kg mass of air at a wind velocity of 10 m s^{-1} (about 22 mph) at $45°$ north latitude, a direct application of eqn (8.12) gives us

$$F_{\text{Coriolis}} = 2m\omega v' \sin \lambda \approx (2)(1)(2\pi \times 10^{-5}) \times (10)(0.707) \approx 10^{-3} \text{ N}$$

If we had considered air flowing in from east or west, the Coriolis forces would not be parallel to the earth's surface, but their components parallel to the surface would be given by the same equation as that used above.

The approximate radius of curvature of the resultant motion may be obtained from

$$F = m\frac{v^2}{R}$$

or

$$R = m\frac{v^2}{F_{\text{Coriolis}}} \approx 1 \times \frac{100}{10^{-3}} = 10^5 \text{ m (about 60 miles)}$$

Occasionally one reads that water draining out of a basin also circulates in a preferred direction because of the Coriolis force. In most cases, the Coriolis force on the flowing water is negligible compared with other larger forces which are present; however, if extremely precise and careful experiments are performed, the effect can be demonstrated.

The tides

As everyone knows, the production of ocean tides is basically the consequence of the gravitational action of the moon—and, to a lesser extent, the sun. Thus we could have discussed this as an example of universal gravitation in Chapter 4. The analysis of the phenomenon is, however, considerably helped by introducing the concept of inertial forces as developed in the present chapter.

The feature that probably causes the most puzzlement when one first learns about the tides is the fact that there are, at most places on the earth's surface, two high tides every day rather than just one. This corresponds to the fact that, at any instant, the general distribution of ocean levels around the earth has two bulges. On the simple model that we shall discuss, these bulges would be highest at the places on the earth's surface nearest to and farthest from the moon (Fig. 8.11a). While the earth performs its rotation during 24 h, the positions of the bulges would remain almost stationary, being defined by the almost constant position of the moon. Thus, if one could imagine the earth completely girdled by water, the depth of the water as measured from a point fixed to the earth's solid surface would pass through two maxima and two minima in each revolution. A better approximation to the observed facts is obtained by considering the bulges to be dragged eastward by friction from the land and the ocean floor, so that their equilibrium positions with respect to the moon are more nearly as indicated in Fig. 8.11(b).

To conclude these preliminary remarks, we may point out that the bulges are, in fact, also being carried slowly eastward all the time by the moon's own

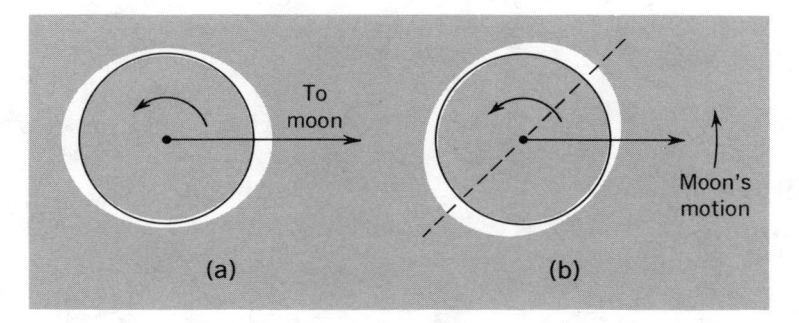

Fig. 8.11 (a) Double tidal bulge as it would be if the earth's rotation did not displace it. The size of the bulge is enormously exaggerated. (b) Approximate true orientation of the tidal bulges, carried eastward by the earth's rotation.

motion around the earth. This motion (one complete orbit relative to the fixed stars every 27.3 days) has the consequence that it takes more than 24 h for a given point on the earth to make successive passages past a particular tidal bulge. Specifically, this causes the theoretical time interval between successive high tides at a given place to be close to 12 h 25 min instead of precisely 12 h.

Now let us consider the dynamical situation. The first point to appreciate is the manner in which the earth as a whole is being accelerated towards the moon by virtue of the gravitational attraction between them. With respect to the CM of the earth–moon system (inside the earth, at about 3000 miles from the earth's centre), the earth's centre of mass has an acceleration of magnitude a_C given by Newton's laws:

$$M_E a_C = \frac{GM_E M_m}{r_m^2}$$

i.e.

$$a_C = \frac{GM_m}{r_m^2} \tag{8.15}$$

where M_m and r_m are the moon's mass and distance. What may not be immediately apparent is that every point in the earth receives this *same* acceleration from the moon's attraction. If one draws a sketch, as shown in Fig. 8.12, of the arcs along which the earth's centre and the moon travel in a certain span of time, one is tempted to think of the earth–moon system as a kind of rigid dumbbell that rotates as a unit about the centre of mass, O. It is true that the moon, for its part, does move so that it presents always the same face towards the earth, but with the earth itself things are different. If the earth were not rotating on its axis, every point on it would follow a circular arc identical in size and direction to the arc $C_1 C_2$ traced out by the earth's centre. The line

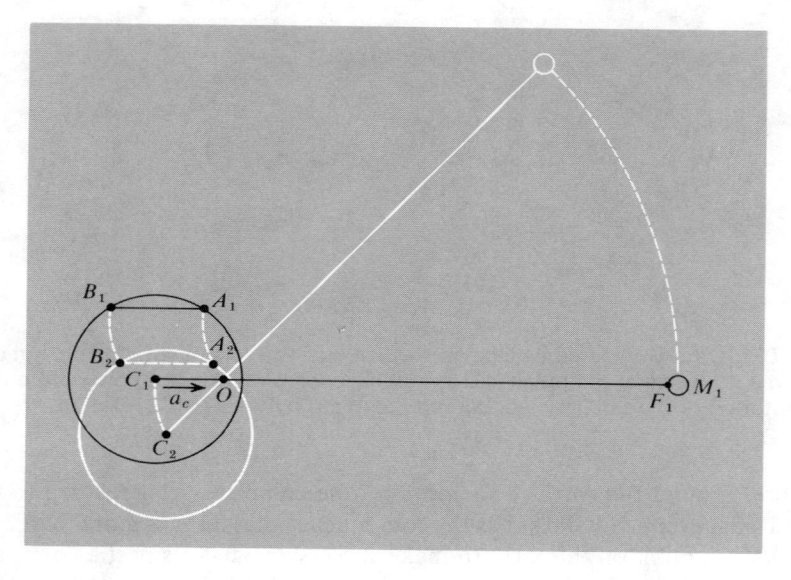

Fig. 8.12 The orbital motion of the earth about the moon does not by itself involve any rotation of the earth; the line A_1B_1 is carried into the parallel configuration A_2B_2.

A_1B_1 would be translated into the parallel line A_2B_2. The earth's intrinsic rotation about its axis is simply superposed on this general displacement and the associated acceleration.

This is where non-inertial frames come into the picture. The dynamical consequences of the earth's orbital motion around the CM of the earth–moon system can be correctly described in terms of an inertial force, $-ma_C$, experienced by a particle of mass m wherever it may be, in or on the earth. This force is then added to all the other forces that may be acting on the particle.

In the model that we are using—corresponding to what is called the *equilibrium theory* of the tides—the water around the earth simply moves until it attains an equilibrium configuration that remains stationary from the viewpoint of an observer on the moon. Now we know that for a particle at the earth's centre, the centrifugal force and the moon's gravitational attraction are equal and opposite. If, however, we consider a particle on the earth's surface at the nearest point to the moon (point A in Fig. 8.13a), the gravitational force on it is greater than the centrifugal force by an amount that we shall call f_0:

$$f_0 = \frac{GM_m m}{(r_m - R_E)^2} - \frac{GM_m m}{r_m^2}$$

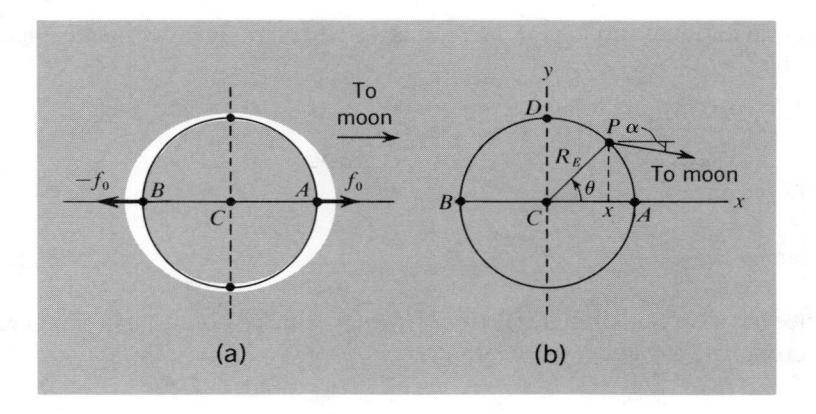

Fig. 8.13 (a) Difference between centrifugal force and the earth's gravity at the points nearest to and farthest from the moon. (b) Tide-producing force at an arbitrary point P, showing existence of a transverse component.

Since $R_E \ll r_m$ ($R_E \approx r_m/60$), we can approximate this expression as follows:

$$f_0 = \frac{GM_m m}{r_m^2}\left[\left(1 - \frac{R_E}{r_m}\right)^{-2} - 1\right]$$

i.e.

$$f_0 \approx \frac{2GM_m m}{r_m^3} R_E \qquad (8.16)$$

By an exactly similar calculation, we find that the tide-producing force on a particle of mass m at the farthest point from the moon (point B in Fig. 8.13a) is equal to $-f_0$; hence we recognize the tendency for the water to be pulled or pushed away from a midplane drawn through the earth's centre (see the figure).

By going just a little further we can get a much better insight into the problem. Consider now a particle of water at an arbitrary point P (Fig. 8.13b). Relative to the earth's centre, C, it has coordinates (x, y), with $x = R_E \cos\theta$, $y = R_E \sin\theta$. The tidal force on it in the x direction is given by a calculation just like those above:

$$f_x \approx \frac{2GM_m m}{r_m^3} x = \frac{2GM_m m}{r_m^3} R_E \cos\theta \qquad (8.17)$$

This yields the results already obtained for the points A and B if we put $\theta = 0$ or π. In addition to this force parallel to the line joining the centres of the earth and the moon there is also, however, a transverse force, because the line from P to the moon's centre makes a small angle, α, with the x axis, and the net gravitational force, $GM_m m/r^2$, has a small component perpendicular to x, given by

$$f_y = -\frac{GM_m m}{r^2} \sin \alpha \quad (\text{with } r \approx r_m)$$

Now we have

$$\tan \alpha = \frac{y}{r_m - x}$$

Since α is a very small angle ($\leq \tan^{-1}[R_E/r_m]$, which is about 1°) we can safely approximate the above expression:

$$\alpha \approx \frac{y}{r_m} = \frac{R_E \sin \theta}{r_m}$$

The component f_y of the tidal force is then given by

$$f_y \approx -\frac{GM_m m}{r_m^3} y = -\frac{GM_m m}{r_m^3} R_E \sin \theta \tag{8.18}$$

We see that this transverse force is greatest at $\theta = \pi/2$, at which point it is equal to half the maximum value (f_0) of f_x. Using eqns (8.17) and (8.18) together, we can develop an over-all picture of the tide-producing forces, as shown in Fig. 8.14. This shows much more convincingly how the forces act in such directions as to cause the water to flow and redistribute itself in the manner already qualitatively described.

Tidal heights; effect of the sun

How high ought the equilibrium tidal bulge to be? If you are familiar with actual tidal variations you may be surprised at the result. The equilibrium tide would be a rise and fall of less than 2 ft. We can calculate this by considering that the work done by the tidal force in moving a particle of water from D to A (Fig. 8.14) is equivalent to the increase of gravitational potential energy needed to raise the water through a height h against the earth's normal gravitational pull. The distance h is the difference of water levels between A and D. Now, using eqs (8.17) and (8.18) we have

$$dW = f_x \, dx + f_y \, dy$$

$$= \frac{GM_m m}{r_m^3} (2x \, dx - y \, dy)$$

$$W_{DA} = \frac{GM_m m}{r_m^3} \left[\int_0^{R_E} 2x \, dx - \int_{R_E}^0 y \, dy \right]$$

TIDAL HEIGHTS; EFFECT OF THE SUN

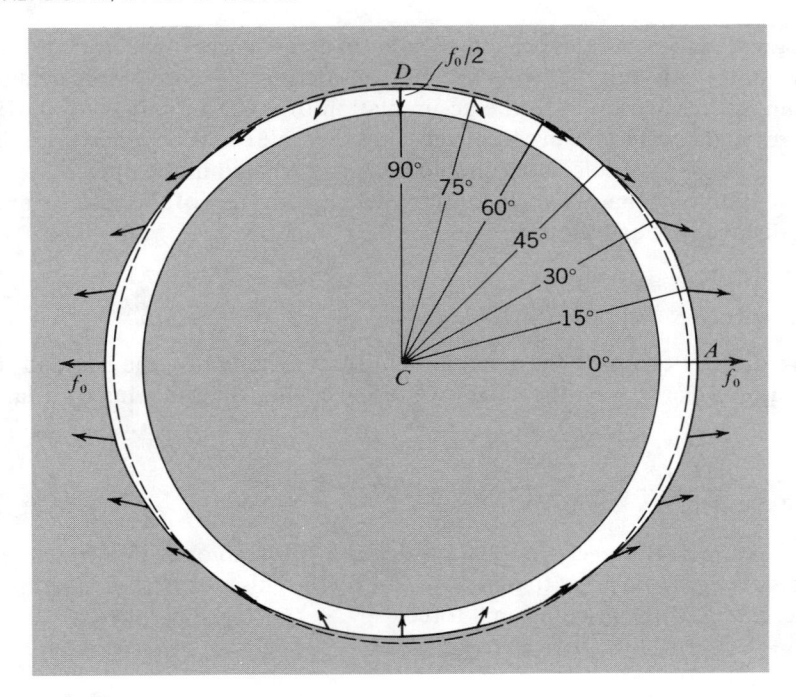

Fig. 8.14 Pattern of tide-producing forces around the earth. The circular dashed line shows where the undisturbed water surface would be.

Hence

$$W_{DA} = \frac{3GM_m m}{2r_m^3} R_E^2$$

Setting this amount of work equal to the gain of gravitational potential energy, *mgh*, we have

$$h = \frac{3GM_m R_E^2}{2gr_m^3} \tag{8.19}$$

The numerical values of the relevant quantities are as follows:

$G = 6.67 \times 10^{-11}$ N m² kg^{-2}

$M_m = 7.34 \times 10^{22}$ kg

$r_m = 3.84 \times 10^8$ m

$R_E = 6.37 \times 10^6$ m

$g = 9.80$ m s^{-2}

Substituting these in eqn (8.19) we find

$h \approx 0.54$ m ≈ 21 in.

The great excess over this calculated value in many places (by factors of 10 or even more) can only be explained by considering the problem in detailed dynamical terms, in which the accumulation of water in narrow estuaries, and resonance effects, can completely alter the scale of the phenomenon. The value that we have calculated should be approximated in the open sea.

The last point that we shall consider here is the effect of the sun. Its mass and distance are as follows:

$M_s = 1.99 \times 10^{30}$ kg

$r_s = 1.49 \times 10^{11}$ m

If we directly compare the gravitational forces exerted by the sun and the moon on a particle on the earth, we discover that the sun wins by a large factor:

$$\frac{F_s}{F_m} = \frac{M_s/r_s^2}{M_m/r_m^2} = \frac{M_s}{M_m}\left(\frac{r_m}{r_s}\right)^2 \approx 180$$

What matters, however, for tide production is the amount by which these forces *change* from point to point over the earth. This is expressed in terms of the *gradient* of the gravitational force:

$$F(r) = \frac{GMm}{r^2}$$

$$f = \Delta F = -\frac{2GMm}{r^3}\Delta r \tag{8.20}$$

Putting $M = M_m$, $r = r_m$ and $\Delta r = \pm R_E$, we obtain the forces $\pm f_0$ corresponding to eqn (8.16).

We now see that the comparative tide-producing forces due to the sun and the moon are given, according to eqn (8.20), by the following ratio:

$$\frac{f_s}{f_m} = \frac{M_s/r_s^3}{M_m/r_m^3} = \frac{M_s}{M_m}\left(\frac{r_m}{r_s}\right)^3 \tag{8.21}$$

Substituting the numerical values, one finds

$$\frac{f_s}{f_m} \approx 0.465$$

This means that the tide-raising ability of the moon exceeds that of the sun by a factor of about 2.15. The effects of the two combine linearly—and, of course, vectorially, depending on the relative angular positions of the moon and the sun. When they are on the same line through the earth (whether on the same side or on opposite sides) there should be a maximum tide equal to 1.465 times that due to the moon alone. This should happen once every 2 weeks, approximately, when the moon is new or full. At intermediate times (half-moon) when the angular positions of sun and moon are separated by 90°, the tidal amplitude should fall to a minimum value equal to 0.535 times that of the moon. The ratio of maximum to minimum values is thus about 2.7.

The search for a fundamental inertial frame

The phenomena that we have discussed in this chapter seem to leave us in no doubt that the acceleration of one's frame of reference can be detected by dynamical means. They suggest that a very special status does indeed attach to inertial frames. But how can we be sure that we have identified a true inertial frame in which Galileo's law of inertia holds exactly?

We saw at the very beginning of our discussion of dynamics that the earth itself represents a good approximation to such a frame for many purposes, especially for dynamical phenomena whose scale in distance and time is small. But we have now seen abundant evidence that a laboratory on the earth's surface is accelerated.

Could a reference frame attached to the so called 'fixed' stars be taken as a true inertial system? Today, thanks to the work of astronomers, we know a good deal about the motions of some of those 'fixed' stars. We have come to be aware of our involvement in a general rotation of our Galaxy. The sun would appear to be making a complete circuit of the Galaxy in about 2.5×10^8 years at a radial distance of about 2.5×10^4 light-years from the centre. For this motion we would have

$\omega \approx 2\pi/(8 \times 10^{15})$ s^{-1}

$R \approx 2.4 \times 10^{20}$ m

$a_3 \approx 10^{-10}$ m s^{-2}

It looks as though this acceleration can be reasonably accounted for by means of Newton's law of universal gravitation, if we regard the solar system as having a centripetal acceleration under the attraction of all the stars lying within its orbit. But no dynamical experiments that we do on earth require us to take into account this extremely minute effect—or, even, for most purposes, the revolution of the earth about the sun. (The rotation of the earth on its own axis is, however, an important consideration—and indeed an important aid in such matters as gyroscopic navigation.) Figure 8.15 schematizes the three rotating frames in which we find ourselves (we ignore here the acceleration caused by the moon).

But we still have not found an unaccelerated object to which we can attach our inertial frame of reference. In fact, we could extend this tantalizing search even further. There is some evidence that galaxies themselves tend to cluster together in groups containing a few galaxies to perhaps thousands. Our local group consists of about 10 galaxies. Although individual galaxies

(a) Acceleration toward earth's axis of rotation

$a_1 \approx 3.4 \times 10^{-2} \cos \lambda$ m s^{-2}

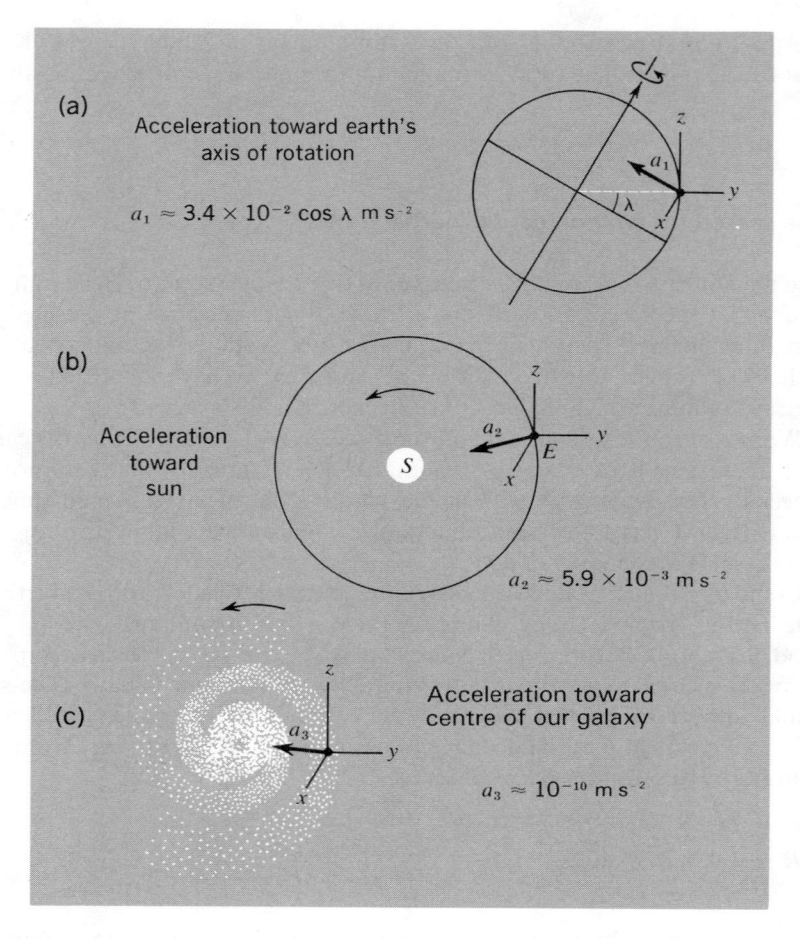

(b) Acceleration toward sun

$a_2 \approx 5.9 \times 10^{-3}$ m s^{-2}

(c) Acceleration toward centre of our galaxy

$a_3 \approx 10^{-10}$ m s^{-2}

Fig. 8.15 Accelerations of any laboratory reference frame attached to the earth's surface.

could have rather complex motions with respect to each other, this group is believed to have a more or less common motion through space.

So where are the 'fixed' stars or other astronomical objects to which we can attach our inertial frame of reference? It appears that referring to the 'fixed stars' is not a solution and contains an uncomfortable element of metaphysics (although we frequently use this phrase as a shorthand designation for the establishment of an inertial frame). This does not mean that the astronomical search for an inertial frame has been without value. For, at least up to the galactic level, it would seem that apparent departures from the law of inertia can be traced to identifiable accelerations of the reference frame in which motions are observed. However, the quest is incomplete, and so it seems likely to remain. Ultimately, therefore, we rely on an operational definition based upon local dynamical experiments and observation. We *define* an inertial frame to be one in which, experimentally, Galileo's law of inertia holds.

Problems

8.1 A man weighs himself on a spring balance calibrated in newtons which indicates his weight as $mg = 700$ N. What will he read if he repeats the observation while travelling in a lift from the first to the twelfth floors in the following manner?
  (a) Between the first and third floors the lift accelerates at the rate of 2 m/s².
  (b) Between the third and tenth floors the lift travels with the constant velocity of 7 m/s.
  (c) Between the tenth and twelfth floors the lift decelerates at the rate of 2 m/s².
  (d) He then makes a similar trip down again.
  (e) If on another trip the balance reads 500 N, what can you say of his motion? Which way is he moving?

8.2 A block of mass 2 kg rests on a frictionless platform. It is attached to a horizontal spring of spring constant 8 N/m, as shown in the figure. Initially the whole system is stationary, but at $t = 0$ the platform begins to move to the right with a constant acceleration of 2 m/s². As a result the block begins to oscillate horizontally relative to the platform.

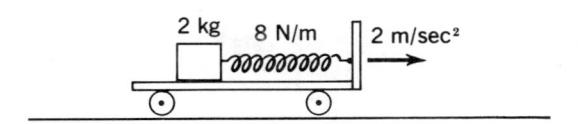

  (a) What is the amplitude of the oscillation?
  (b) At $t = 2\pi/3$ s, by what amount is the spring longer than it was in its initial unstretched condition?

8.3 A plane surface inclined 37° ($\sin^{-1} \frac{3}{5}$) from the horizontal is accelerated horizontally to the left (see the figure). The magnitude of the acceleration is gradually increased until a block of mass m, originally at rest with respect to the plane, just starts to slip up the plane. The static friction force at the block-plane surface is characterized by $\mu = \frac{4}{5}$.
  (a) Draw a diagram showing the forces acting on the block, just before it slips, in an inertial frame fixed to the floor.

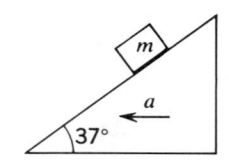

(b) Find the acceleration at which the block begins to slip.

(c) Repeat part (a) in the non-inertial frame moving along with the block.

8.4 A nervous passenger in an airplane at takeoff removes his tie and lets it hang loosely from his fingers. He observes that during the takeoff run, which lasts 30 sec, the tie makes an angle of 15° with the vertical. What is the speed of the plane at take-off, and how much runway is needed? Assume that the runway is level.

8.5 A uniform steel rod (density = 7500 kg/m^3, ultimate tensile strength 5 × 10^8 N/m^2) of length 1 m is accelerated along the direction of its length by a constant force applied to one end and directed away from the centre of mass of the rod. What is the maximum allowable acceleration if the rod is not to break? If this acceleration is exceeded, where will the rod break?

8.6 (a) A train slowed with deceleration a. What angle would the liquid level of a bowl of soup in the restaurant car have made with the horizontal? A child dropped an apple from a height h and a distance d from the front wall of the restaurant car. What path did the apple take as observed by the child? Under what conditions would the apple have hit the ground?; the front wall?

(b) As a reward for making the above observations, the parents bought the child a helium-filled balloon at the next stop. For fun, they asked her what would happen to the balloon if the train left the station with acceleration a'. Subsequently, they were surprised to find her predictions correct. What did the precocious child answer?

8.7 A lift has a *downward* acceleration equal to $g/3$. Inside the lift is mounted a pulley, of negligible friction and inertia, over which passes a string carrying two objects, of masses m and $3m$, respectively (see the figure).

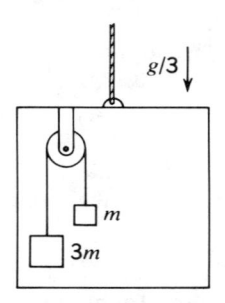

(a) Calculate the acceleration of the object of mass $3m$ *relative to the lift*.

(b) Calculate the force exerted on the pulley by the rod that joins it to the roof of the lift.

(c) How could an observer, completely isolated inside the lift, explain the acceleration of m in terms of forces that he himself could measure with the help of a spring balance?

8.8 In each of the following cases, find the equilibrium position as well as the period of small oscillations of a pendulum of length L:

(1) In a train moving with acceleration a on level tracks.
(2) In a train free-wheeling on tracks making an angle θ with the horizontal.
(3) In a lift falling with acceleration a.

8.9 (a) A man travels in a lift with vertical acceleration a. He swings a bucket of water in a vertical circle of radius R. With what angular velocity must he swing the bucket so that no water spills?

(b) With what angular frequency must the bucket be swung if the man is on a train with horizontal acceleration a? (The plane of the circle is again vertical and contains the direction of the train's acceleration.)

8.10 Consider a thin rod of material of density ρ rotating with constant angular velocity ω about an axis perpendicular to its length.

(a) Show that if the rod is to have a constant stress S (tensile force per unit area of cross-section) along its length, the cross-sectional area must decrease exponentially with the square of the distance from the axis:

$$A = A_0 e^{-kr^2} \quad \text{where } k = \rho\omega^2/2S$$

[Consider a small segment of the rod between r and $r + \Delta r$, having a mass $\Delta m = \rho A(r) \Delta r$, and notice that the difference in tensions at its ends is $\Delta T = \Delta(SA)$.]

(b) What is the maximum angular velocity ω_{max} in terms of ρ, S_{max} and k?

(c) The ultimate tensile strength of steel is about 10^9 N/m². Estimate the maximum possible number of rpm of a steel rotor for which the 'taper constant' $k = 100$ m^{-2} ($\rho = 7500$ kg/m³).

8.11 A spherically shaped influenza virus particle, of mass 6×10^{-16} g and diameter 10^{-5} cm, is in a water suspension in an ultracentrifuge. It is 4 cm from the vertical axis of rotation, and the speed of rotation is 10^3 rps. The density of the virus particle is 1.1 times that of water.

(a) From the standpoint of a reference frame rotating with the centrifuge, what is the effective value of 'g'?

(b) Again from the standpoint of the rotating reference frame, what is the net centrifugal force acting on the virus particle?

(c) Because of this centrifugal force, the particle moves radially outward at a small speed v. The motion is resisted by a viscous force given by $F_{res} = 3\pi\eta v d$, where d is the diameter of the particle and η is the viscosity of water, equal to 10^{-2} cgs units (g/cm/s). What is v?

(d) Describe the situation from the standpoint of an inertial frame attached to the laboratory.

[In (b) and (c), account must be taken of buoyancy effects. Think of the ordinary hydrostatics problem of a body completely immersed in a fluid of different density.]

8.12 (a) Show that the equilibrium form of the surface of a rotating body of liquid is parabolic (or, strictly, a paraboloid of revolution). This problem is most simply considered from the standpoint of the rotating frame, given that a liquid cannot withstand forces tangential to its surface and will tend towards a configuration in which such forces disappear. It is instructive to consider the situation from the standpoint of an inertial frame also.

(b) It has been proposed that a parabolic mirror for an astronomical telescope might be formed from a rotating pool of mercury. What rate of rotation (rpm) would make a mirror of focal length 20 m?

8.13 To a first approximation, an object released anywhere within an orbiting spacecraft will remain in the same place relative to the spacecraft. More accurately, however, it experiences a net force proportional to its distance from the centre of mass of the spacecraft. This force, as measured in the non-inertial frame of the craft, arises from the small variations in both the gravitational force and the centrifugal force due to the change of distance from the earth's centre. Obtain an expression for this force as a function of the mass m of the object, its distance ΔR from the centre of the spacecraft, the radius R of the spacecraft's orbit around the earth, and the gravitational acceleration g_R at the distance R from the earth's centre.

8.14 On a long-playing record (33 rpm, 12 in.) an insect starts to crawl towards the rim. Assume that the coefficient of friction between its legs and the record is 0.1. Does it reach the edge by crawling or otherwise?

8.15 A child sits on the ground near a rotating merry-go-round. With respect to a reference frame attached to the earth the child has no acceleration (accept this as being approximately true) and experiences no force. With respect to polar coordinates fixed to the merry-go-round, with origin at its centre:
  (a) What is the motion of the child?
  (b) What is his acceleration?
  (c) Account for this acceleration, as measured in the rotating frame, in terms of the centrifugal and Coriolis forces judged to be acting on the child.

8.16 Calculate the Coriolis acceleration of a satellite in a circular polar orbit as observed by someone on the rotating earth. Obtain the direction of this acceleration throughout the orbit, thereby explaining why the satellite always passes through the poles even though it is subjected to the Coriolis force. Is there a similar force on a satellite in an equatorial orbit?

9 Motion under central forces

Basic features of the problem

As we saw in Chapter 7, a central force field that is also conservative must be spherically symmetric, and some of the most important fields in nature (notably electrical and gravitational) are precisely of this type. The frequent occurrence of spherically symmetric models to describe physical reality is closely linked to the basic assumption that space is isotropic and is the intuitively natural starting point in building theoretical models of various kinds of dynamical systems.

We shall begin with the specific problem of the motion of a single particle of mass m in a spherically symmetric central field of force. Initially, at least, we shall assume that the object responsible for this central field is so massive that it can be regarded as a fixed centre that defines a convenient origin of coordinates for the analysis of the motion.

The first thing to notice is that the path of the moving particle will lie in a fixed plane that passes through the centre of force. This plane is defined by the initial velocity vector v_0 of the particle and the initial vector position r_0 of the particle with respect to the centre of force. Since the force acting on the

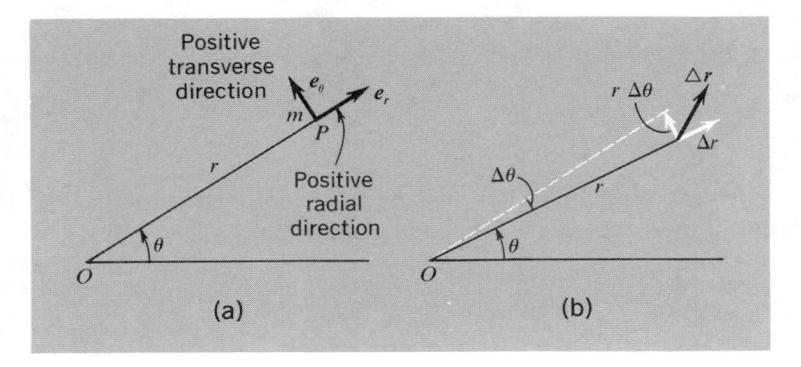

Fig. 9.1 (a) Unit vectors associated with radial and transverse directions in a plane polar coordinate system. (b) Radial and transverse components of an elementary vector displacement Δr.

particle is in this plane, and since there is no component of initial velocity perpendicular to it, the motion must remain confined to this plane of r_0 and v_0. To analyse the motion we must first pick an appropriate coordinate system. Because the force F is a function of the scalar distance r only and is along the line of the vector r (positively or negatively), it is clearly most convenient to work with the plane polar coordinates (r, θ), as indicated in Fig. 9.1(a). This means that we shall be making use of the acceleration vector expressed in these coordinates. In Chapter 1 we calculated this vector for the particular case of circular motion ($r = $ constant). Now we shall develop the more general expression that embraces changes of both r and θ.

Using the unit vectors e_r and e_θ as indicated in Fig. 9.1(a) we have

$$r = re_r \tag{9.1}$$

$$v = \frac{dr}{dt} = \frac{dr}{dt} e_r + r \frac{d\theta}{dt} e_\theta \tag{9.2}$$

We now proceed to differentiate both sides of eqn (9.2) with respect to t:

$$a = \frac{dv}{dt} = \frac{d^2r}{dt^2} e_r + \frac{dr}{dt} \frac{d}{dt} e_r + \frac{dr}{dt} \frac{d\theta}{dt} e_\theta + r \frac{d^2\theta}{dt^2} e_\theta$$
$$+ r \frac{d\theta}{dt} \frac{d}{dt} e_\theta$$

Substituting $d(e_r)/dt = (d\theta/dt)e_\theta$, and $d(e_\theta)/dt = -(d\theta/dt)e_r$, the expression for a can be rewritten as follows:

$$a = \left[\frac{d^2r}{dt^2} - r \left(\frac{d\theta}{dt} \right)^2 \right] e_r + \left[r \frac{d^2\theta}{dt^2} + 2 \frac{dr}{dt} \frac{d\theta}{dt} \right] e_\theta \tag{9.3}$$

It will be convenient to extract from this the separate radial and transverse components of the total acceleration:

$$a_r = \frac{d^2r}{dt^2} - r \left(\frac{d\theta}{dt} \right)^2 \tag{9.4}$$

$$a_\theta = r \frac{d^2\theta}{dt^2} + 2 \frac{dr}{dt} \frac{d\theta}{dt} \tag{9.5}$$

The statement of Newton's law in plane polar coordinates can then be made in terms of these separate acceleration components:

$$F_r = m \left[\frac{d^2r}{dt^2} - r \left(\frac{d\theta}{dt} \right)^2 \right] \tag{9.6}$$

$$F_\theta = m \left[r \frac{d^2\theta}{dt^2} + 2 \frac{dr}{dt} \frac{d\theta}{dt} \right] \tag{9.7}$$

The above two equations provide a basis for the solution of any problem of motion in a plane, referred to an origin of polar coordinates. We shall, however, consider their application to central forces in particular.

The conservation of angular momentum

In the case of any kind of conservative central force, we have $F_r = F(r)$ simply, and $F_\theta = 0$. The second of these immediately implies that $a_\theta = 0$. Substituting the specific expression for a_θ from eqn (9.5), we have

$$r \frac{d^2\theta}{dt^2} + 2 \frac{dr}{dt} \frac{d\theta}{dt} = 0$$

If we integrate this expression, we have

$$r^2 \frac{d\theta}{dt} = \text{const.}$$

But $r^2 (d\theta/dt)$ is just equal to twice the rate at which area is being swept out by the radius vector r, and we therefore have

(any central force) $$\frac{dA}{dt} = \frac{1}{2} r^2 \frac{d\theta}{dt} = \text{const.} \qquad (9.8)$$

The result expressed by eqn (9.8) was first discovered by Kepler in his analysis of planetary motions. It was stated by him in what is known as his second law (although it was actually the first chronologically). Newton understood it on the same dynamical grounds as we have discussed above, i.e. as a feature of the motion of an object acted on by any kind of force that is directed always to the same point.

This result can be expressed in terms of the conservation of *orbital angular momentum*. If a particle at P (Fig. 9.2a) is acted on by a force F, we have

$$F = ma = m \frac{dv}{dt}$$

Let us now form the vector (cross) product of the position vector r with both sides of this equation:

$$r \times F = r \times m \frac{dv}{dt} \qquad (9.9)$$

The left-hand side is the moment M of F about O.

We now introduce the *moment of the momentum*, L, of the particle with respect to O:

$$L = r \times mv = r \times p \qquad (9.10)$$

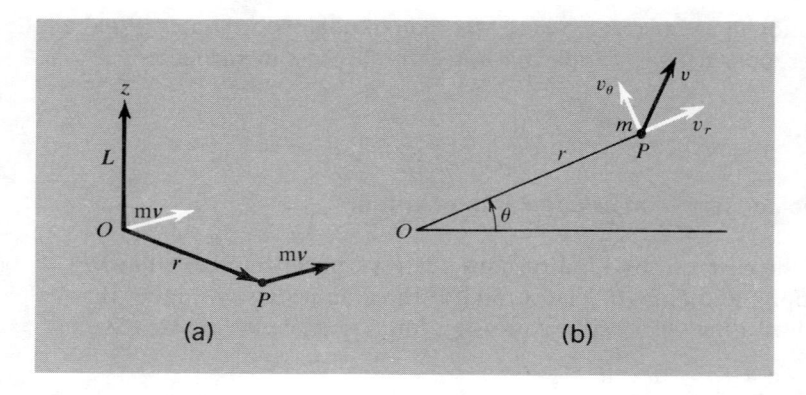

Fig. 9.2 (a) Vector relationship of position, linear momentum and orbital momentum. (b) Analysis of the velocity into radial and transverse components.

Thus L bears the same relation to the linear momentum, p, as the moment of the force, M, does to F. It is a vector, as shown in Fig. 9.2(a), perpendicular to the plane defined by r and v. If we calculate the rate of change of L—i.e. the derivative of L with respect to t—we have

$$\frac{dL}{dt} = \frac{dr}{dt} \times mv + r \times m\frac{dv}{dt}$$

$$= v \times mv + r \times m\frac{dv}{dt}$$

However, the value of $v \times mv$ is zero, because it is the cross product of two parallel vectors. The second term on the right of the above equation is, by eqn (9.9), equal to the moment of the applied force about O. Thus we have the simple result

$$M = r \times F = \frac{dL}{dt} \tag{9.11}$$

This, then, relates the moment of F to the rate of change of the moment of momentum, or orbital angular momentum, of the particle about the origin O. If, now, we take F to be a central force directed radially toward or away from O, the value of M is zero. In this case, therefore, we have

$$\frac{dL}{dt} = 0$$

and

$$L = r \times p = \text{const.} \tag{9.12}$$

This, then, is the statement of the conservation of orbital angular momentum

THE CONSERVATION OF ANGULAR MOMENTUM

for motion under a central force—or, of course, under zero force.

If we look at the situation in the plane of the motion (Fig. 9.2b) the vector L points upward from the page, and its scalar magnitude L is given by

$$L = rmv_\theta = mr^2 \frac{d\theta}{dt} \tag{9.13a}$$

The constant rate of sweeping out area [eqn (9.8)] is thus given by

$$\frac{dA}{dt} = \frac{1}{2} r^2 \frac{d\theta}{dt} = \frac{L}{2m} \tag{9.13b}$$

The result expressed by eqns (9.12) and (9.13) is a valuable one. As we have pointed out before, if we find quantities that remain constant throughout some process, i.e. they are 'conserved', these become powerful tools in the analysis of phenomena. Angular momentum is a conserved quantity of this sort which is particularly valuable in the analysis of central field problems. It should be noted that the conservation of angular momentum depends only on the absence of a torque and is independent of the conservation or non-conservation of mechanical energy. A nice example of this last feature is the speeding up of an object that whirls around at the end of a gradually shortening string (Fig. 9.3). If the decrease of r in one revolution is small compared to r itself, the velocity v is almost perpendicular to r at each instant, and the angular momentum (rmv_θ) is very nearly equal to mvr. The tension T in the string that pulls the object inward exerts no torque about O, and so we have $L = $ const. In a change of r from r_1 to r_2 we thus have

$$mv_1 r_1 = mv_2 r_2$$

or

$$v_2 = \frac{r_1}{r_2} v_1$$

Thus, if r_2 is less than r_1, there is a gain of kinetic energy given by

$$K_2 - K_1 = \tfrac{1}{2} m v_1^2 \left(\frac{r_1^2}{r_2^2} - 1 \right)$$

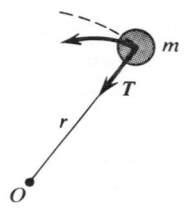

Fig. 9.3 Small portion of the path of an object that is being whirled around O at the end of a string the length of which is being steadily shortened.

The work equivalent to this gain of kinetic energy must be provided by the tension in the string. In an equilibrium orbit we would have

$$T = \frac{mv^2}{r}$$

Substituting $v = L/mr$, this becomes

$$T = \frac{L^2}{mr^3} = \frac{mv_1^2 r_1^2}{r^3}$$

The work done by T in a change from r to $r + dr$ is $-T\,dr$ (since T acts radially inward). Thus the total work between r_1 and r_2 is given by

$$W = -mv_1^2 r_1^2 \int_{r_1}^{r_2} \frac{dr}{r^3} = \tfrac{1}{2} mv_1^2 r_1^2 \left(\frac{1}{r_2^2} - \frac{1}{r_1^2} \right)$$

It may be seen that this equals the value of $K_2 - K_1$ already calculated. The problem has a good deal in common with that of the earth satellite spiralling inward, although in that case conservation of orbital angular momentum does not apply, because the air resistance represents a transverse force providing a negative torque (see discussion on pp. 184–186).

Energy conservation in central force motions

If we are dealing with a conservative central force (which was not the case in the example analysed above) we can write a statement of the conservation of total mechanical energy in the following form:

$$\frac{m}{2}(v_r^2 + v_\theta^2) + V(r) = E \tag{9.14}$$

In addition, we have the law of conservation of angular momentum as given by eqn (9.13a):

$$mrv_\theta = L$$

The quantities E and L are thus 'constants of the motion'. $V(r)$ is the potential energy of the particle m in the central field. The orbital-momentum equation allows us to reduce eqn (9.14) to the *form* of a problem of a particle moving in one dimension only under the action of a conservative force. This is the key to the method of handling central field problems. From eqn (9.13a) we take the value of v_θ and use this value in eqn (9.14). There follows

$$\frac{mv_r^2}{2} + \frac{L^2}{2mr^2} + V(r) = E \tag{9.15}$$

which describes the radial part of the motion. This is of the form

$$\frac{mv_r^2}{2} + V'(r) = E \tag{9.16}$$

where

$$V'(r) = V(r) + \frac{L^2}{2mr^2}$$

The quantity $V'(r)$ plays the role of an *equivalent* potential energy for the one-dimensional radial problem. The additional term $L^2/2mr^2$ takes complete account, as far as the radial motion is concerned, of the fact that the position vector r in the actual motion is continuously changing its direction. But it is to be noted that nothing that depends explicitly on the angular coordinate θ or its time derivatives appears in eqn (9.16).

The term $L^2/2mr^2$ is often referred to as the 'centrifugal' potential energy, because the force represented by the negative gradient of this potential energy is given by

$$F_{\text{centrifugal}} = -\frac{d}{dr}\left(\frac{L^2}{2mr^2}\right) = \frac{L^2}{mr^3}$$

Putting $L = mr^2(d\theta/dt)$, this becomes

$$F_{\text{centrifugal}} = mr\left(\frac{d\theta}{dt}\right)^2$$

which is identical with the centrifugal force $m\omega^2 r$ in a frame rotating at an angular velocity ω equal to the instantaneous value of $d\theta/dt$.

It must be remembered, of course, that the 'centrifugal potential energy' is in fact a portion of the kinetic energy of the particle: that part due to the component of its motion transverse to the instantaneous direction of the radius vector. The circumstance that this kinetic-energy contribution can be expressed as a function of radial position alone enables us to treat the radial motion as an independent one-dimensional problem.

The potential-energy function $V(r)$ in eqn (9.15) can be any function of radius. In developing this discussion we shall first consider the general properties of motion in any central field for which $V(r)$ approaches zero as r increases. Subsequently we shall take up the important special case of an inverse-square force field in which the potential energy of a particle is inversely proportional to r.

Use of the effective potential-energy curves

The general properties of the motion of a particle in a central field can be most readily obtained by using the energy-diagram method of Chapter 6 applied to eqn (9.15) or (9.16). There are, however, two significant differences between the use of the energy method for one-dimensional motion and its use for any two-dimensional motion which can be reduced to two one-dimensional motions, as in the case of central fields. First, in one dimension, the energy alone determines the general character of the motion in a conservative field. In the central-field case, however, specification of the energy is not enough. The angular momentum L of the particle must also be specified, and the motion depends on both parameters E and L, instead of on E alone. This shows up clearly in the energy diagram, because the plot of $V' = V + (L^2/2mr^2)$ will be different for different values of L. We have, then, a whole family of curves of equivalent potential energy, corresponding to different values of the angular momentum, and must consider what happens to the motion with a given energy for the different values of L.

In addition, to translate the results obtained for the radial motion by this scheme into the actual motion of the particle, we must remember that while the radial coordinate r is changing with time, so is θ. The changes in r are accompanied by a simultaneous rotation of the vector r, and the actual orbit of the particle depends on both. The rotation of the vector r is not uniform except in the special case of circular motion, for which the length of the position vector r does not change with time.

The angular momentum L and the energy E define the basic dynamical conditions and are related to the position r_0 and velocity v_0 at an arbitrary time t_0 by

$$E = \frac{mv_0^2}{2} + V(r_0)$$
$$L = mr_0(v_\theta)_0$$
(9.17)

Since we are dealing with a two-dimensional problem, we need a total of four initial conditions to describe the situation completely. A specification of the vectors r_0 and v_0 (two components each) fulfils this purpose.

Now let us look at a typical energy diagram. In Fig. 9.4(a) are plotted the curves of $L^2/2mr^2$ against r for several different values of the angular momentum, and $V(r)$ against r for an attractive potential that rises with increasing distance from $r = 0$. $V(r) \to 0$ as $r \to \infty$. Figure 9.4(b) shows the effective potential-energy curves $V'(r)$ obtained by combining the single function $V(r)$ with each of the separate centrifugal potential curves. (To obtain curves with minima as shown, it is necessary for the variation of V with r to be less rapid than $1/r^2$.) Let us now consider how to use such curves to draw valuable qualitative conclusions about the possible motions.

In Figure 9.5 we show an effective potential-energy curve drawn for a particular value of L. We then see that, given this value of L, no physically meaningful situation can exist for any value of the total energy less than the energy E_m equal to the minimum value of $V'(r)$. At this minimum possible value

BOUNDED ORBITS   231

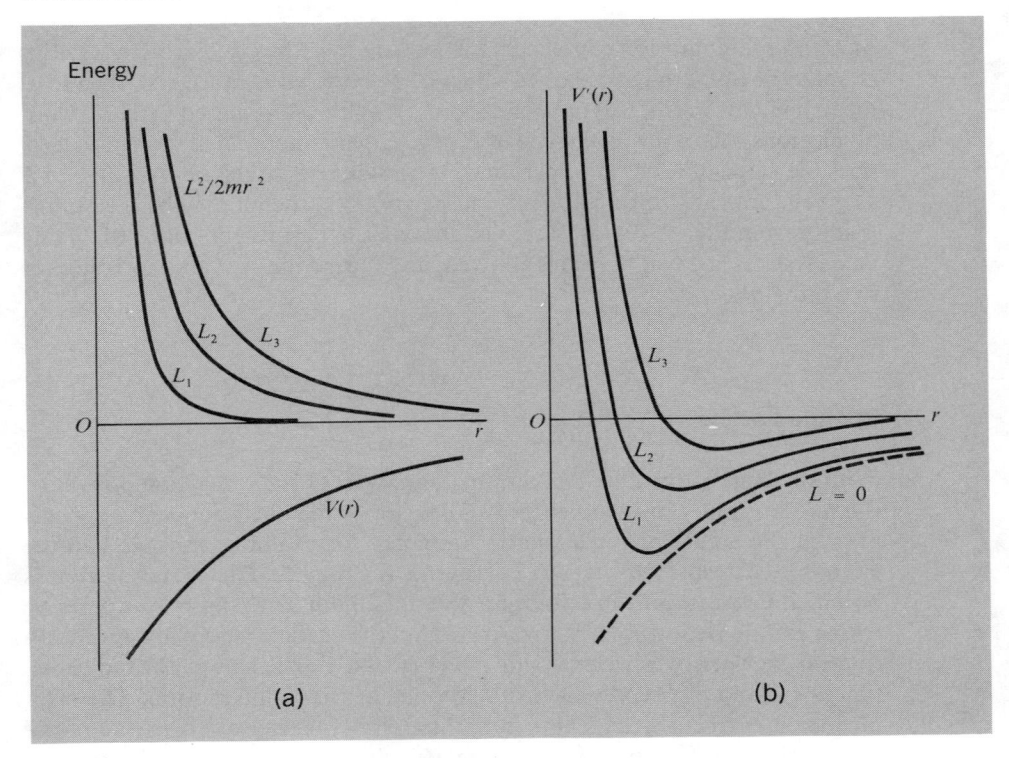

Fig. 9.4 (a) True potential energy $V(r)$, and several 'centrifugal potential energy' curves belonging to different values of L. (b) Resultant effective potential energy $V'(r)$ for different orbital momenta.

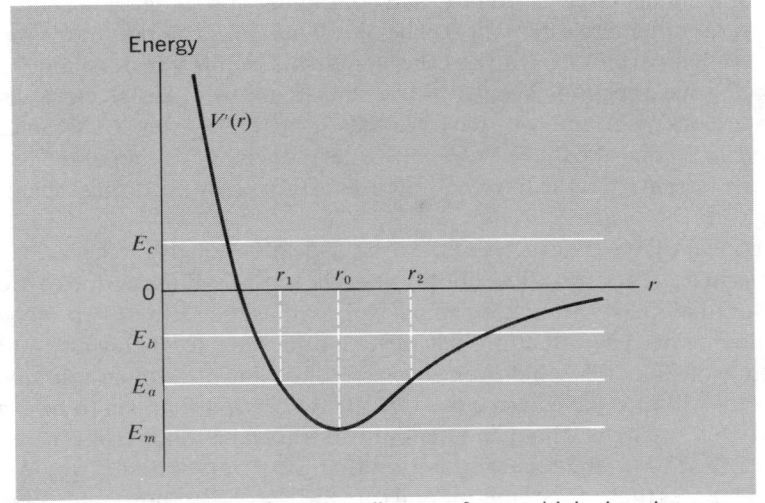

Fig. 9.5 Effective potential-energy diagram from which the character of the radial motion for different values of the total energy can be inferred.

of E, no radial motion can occur; the motion must be circular with a radius equal to r_0. For a range of larger values of E, between E_m and zero, the radial motion will be periodic (e.g. $E = E_a$ or E_b). With the assumed form of $V(r)$, all motions with a positive value of E (e.g. E_c) are unbounded; there is a least possible value of r but no upper limit. If one chooses to regard the value of E as given, but the orbital momentum as a variable, then it may be seen (most readily from Fig. 9.4b) that there is a maximum permissible value of L, any value from this down to zero is allowed, and the maximum value corresponds to a circular orbit.

Bounded orbits

With the help of the above preliminary analysis of the radial component of the motion, one can then proceed to develop some ideas about the appearance of the actual orbits in space. Suppose, for example, that we had the situation corresponding to the energy E_b in Fig. 9.5. The radial motion is bounded between certain minimum and maximum values of r. It is periodic with a certain period T_r. Thus we know that the particle moves always within the area between two circles as shown in Fig. 9.6. Furthermore, the radial distances r_{min} and r_{max} represent turning points of the radial motion. The orbit must be such that it is tangent to both these circles, because at these points the radial velocity is zero but the tangential velocity cannot be zero, given that the particle has angular momentum. Consider the particle after it has reached point A of the figure. It moves in as indicated, its trajectory becoming tangent to the inner circle at point C and continuing until it again becomes tangent to the outer circle at point B. The time it takes for this part of the motion is the radial period T_r. On the other hand, the radius vector is continually changing its direction, always in the same sense (i.e. either clockwise or counterclockwise) and will have turned through 2π after a characteristic period T_θ. In Fig. 9.6 the line OA represents the vector position of the particle at some instant, and the line OA' represents the vector position at the tim T_θ later.

It is clear that the character of the orbit depends strongly on the ratio of the two periods, T_r and T_θ, of the doubly periodic motion. If the periods are commensurable (i.e. if their ratio can be expressed as the ratio of two integers), the moving particle will ultimately (after a time equal to the lowest common multiple of T_r and T_θ) find itself in exactly the same position as initially and the orbit will thus have been closed. If the two periods happen to be exactly equal, this closure will happen after only one radial period and one increment of 2π in θ. In Fig. 9.6 this would mean that the points A and B would coincide. One sees that this is a unique and (on the face of it) a thoroughly improbable situation, yet it is precisely what one has if the force is one of attraction

UNBOUNDED ORBITS

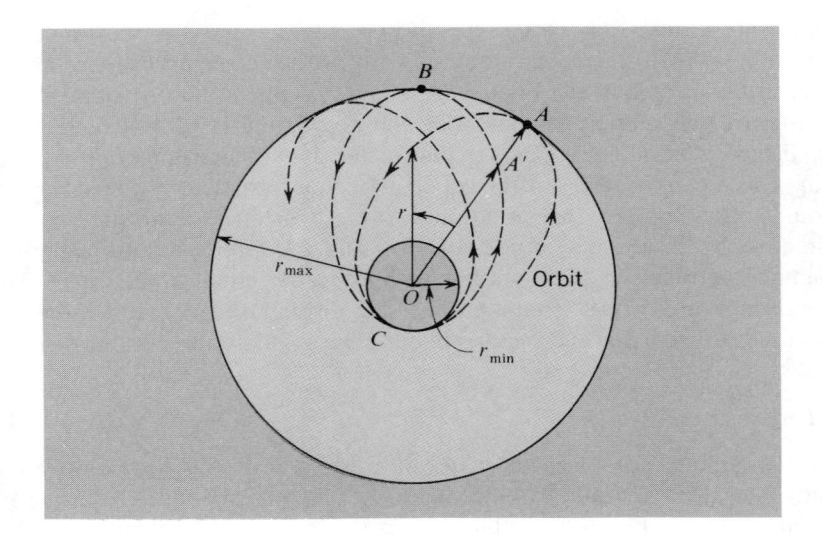

Fig. 9.6 Plan view of an orbital motion for a case in which the radial and angular periods differ. The effect is that of an approximately elliptic orbit that keeps turning (precessing) in its own plane.

proportional to r^{-2}. Thus the most important forces in nature (gravitational and electrostatic) yield orbits of this remarkable character, as we shall show shortly.

If the radial and angular periods are comparable but definitely different, then one has just the kind of situation shown in Fig. 9.6. This corresponds to a case in which T_r is somewhat greater than T_θ, so that the radius vector rotates through rather more than 2π before r completes its variation from r_{max} to r_{min} and back again. In a situation like this, where the path is near to being a closed curve but is, in effect, also turning (either forward or backward), one says that the orbit is *precessing*. The study of orbital precession is important in astronomical systems.

If the radial and angular periods are incommensurable, the orbit will never close and will eventually fill up the whole region between r_{min} and r_{max}.

Unbounded orbits

We have already pointed out that a positive total energy in the effective potential of Fig. 9.5 leads to a lower limit of r but to no upper limit. This is, in fact, a rather general result, applying to any potential that tends to zero at $r = \infty$, because it corresponds to the fact that the particle has a positive kinetic energy at infinity. If $V(r)$ is a repulsive potential, everywhere positive, then (assuming that it decreases monotonically with increasing r) there are no bounded motions at all. Figure 9.7 compares the situations obtained by taking a given centrifugal potential and combining it with attractive and repulsive potential-energy curves that differ only in sign. Figure 9.8 shows what this means in terms of the trajectories that a particle of a given energy would follow in these two situations. At large distances from the centre of force, such that the magnitudes of $V(r)$ and $L^2/2mr^2$ are both negligible, the particle travels in a straight line with a speed v_0 equal to $(2E/m)^{1/2}$. This line of motion is offset from a parallel line through the centre of force by a certain distance b that is directly related to the angular momentum; we have, in fact,

$$L = mv_0 b \qquad (9.18)$$

Thus an assumption of given values of E and L in Fig. 9.7 is expressed by using the values of v_0 and b in Fig. 9.8(a), which corresponds to an attractive potential, and in Fig. 9.8(b) which corresponds to a repulsive potential. The distance b is called the *impact parameter*, and it is a very useful quantity for characterizing situations in which a particle, in a unbounded orbit, approaches a centre of force from a large distance away. For a given value of

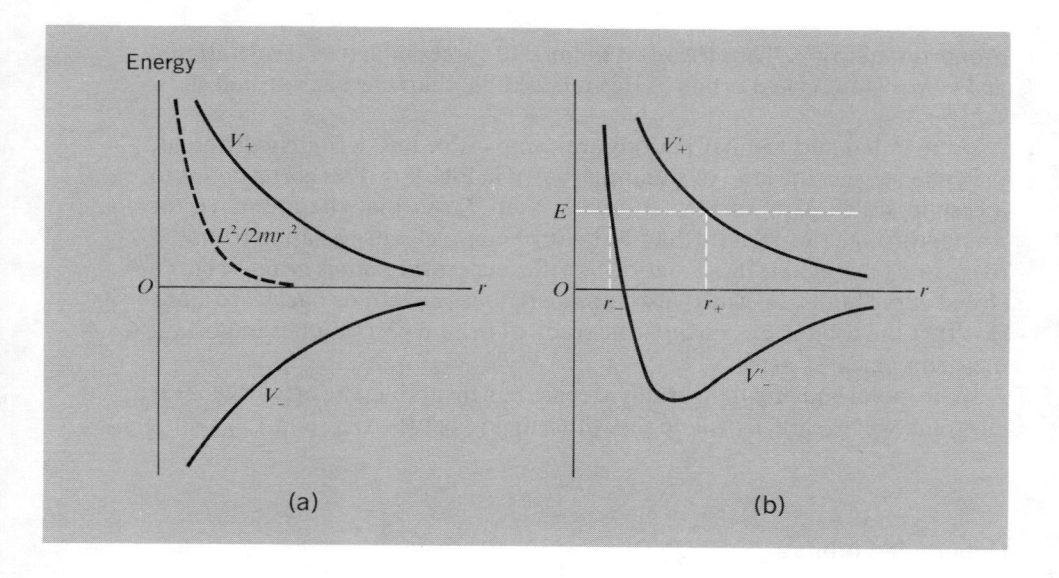

Fig. 9.7 (a) Centrifugal potential-energy curve, and two potential-energy curves, differing only in sign, that might arise from electrical interactions of like and unlike charges. (b) Effective potential-energy curves corresponding to the two cases shown in (a), indicating different distances of closest approach for a given positive total energy.

UNBOUNDED ORBITS

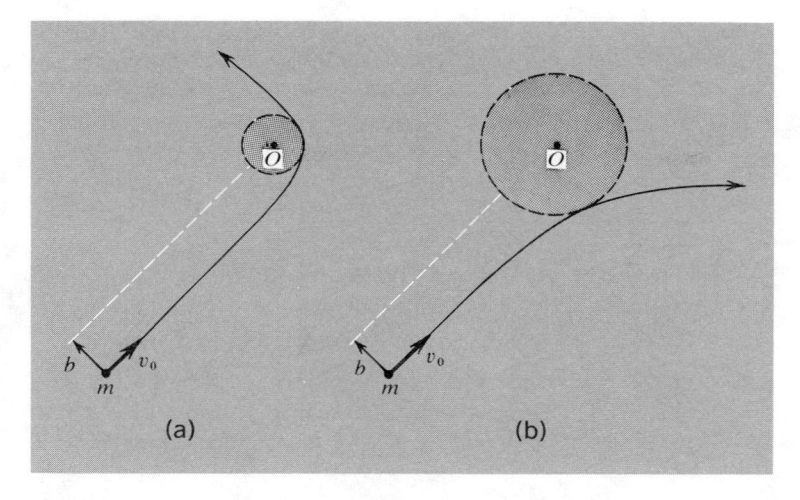

Fig. 9.8 (a) Plan view of trajectory of a particle moving around a centre of attraction. The angular momentum is defined by the 'impact parameter' b. (b) Corresponding trajectory with the same impact parameter, but with a repulsive centre of force.

v_0, the value of b completely defines the orbital angular momentum, via eqn (9.18).

It is clear that these unbounded trajectories represent single encounters of the particle with the centre of force; there is no possibility of successive returns as we have with the bounded orbits of the kind shown in Fig. 9.6. The situations shown in Fig. 9.8 thus represent individual collisions or scattering processes, of the kind so important in atomic and nuclear physics. We shall return to them later.

Certain potentials may lead to the possibility of having both bounded and unbounded motions at the same energy. This possibility exists for any attractive potential curve $V(r)$ that falls off *more* rapidly than $1/r^2$ with increasing r. An example is shown in Fig. 9.9. As long as L is not zero, the combination of $V(r)$ with the centrifugal potential gives an effective potential-energy curve that looks very much like an upside-down version of Fig. 9.5. For a positive total energy E that is less than the maximum of $V'(r)$, there are now two distinct regions of possible motion:

$0 \le r \le r_1$ bounded orbits

$r_2 \le r \le \infty$ unbounded orbits

This is not just an academic curiosity. The effective potential-energy curve representing the interaction between, let us say, a proton and an atomic nucleus resembles Fig. 9.9 and suggests that the particle may be either trapped within r_1, or free outside r_2, with no possibility of going from one

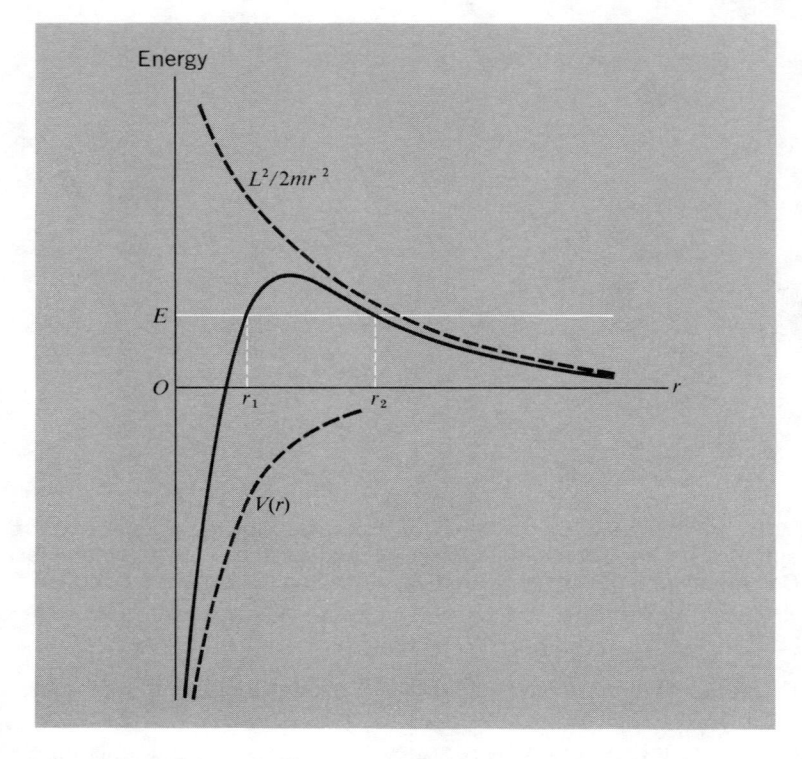

Fig. 9.9 Construction of effective potential-energy diagram for an attractive potential that varies more strongly than $1/r^2$.

state to the other. It is, as we first mentioned in Chapter 6, one of the fascinating results of quantum mechanics that a transition through the classically forbidden region can in fact occur with a certain probability—a probability that is far too small to be significant in most cases but can become very important in atomic and nuclear systems.

Circular orbits in an inverse-square force field

As a quantitative example of the general approach discussed in the preceding sections, we shall take the case of circular orbits under an attractive central force varying as $1/r^2$. This means that we shall mainly be discussing, from a diferent point of view, a situation that we have already analysed in some detail in earlier chapters. There is merit in this, because it enables one to see more readily the relationship between the different approaches.

To be specific, let us consider the motion of a satellite of mass m in the

CIRCULAR ORBITS IN AN INVERSE-SQUARE FORCE FIELD

gravitational field of a vastly more massive planet. For this case, the potential energy in the field of a planet of mass M takes the special form (see eqn 7.28)

$$V(r) = -\frac{GMm}{r} \tag{9.19}$$

where G is the universal gravitational constant and M the mass of the attracting object. Equation (9.15) becomes for this case

$$\frac{mv_r^2}{2} + \frac{L^2}{2mr^2} - \frac{GMm}{r} = E \tag{9.20}$$

In Fig. 9.10 are plotted the equivalent potential-energy curves

$$V'(r) = \frac{L^2}{2mr^2} - \frac{GMm}{r}$$

for two values of the angular momentum L.

A circular orbit corresponds, as we have seen, to a total energy equal to the minimum value of $V'(r)$ for a given value of L. Now from the above equation for $V'(r)$, taking L as fixed, we have

$$\frac{dV'}{dr} = -\frac{L^2}{mr^3} + \frac{GMm}{r^2}$$

Putting $dV'/dr = 0$, we thus find that

$$\frac{L^2}{mr} = GMm = \text{const.} \tag{9.21}$$

The orbital radius is therefore proportional to the square of the orbital angular momentum. This is indicated qualitatively in Fig. 9.10.

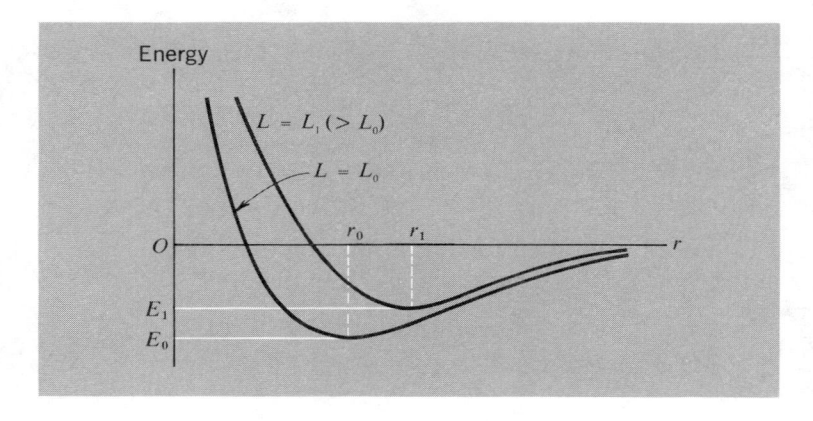

Fig. 9.10 Energy diagrams for orbits of different orbital angular momentum around a given centre of force.

Let us now consider the energy of the motion. A circular orbit is characterized by having the radial velocity component, v_r, always equal to zero. Thus in eqn (9.20) we can put

$$\text{(circular orbit)} \quad E = \frac{L^2}{2mr^2} - \frac{GMm}{r}$$

From eqn (9.21), however, we have

$$\frac{GMm}{r} = \frac{L^2}{mr^2}$$

Hence the energy can be expressed in either of the following ways:

$$\text{(circular orbit)} \quad E = -\frac{L^2}{2mr^2} = -\frac{GMm}{2r} \tag{9.22}$$

The second form shows that the orbital radius is inversely proportional to $|E|$; the first, taken together with eqn (9.21) shows that $|E|$ is inversely proportional to L^2. These properties, too, are qualitatively indicated in Fig. 9.10.

The period of the orbit can be obtained, in this approach, with the help of the law of areas. Earlier in this chapter we established the following result (eqn 9.13b):

$$\frac{dA}{dt} = \frac{L}{2m}$$

For a circular orbit, the total area A is πr^2 and the value of L is simply mvr. Thus we have $dA/dt = vr/2$ and hence $T_\theta = \pi r^2/(vr/2) = 2\pi r/v$—hardly a surprising result! To express T_θ in more interesting terms we can make use of eqn (9.21), putting $L = mvr$:

$$\frac{(mvr)^2}{mr} = GMm$$

giving

$$v = \left(\frac{GM}{r}\right)^{1/2}$$

Using this explicit expression for v as a function of r, we then arrive once again at the equation that expresses Kepler's third law:

$$T_\theta = \frac{2\pi}{(GM)^{1/2}} r^{3/2} \tag{9.23}$$

Elliptic orbits: analytical treatment

We know that the orbits of the planets are well described as ellipses with the sun (acting as the centre of force) at one focus. In this section we shall show that such orbits (conforming also to Kepler's Second Law—the law of equal areas) necessitate an attractive force varying as the inverse square of the distance.

We start with one form of the equation of an ellipse in polar coordinates:

$$\frac{1}{r} = \frac{a}{b^2}(1 - \epsilon \cos \theta) \tag{9.24}$$

where a and b are the semi-major and semi-minor axes and ϵ is the eccentricity. Our first and most essential step is to calculate the radial component of acceleration. By eqn (9.4) we have

$$a_r = \frac{d^2 r}{dt^2} - r\left(\frac{d\theta}{dt}\right)^2 \tag{9.25}$$

Now by differentiating both sides of eqn (9.24) with respect to t, we get

$$-\frac{1}{r^2}\frac{dr}{dt} = \frac{\epsilon a}{b^2} \sin \theta \frac{d\theta}{dt}$$

However, by the law of areas we have

$$r^2 \frac{d\theta}{dt} = C \tag{9.26}$$

where C is a constant (equal to twice the areal velocity). Using this, the previous equation gives us

$$\frac{dr}{dt} = -\frac{C \epsilon a}{b^2} \sin \theta$$

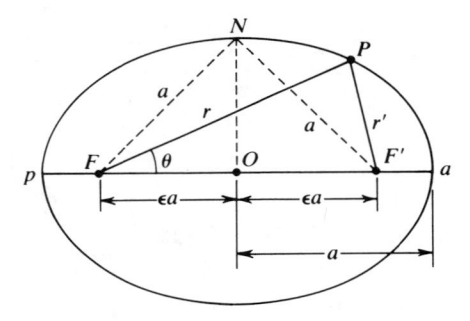

Fig. 9.11 Basic geometrical features of an elliptic orbit with a centre of force at the focus F.

Differentiating a second time, we get

$$\frac{d^2 r}{dt^2} = -\frac{C\epsilon a}{b^2} \cos\theta \frac{d\theta}{dt}$$

and hence

$$a_r = -\frac{C^2}{r^2}\left(\frac{\epsilon a \cos\theta}{b^2} + \frac{1}{r}\right)$$

This looks complicated, but if we look back at the original equation for $1/r$ (eqn 9.24) we see that

$$\frac{\epsilon a \cos\theta}{b^2} + \frac{1}{r} = \frac{a}{b^2}$$

Hence the term in parentheses in the above equation for a_r is just a geometrical constant of the ellipse, and we have

$$a_r = -\frac{C^2 a}{b^2}\frac{1}{r^2} \qquad (9.27)$$

Thus the operation of an inverse-square law is mathematically verified for any elliptic orbit known to be taking place under the action of a central force directed toward one focus. (But note the last qualification. It is perfectly possible, for example, to have an elliptic path under the action of a force directed toward the geometric centre of the ellipse, but the law of force is no longer the inverse square. Can you guess what it is?)

We can at once use eqn (9.27) to develop another important result—Kepler's third law in a rigorous form. The parameter C is, as we have said, equal to twice the constant rate of sweeping out area. But the total area of an ellipse is given by the equation

$$A = \pi a b$$

Hence the period T of the orbit is given by

$$T = \frac{\pi a b}{C/2}$$

i.e.

$$C = \frac{2\pi a b}{T}$$

Substituting this value in eqn (9.27) we find

$$a_r = -\frac{4\pi^2 a^3}{T^2}\frac{1}{r^2} \qquad (9.28)$$

However, by introducing the specific law of gravitation between the

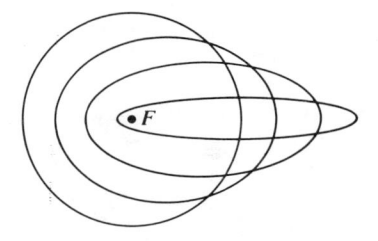

Fig. 9.12 Family of elliptic orbits of the same total energy, sharing the focus F, where the centre of force is located. The major axes of the ellipses are all equal.

orbiting mass m and a mass M fixed at F, we obtain another expression for the radial acceleration:

$$F_r = -\frac{GMm}{r^2}$$

Therefore,

$$a_r = -\frac{GM}{r^2} \tag{9.29}$$

Equating the right-hand sides of eqns (9.28) and (9.29), we have the identity

$$\frac{4\pi^2 a^3}{T^2} = GM$$

whence

$$T^2 = \frac{4\pi^2 a^3}{GM} \tag{9.30}$$

This result, although identical in form with what we identified in Chapter 4 as Kepler's third law, contains an important new feature. Previously we considered only circular orbits. Now we see that, according to eqn (9.30), all orbits having the same major axis have the same period (for a given value of M), whether they are circular or strongly flattened. If one takes into account the fact that the gravitating mass must be at one focus, a group of elliptic orbits of the same period but different eccentricities might be as shown in Fig. 9.12.

Kepler, in stating his third law, said that the squares of the planetary periods are proportional to the cubes of the mean distances from the sun. We see now that the rather vague phrase 'mean distance' takes on a very sharp and precise meaning; it is just the average of the maximum and minimum distances of a planet from the sun. This is identical with the semimajor axis, a, for from Fig. 9.11 we have

$$r_{max} = a(1 + \epsilon)$$
$$r_{min} = a(1 - \epsilon)$$
$$a = \frac{r_{max} + r_{min}}{2}$$

Energy in an elliptic orbit

The purpose of this section is to show that, as with the orbital period, the only geometrical parameter entering into the total energy is the length of the major axis of the elliptical path.

Since the total energy is constant, we are free to evaluate it at any convenient point in the orbit. We shall choose the *apogee* (or *aphelion*) point a on the major axis (Fig. 9.11), for which we have

$$r = r_{max} = a(1 + \epsilon)$$

The potential energy can be stated directly:

$$V(r_{max}) = -\frac{GMm}{a(1 + \epsilon)} \qquad (9.31)$$

The kinetic energy is a little harder to come by. Since the velocity at a is purely transverse, we have

$$K = \tfrac{1}{2}mv_\theta^2 = \tfrac{1}{2}mr^2\left(\frac{d\theta}{dt}\right)^2$$

With the help of eqn (9.26) this becomes

$$K = \frac{mC^2}{2r^2}$$

Specifically,

$$K(r_{max}) = \frac{mC^2}{2a^2(1 + \epsilon)^2}$$

Now the constant C is twice the rate of sweeping out area, which means, as we saw before, that $C = 2\pi ab/T$. Hence

$$C^2 = \frac{4\pi^2 a^2 b^2}{T^2} = \frac{4\pi^2 a^4(1 - \epsilon^2)}{T^2}$$

Substituting now for T^2 from eqn (9.30) we have

$$C^2 = GMa(1 - \epsilon^2)$$

Using this expression for C^2 in the equation for $K(r_{max})$ we find that

$$K(r_{max}) = \frac{GMm(1-\epsilon)}{2a(1+\epsilon)} \tag{9.32}$$

Combining the results expressed in eqns (9.31) and (9.32), we obtain the following formula for the total energy of the motion:

$$E = -\frac{GMm}{2a} \tag{9.33}$$

The total energy of any elliptic orbit is thus the same as that of a circular orbit whose diameter is equal in length to the major axis of the ellipse.

It may be seen from eqn (9.33) that any increase of E implies an increase in length of the major axis; the total energy remains negative but becomes numerically smaller. The negative total energy of an elliptic orbit expresses the fact that the orbiting object is bound to the centre of force and cannot escape unless a positive amount of energy at least equal to $GMm/2a$ is supplied.

If we consider an object at an arbitrary point in its orbit, its gravitational potential energy is $-GMm/r$ and the total energy must, by eqn (9.33), be equal to $-GMm/2a$. Thus we have

$$E = \tfrac{1}{2}mv^2 - \frac{GMm}{r} = -\frac{GMm}{2a} \tag{9.34}$$

Possible orbits under a $1/r^2$ force

The basic equations governing orbital motions are, as we have seen, the following:

$$\begin{aligned} F_r = ma_r &= m\left[\frac{d^2r}{dt^2} - r\left(\frac{d\theta}{dt}\right)^2\right] \\ r^2\frac{d\theta}{dt} &= C = \frac{L}{m} \end{aligned} \tag{9.35}$$

The *shape* of the orbit is something that can be described without reference to the time; it is just the spatial description of the curve—given, in these problems, by r as a function of θ. Thus we shall be interested in suppressing the explicit time dependence that is represented by the derivatives d^2r/dt^2 and $d\theta/dt$. The clue to doing this is given by the second equation of (9.35): we can put

$$\frac{d\theta}{dt} = \frac{C}{r^2}$$

It proves to be very advantageous in the analysis to introduce the reciprocal of r as a variable. Calling this u we have

$$r = \frac{1}{u}$$

$$\frac{d\theta}{dt} = Cu^2 \qquad (9.36)$$

Also, taking the first derivative of r with respect to t, we have

$$\frac{dr}{dt} = -\frac{1}{u^2}\frac{du}{dt}$$

Using the chain rule, we can rewrite this as follows:

$$\frac{dr}{dt} = -\frac{1}{u^2}\frac{du}{d\theta}\frac{d\theta}{dt} = -C\frac{du}{d\theta}$$

Differentiating again, we get

$$\frac{d^2r}{dt^2} = -C\frac{d^2u}{d\theta^2}\frac{d\theta}{dt} = -C^2u^2\frac{d^2u}{d\theta^2} \qquad (9.37)$$

Using eqns (9.36) and (9.37) gives us the following expression of Newton's law as applied to the radial component of the motion:

$$F_r = ma_r = -mC^2u^2\left(\frac{d^2u}{d\theta^2} + u\right) \qquad (9.38)$$

The value of developing this particular formulation of the radial equation of motion shows up at once when we designate F_r as a specific function of r. In particular, for the case of motion under gravity with a mass M fixed at the origin, we have

$$F_r = -\frac{GMm}{r^2} = -GMmu^2$$

Substituting this in eqn (9.38) then leads at once to the following simple equation:

$$\frac{d^2u}{d\theta^2} + u = \frac{GM}{C^2} = A \qquad (9.39)$$

where A is a constant of the motion. If we rewrite this in the form

$$\frac{d^2}{d\theta^2}(u - A) = -(u - A)$$

it is easy to see that the integrated solution can be written (with a suitable choice of the zero of θ) in the form

$$u - A = B\cos\theta$$

POSSIBLE ORBITS UNDER A $1/r^2$ FORCE

where B is another constant. Returning now to r as a variable, we have the following equation for the orbit:

$$\frac{1}{r} = A + B \cos \theta \tag{9.40}$$

We shall point at once to one feature of eqn (9.40) resulting from our particular choice of the zero of θ. This is that as θ increases from zero, in either the positive or the negative sense, the value of $1/r$ decreases and so r increases (provided that B is positive). Thus the point corresponding to $\theta = 0$ is the perigee point of the orbit; r is passing through its minimum value, which we shall call r_1 (see Fig. 9.13a).

Equation (9.40) has a geometrical interpretation that can be described with reference to Fig. 9.13. Rewriting the equation slightly, we have

$$\frac{1}{B} = d = \alpha r + r \cos \theta$$

where $d = 1/B$ and $\alpha = A/B$. If we take a line FD of length d passing through the perigee and draw an axis at right angles to this at D, then the orbit is the locus of a point P that moves so that its perpendicular distance $(d - r \cos \theta)$ from the line DN is a constant multiple, α, of its distance r from the focus F. This corresponds to a general prescription for generating the various conic sections.

To interpret the result more fully, we must consider the values of the constants A and B in eqn (9.40). The value of A is defined in eqn (9.39):

$$A = \frac{GM}{C^2}$$

Now C is the constant value of $r^2 \, d\theta/dt$. We can express this in terms of the radial distance and the speed at perigee:

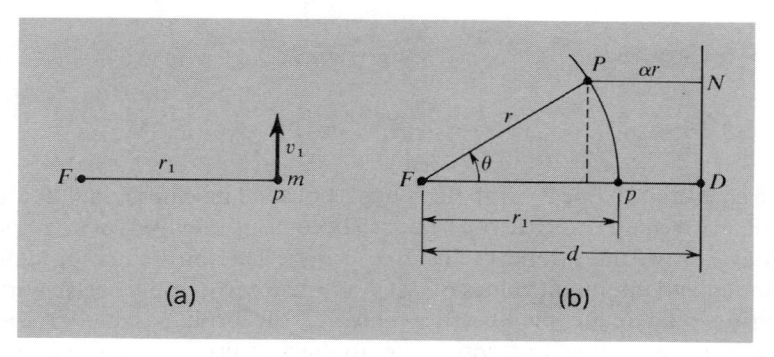

Fig. 9.13 (a) Particle at perigee in an orbit under an inverse-square force. (b) Portion of the orbit, showing the geometrical relationships of the focus, the particle's position, and the directrix (the line DN).

$$C = r_1 v_1$$

Hence

$$A = \frac{GM}{r_1^2 v_1^2}$$

Now the potential energy and the kinetic energy at perigee are given by the following expressions:

$$V_1 = -\frac{GMm}{r_1}; \quad K_1 = \tfrac{1}{2}mv_1^2$$

This permits us to write the constant A as follows:

$$A = -\frac{V_1}{2r_1 K_1}$$

The value of B follows immediately from putting $\theta = 0$ in eqn (9.40) itself:

$$B = \frac{1}{r_1} - A = \frac{1}{r_1}\left(1 + \frac{V_1}{2K_1}\right)$$

Now consider the difference between B and A:

$$B - A = \frac{1}{r_1}\left(1 + \frac{V_1}{K_1}\right) = \frac{E}{r_1 K_1}$$

where E is the *total* energy of the motion at every point in the orbit. This is the key to the problem, for we can now recognize three situations that correspond to three different types of orbit, according to whether E is zero, negative or positive:

$$E = 0 \ (B = A, \alpha = 1): \quad \frac{1}{r} = A(1 + \cos\theta)$$

$$E > 0 \ (B > A, \alpha < 1): \quad \frac{1}{r} = A(\alpha_h + \cos\theta) \quad (9.41)$$

$$E < 0 \ (B < A, \alpha > 1): \quad \frac{1}{r} = A(\alpha_e + \cos\theta)$$

These equations define a parabola, a hyperbola and an ellipse, in that order. The first two equations clearly permit r to become infinitely great (the first of them as $\theta \to \pi$, the second as $\cos\theta \to -\alpha_h$). The third equation defines maximum and minimum values of r at $\theta = \pi$ and zero, respectively. Further analysis would relate the specific values of the orbit parameters to the dynamical constants of the motion, i.e. to the magnitudes of the energy and the orbital momentum.

Many problems involving force laws other than the inverse square can also

Rutherford scattering

As another example of motion in an inverse-square central field of force, we shall consider the deflection of an electrically charged particle in the electric field of a much more massive object carrying an electric charge of the same sign. The field is repulsive, obeys Coulomb's law, and has the equivalent potential energy

$$V'(r) = \frac{kQ_1Q_2}{r} + \frac{L^2}{2mr^2} \qquad (9.42)$$

as shown plotted in Fig. 9.14(a). k is the proportionality constant in

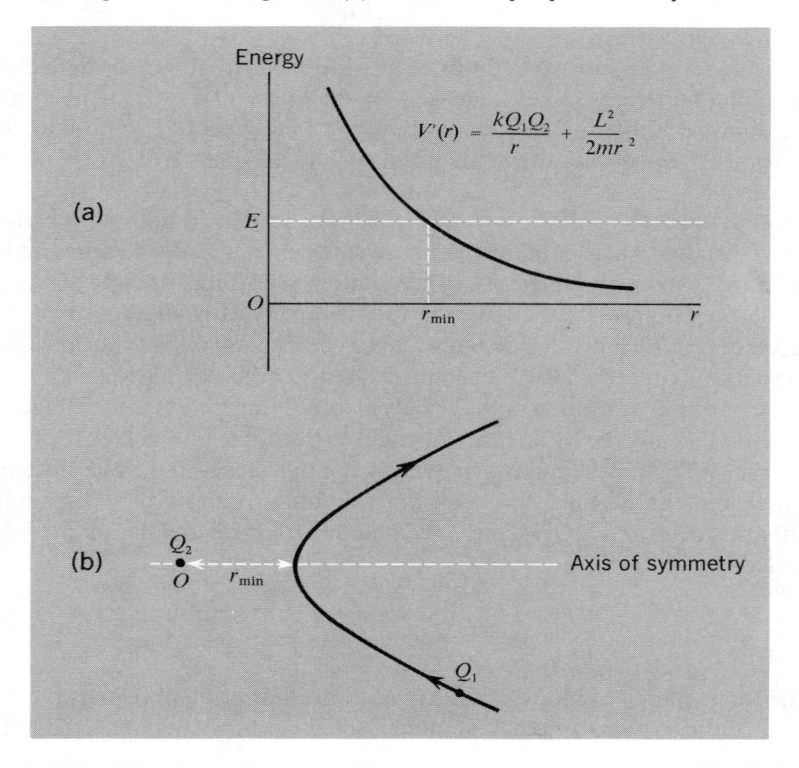

Fig. 9.14(a) Effective radial potential-energy diagram for a particle with orbital angular momentum L in a repulsive Coulomb field. (b) Plan view of the trajectory of an alpha particle (Q_1) in the neighbourhood of a massive nuclear charge Q_2).

Coulomb's law and Q_1 and Q_2 are the electrical charges on the two particles.

Motion is possible only for positive energies (E) and all such motions are unbounded, characterized by a distance of nearest approach, r_{min}, which depends on the energy of the moving particle. Because the particle retraces all the values of radial speed on the way out that it had on the way in, and because the angular velocity ($d\theta/dt$) of the particle depends only on its distance r from O, the trajectory will be symmetrical as shown in Fig. 9.14(b).

Historically, the understanding of this type of motion played a basic role in one of the most important experiments of this century. About 1910 Lord Rutherford and his students, especially Geiger and Marsden, performed a series of experiments on the scattering of a beam of alpha particles by thin metallic foils. These experiments showed that most of the mass of atoms is concentrated in a small positively charged nucleus. Presumably the electrons in the atom surrounded this nucleus like a cloud. This nuclear model of an atom was in sharp contrast to that previously proposed by J.J. Thomson, which was essentially a ball of distributed positive charge in which the electrons were embedded.

Although we can use the methods of satellite orbits to obtain the deflection of an alpha particle by a gold nucleus, we shall present a much simpler and more direct argument. Consider an alpha particle (charge Q_1 and mass m) moving with a speed v_0 toward a gold nucleus of charge Q_2 as shown in Fig. 9.15.

Clearly, the deflection will be larger, the more nearly v_0 points at the charge Q_2. The distance b in the figure is the impact parameter defined earlier in this chapter (p. 234) and is a measure of the aiming error. Since at large distances from Q_2 the potential energy of m in the field of Q_2 is negligibly small, the kinetic energy $\frac{1}{2}mv_0^2$ at such distances is the total energy of the motion. This total energy is conserved in the encounter. Hence the alpha particle regains its initial speed after scattering, and the only effect of the process is to change the direction of its motion by an amount equal to the angle φ in Fig. 9.15. To be sure, the alpha particle slows down as it approaches the gold nucleus, but it regains its original speed on the way out. In addition, the angular momentum of m about Q_2 is conserved and this constant angular momentum L is given by

$$L = mv_0 b = mr^2 \frac{d\theta}{dt} = \text{const.} \tag{9.43}$$

The total change of momentum $\Delta(mv)$ in the scattering process is the difference of the two vectors, each of magnitude mv_0, shown in Fig. 9.16, and is equal in magnitude to

$$\Delta(mv) = 2mv_0 \sin\frac{\varphi}{2} \tag{9.44}$$

RUTHERFORD SCATTERING

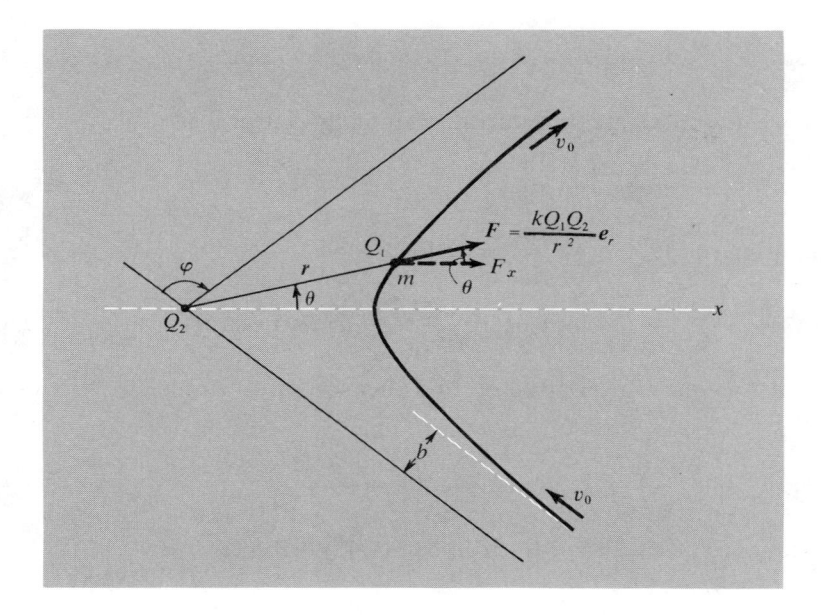

Fig. 9.15 Geometry of a Rutherford-scattering event.

This must be equal to the total impulse supplied by the force F of Fig. 9.15 during the scattering process. This impulse is the vector

$$\Delta p = \int F \, dt$$

From the symmetry of Fig. 9.15, we see that only the component F_x of F contributes to this impulse, because the perpendicular contributions from F_y at points on the trajectory below the x axis just cancel the corresponding contributions at points above the x axis. This makes sense, because $\Delta(mv)$ is parallel to x as indicated in Fig. 9.16. Thus, Newton's law of motion gives us

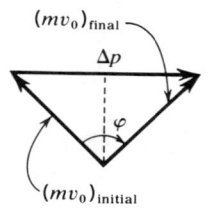

Fig. 9.16 Net dynamical result of a scattering event in terms of the impulse Δp that changes the direction but not the magnitude of the alpha-particle momentum.

$$\Delta p = \int F_x \, dt = \int F \cos\theta \, dt = \Delta(mv) = 2mv_0 \sin\frac{\varphi}{2}$$

and we must evaluate the integral. Writing this integral as

$$\Delta p = \int \frac{kQ_1Q_2 \cos\theta}{r^2} \, dt$$

and using eqn (9.43), we have

$$\Delta p = \frac{kQ_1Q_2}{v_0 b} \int_{\theta_1}^{\theta_2} \cos\theta \, d\theta = \frac{kQ_1Q_2}{v_0 b} (\sin\theta_2 - \sin\theta_1) \tag{9.45}$$

θ_1 and θ_2 are the values of θ before and after scattering. From Fig. 9.15 we see that

$$\theta_1 = -\left(\frac{\pi - \varphi}{2}\right) \quad \text{and} \quad \theta_2 = +\left(\frac{\pi - \varphi}{2}\right)$$

Since $\sin[(\pi - \varphi)/2] = \cos(\varphi/2)$, eqn (9.45) becomes

$$\Delta p = \frac{2kQ_1Q_2}{v_0 b} \cos\frac{\varphi}{2}$$

Equating this expression for Δp to the value of $\Delta(mv)$ according to eqn (9.44), we find

$$\tan\frac{\varphi}{2} = \frac{kQ_1Q_2}{mv_0^2 b} \tag{9.46}$$

This tells us, for each value of the impact parameter b, the angle of scattering of particles of a given energy. If the beam of alpha particles is essentially monoenergetic, we can proceed to use eqn (9.46) to calculate the relative numbers of the incident particles that are scattered through different angles φ. This involves finding the fraction of the incident alpha particles that have impact parameters between b and $b + db$ and from this the fraction scattered into the corresponding range of angles φ. Rutherford found that his experimental results were in complete accord with this theory of scattering by a point-like nucleus.

Problems

9.1 In the Bohr model of the hydrogen atom an electron (mass m) moves in a circular orbit around an effectively stationary proton, under the central Coulomb force $F(r) = -ke^2/r^2$.
   (a) Obtain an expression for the speed v of the electron as a function of r.
   (b) Obtain an expression for the orbital angular momentum L as a function of r.
   (c) Introduce Bohr's postulate (of the so-called 'old quantum theory', now super-

seded) that the angular momentum in a circular orbit is equal to $nh/2\pi$, where h is Planck's constant. Obtain an expression for the permitted orbital radii.

(d) Calculate the potential energy of the system from the equation

$$V(r) = -\int_{\infty}^{r} F(r)\,dr$$

Hence find an expression for the total energy of the quantized system as a function of n.

(e) For the lowest energy state of the atom (corresponding to $n = 1$) calculate the numerical values of the orbital radius and the energy, measured in electron volts, needed to ionize the atom. ($k = 9 \times 10^9$ N-m^2/C^2; $e = 1.6 \times 10^{-19}$ C; $m = 9.1 \times 10^{-31}$ kg; $h = 6.63 \times 10^{-34}$ J s.)

9.2 A mass m is joined to a fixed point O by a string of length L. Iniitially the string is slack and the mass is moving with constant speed v_0 along a straight line. At its closest approach the distance of the mass from O is h. When the mass reaches a distance L from O, the string becomes taut and the mass goes into a circular path around O. Find the ratio of the final kinetic energy of the mass to its initial kinetic energy. (Neglect any effects of gravity.) Where did the energy go?

9.3 A particle A, of mass m, is acted on by the gravitational force from a second particle, B, which remains fixed at the origin. Initially, when A is very far from $B (r = \infty)$, A has a velocity v_0 directed along in line shown in the figure. The perpendicular distance between B and this line is D. The particle A is deflected from its initial course by B and moves along the trajectory shown in the figure. The shortest distance between this trajectory and B is found to be d. Deduce the mass of B in terms of the quantities given and the gravitational constant G.

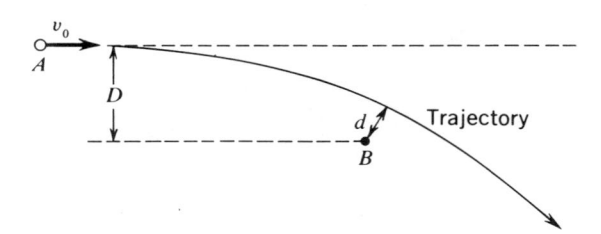

9.4 A particle of mass m moves in the field of a repulsive central force Am/r^3, where A is a constant. At a very large distance from the force centre the particle has speed v_0 and its impact parameter is b. Show that the closest m comes to the centre of force is given by

$$r_{min} = (b^2 + A/v_0^2)^{1/2}$$

9.5 A non-rotating, spherical planet with no atmosphere has mass M and radius R. A particle is fired off from the surface with a speed equal to three-quarters of the escape speed. By considering conservation of total energy and angular momentum, calculate the farthest distance that it reaches (measured from the centre of the planet) if it is fired off (a) radially and (b) tangentially. Sketch the effective potential-energy curve, given by

$$V'(r) = -\frac{GMm}{r} + \frac{L^2}{2mr^2}$$

for case (b). Draw the line representing the total energy of the motion, and thus verify your result.

9.6 Imagine a spherical, non-rotating planet of mass M, radius R, that has no atmosphere. A satellite is fired from the surface of the planet with speed v_0 at 30° to the local vertical. In its subsequent orbit the satellite reaches a maximum distance of $5R/2$ from the centre of the planet. Using the principles of conservation of energy and angular momentum, show that

$$v_0 = (5GM/4R)^{1/2}$$

9.7 A particle moves under the influence of a central *attractive* force, $-k/r^3$. At a very large (effectively infinite) distance away, it has a non-zero velocity that does *not* point toward the centre. Construct the effective potential-energy diagram for the radial component of the motion. What conclusions can you draw about the dependence on r of the radial component of velocity?

9.8 A satellite in a circular orbit around the earth fires a small rocket. Without going into detailed calculations, consider how the orbit is changed according to whether the rocket is fired (a) forwards; (b) backwards; (c) towards the earth; and (d) perpendicular to the plane of the orbit.

9.9 A satellite of mass m is travelling at speed v_0 in a circular orbit of radius r_0 under the gravitational force of a fixed mass at O.

(a) Taking the potential energy to be zero at $r = \infty$, show that the total mechanical energy of the satellite is $-\frac{1}{2}mv_0^2$.

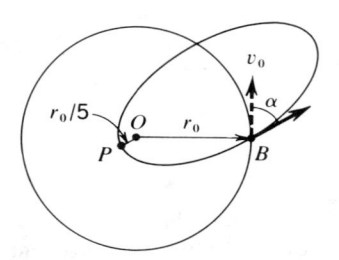

(b) At a certain point B in the orbit (see the figure) the direction of motion of the satellite is suddenly changed *without any change in the magnitude of the velocity*. As a result the satellite goes into an elliptic orbit. Its closest distance of approach to O (at point P) is now $r_0/5$. What is the speed of the satellite at P, expressed as a multiple of v_0?

(c) Through what angle α (see the figure) was the velocity of the satellite turned at the point B?

PROBLEMS

9.10 A small satellite is in a circular orbit of radius r_1 around the earth. The direction of the satellite's velocity is now changed, causing it to move in an elliptical orbit around the earth. The change in velocity is made in such a manner that the satellite loses half its orbital angular momentum, but its total energy remains unchanged. Calculate, in terms of r_1, the perigee and apogee distances of the new orbit (measured with respect to the earth's centre).

9.11 A satellite of mass m is travelling in a perfectly circular orbit of radius r about the earth (mass M). An explosion breaks up the satellite into two equal fragments, each of mass $m/2$. Immediately after the explosion the two fragments have radial components of velocity equal to $\pm v_0/2$, where v_0 is the orbital speed of the satellite prior to the explosion; in the reference frame of the satellite at the instant of the explosion the fragments appear to separate along the line joining the satellite to the centre of the earth.

(a) In terms of G, M, m and r, what are the energy *and* the angular momentum (with respect to the earth's centre) of each fragment?

(b) Make a sketch showing the original circular orbit and the orbits of the two fragments. In making the sketch, use the fact that the major axis of the elliptical orbit of a satellite is inversely proportional to the total energy.

9.12 The commander of a spaceship that has shut down its engines and is coasting near a strange-appearing gas cloud notes that the ship is following a circular path that will lead directly *into* the cloud (see the figure). He also deduces from the ship's motion that its angular momentum with respect to the cloud is not changing. What attractive (central) force could account for such an orbit?

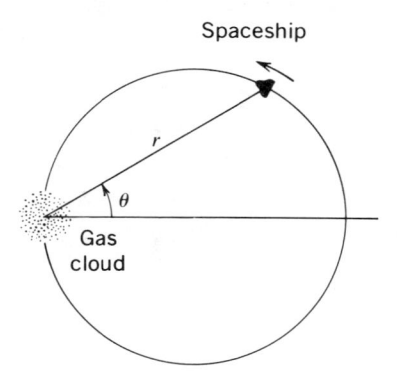

9.13 The problem of dropping a spacecraft into the sun from the earth's orbit with the application of minimum possible impulse (given to the spacecraft by firing a rocket engine) is not solved by firing the rocket in a direction opposite to the earth's orbital motion so as to reduce the velocity of the spacecraft to zero. A two-step process can accomplish the goal with a smaller rocket. Assume the initial orbit to be a circle of radius r_1 with the sun at the centre (see the figure). By means of a brief rocket burn the spacecraft is speeded up tangentially in the direction of the orbit velocity, so that it assumes an elliptical orbit whose perihelion coincides with the firing point. At the aphelion of this orbit the spacecraft is given a backward impulse sufficient to reduce its space velocity to zero, so that it will subsequently fall into the sun. (Neglect the effect of the earth's gravity.)

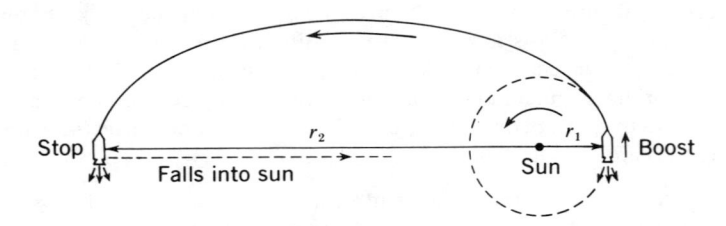

(a) For a given value of the aphelion distance, r_2 of the spacecraft, calculate the required increment of speed given to it at first firing.

(b) Find the speed of the spacecraft at its aphelion distance, and so find the sum of the speed increments that must be given to the spacecraft in the two steps to make it fall into the sun. This sum provides a measure of the total impulse that the rocket engine must be able to supply. Compare this sum with the speed of the spacecraft in its initial earth orbit for the case $r_2 = 10r_1$.

[*Note:* This problem is discussed by E. Feenberg, 'Orbit to the Sun', *Am. J. Phys.*, **28**, 497 (1960).]

9.14 The sun loses mass at the rate of about 4×10^6 tons/s. What change in the length of the year should this have produced within the span of recorded history (~ 5000 yr)? Note that the equation for circular motion can be employed (even though the earth spirals away from the sun) because the fractional yearly change in radius is so small. The other condition needed to describe the gradual shift is the overall conservation of angular momentum about the CM of the system.

10 Extended systems and rotational dynamics

Nearly all our discussion of dynamics so far has been limited to the translational motions of particles regarded as point masses. But every particle has structure; it has a finite size and a greater or lesser degree of rigidity. The full description of the motion of any real physical object must include, in addition to the motion of its centre of mass, a consideration of its rotation and other internal motions. In many instances one may regard a complex, extended object as being an assemblage of the ideal particles of basic mechanics. In this chapter we shall discuss a number of topics, touching upon what are in many respects widely different physical systems, yet having in common the feature that they involve the motions of two or more individual particles. We shall be devoting special attention to those physical systems in which particles interact strongly with one another and, in some instances, we shall treat the interactions as being so strong that the system is effectively rigid.

Momentum and kinetic energy of a many-particle system

In Chapter 5 we analysed the dynamics of two-particle systems as described in an arbitrary frame (the laboratory) and in the unique centre-of-mass (CM) frame defined by the particles themselves. We saw how the introduction of the centre of mass allows one to separate motions relative to the centre of mass from bodily motions of the system as a whole. We shall now show that this possibility exists for any system of particles; it is a result that makes for very important simplifications in our analysis of complete objects of arbitrary shapes and sizes.

We shall suppose that our system is made up of particles of masses m_1, m_2, m_3, ..., located instantaneously at the positions r_1, r_2, r_3, ... and moving instantaneously with the velocities v_1, v_2, v_3, ... The position and velocity of the centre of mass, C, are then defined by the following equations:

$$Mr_c = m_1r_1 + m_2r_2 + m_3r_3 + \ldots$$
$$Mv_c = m_1v_1 + m_2v_2 + m_3v_3 + \ldots$$

where

$$M = m_1 + m_2 + m_3 + \ldots$$

We can express these results more compactly as follows:

$$Mr_c = \sum_i m_i r_i$$
$$Mv_c = \sum_i m_i v_i \tag{10.1}$$

where the suffix i runs from 1 to N (N being the total number of particles in the system).

Consider now the statement of $F = ma$ as it applies to any one particle, i, in the system. In general it may be subjected to an external force, F_i, and also to internal interactions from all the other particles of the system. We shall denote these latter by symbols such as f_{ik}, to be read as the force exerted on particle i by particle k. Then the specific statement of Newton's law for particle i is as follows:

$$F_i + \sum_{k \neq i} f_{ik} = \frac{d}{dt}(m_i v_i)$$

We can now proceed to write a similar equation for every other particle in the system and add them all up. When we do this, the right-hand side is, by eqn (10.1), just the rate of change of the momentum of a single particle of mass M travelling at the centre-of-mass velocity v_c. What about the left-hand side? The first part of it is the sum (F_{ext}) of all the external forces, regardless of which particles they are applied to. The second part—let us call it f_{int}—is a double summation over all the interactions that can occur between the particles in pairs:

$$f_{int} = \sum_i \sum_{k \neq i} f_{ik}$$

Now if you consider what this summation entails, you will see that it can be broken down into a set of pairs of contributions of the type $f_{ik} + f_{ki}$. (It may help you to take the simplest specific case, $N = 3$, and write out all the terms in detail.) In other words, it is made up of a set of terms, each of which is the sum of the forces of action and reaction between two particles. Since, however, Newtonian mechanics has it as a basic tenet that the forces of action and reaction are equal and opposite, each one of these pairs gives zero, and it follows that the resultant of all the internal forces, f_{int}, is itself zero. Thus, for

MOMENTUM AND KINETIC ENERGY OF A MANY-PARTICLE SYSTEM

any system of particles whatever, we have a statement of Newton's law exactly like that for a single particle of the total mass M:

$$F_{\text{ext}} = \frac{\mathrm{d}}{\mathrm{d}t}(Mv_c) = M\frac{\mathrm{d}v_c}{\mathrm{d}t} \qquad (10.2)$$

Figure 10.1 shows an example of this result in action. The centre of mass of a complex object, with innumerable internal interactions, follows a simple parabolic path under gravity.

The total kinetic energy of the system is also amenable to a simple analysis. In this case we shall introduce the velocities $v'_1, v'_2, v'_3, \ldots$ of the particles relative to the centre of mass. Thus the velocity, v_i, of any particle as measured in

Fig. 10.1 The centre of mass of a complicated object follows a simple parabolic path under the net gravitational force. Photograph by Prof. Harold E. Edgerton, MIT, of a drum majorette tossing a baton. Time between flashes was 1/60 s. Dashed lines show path of CM before the baton was released and after it was caught again.

the laboratory can be written as $v_i + v_c$. The kinetic energy K_i of this particle can thus be written as follows:

$$K_i = \tfrac{1}{2}m_i v_i^2 = \tfrac{1}{2}m_i(v_i \cdot v_i)$$
$$= \tfrac{1}{2}m_i(v_i' + v_c) \cdot (v_i' + v_c)$$
$$= \tfrac{1}{2}m_i(v_i')^2 + m_i(v_i' \cdot v_c) + \tfrac{1}{2}m_i v_c^2$$

Thus we have

$$K_i = K_i' + (m_i v_i') \cdot v_c + \tfrac{1}{2}m_i v_c^2$$

Let us now consider the result of summing the individual kinetic energies such as K_i for all the particles in the system. The first term on the right gives us the total kinetic energy, K', of all the particles relative to the centre of mass. The last term gives us the kinetic energy of a particle of the total mass, M, moving with the speed v_c of the centre of mass. And the *middle* term vanishes, because by the definition of the centre of mass we have

$$\sum_i m_i r_i' = 0$$

and hence

$$\left(\sum_i m_i v_i'\right) \cdot v_c = 0$$

Thus, for any system of particles, we can put

$$K = K' + \tfrac{1}{2}M v_c^2 \tag{10.3}$$

Angular momentum

In Chapter 9 we recognized that the orbital angular momentum, L, of a particle with respect to a centre of force is an important dynamical quantity. You will recall that L (the 'moment of momentum') is defined through the following equation:

$$L = r \times (mv) = r \times p \tag{10.4}$$

Thus the actual magnitude of L depends on the particular choice of origin from which the position vector r of the particle is measured. If, as in the situations we considered, there is a well-defined fixed centre of force, the appropriate choice of origin is clear. In general, however, the angular momentum of an individual moving particle is not a uniquely definable quantity. But as soon as one has two or more particles, or a single object that cannot be approximated as a point particle, it does become possible to speak unambiguously of the *internal* angular momentum of the complete system. Let us see how.

To introduce the discussion, consider first a very simple and specific situation. Two particles, of masses m_1 and m_2, are joined by a very light rigid bar that is pivoted at the centre of mass, C, of the two particles (see Fig. 10.2a). The system rotates with angular velocity ω about an axis through C perpendicular to the plane of the diagram. We shall calculate the total angular momentum of the two particles about a parallel axis through an arbitrary origin, O. With respect to O, the orbital angular momentum of m_1 is counterclockwise and that of m_2 is clockwise. To calculate the actual magnitude of the combined angular momentum, we can draw a line OA parallel to the line joining the masses. The velocity vectors are perpendicular to this line and intersect it at the points A and B. A line through C parallel to these velocity vectors intersects OA at a point D. Let $OD = d$. Then the total angular momentum of the two particles with respect to O is given by the following expression:

$$L_O = m_1 v_1 (d + r_1') - m_2 v_2 (d - r_2')$$
$$= (m_1 v_1 - m_2 v_2)d + (m_1 v_1 r_1' + m_2 v_2 r_2')$$

where r_1' and r_2' are the distances of the particles from their common centre of mass. The second term in parentheses is the total angular momentum, L_c, about the axis through C. The first term in parentheses is zero; this follows at once from the fact that we have taken C to be at rest and therefore we are in the zero-momentum frame (or we could write $v_1 = \omega r_1'$ and $v_2 = \omega r_2'$ and invoke the definition of the centre of mass as such). Thus in this case we have

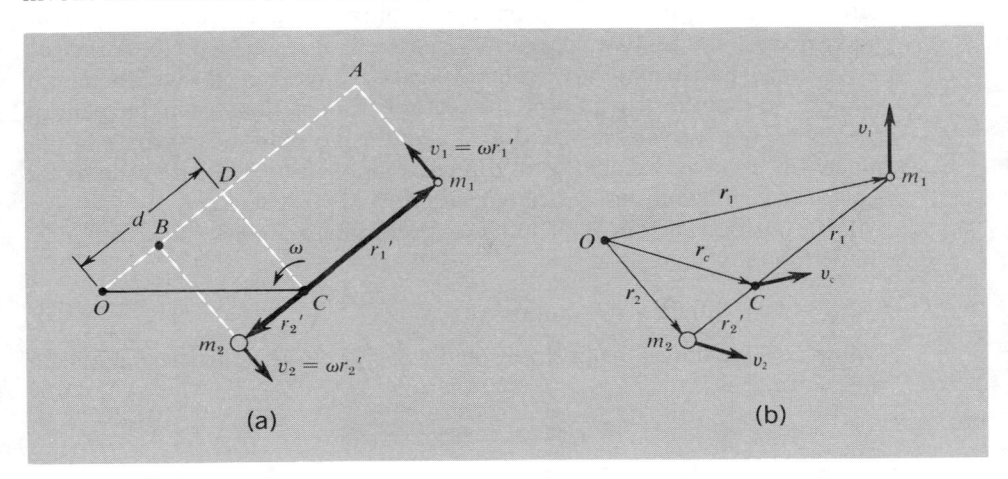

Fig. 10.2 (a) A rigid two-body system rotating about its centre of mass, C. The angular momentum can be calculated with respect to C or to an arbitrary point, O. (b) Individual and centre-of-mass motions in an arbitrary two-particle system.

$L_O = L_c$—the rotational angular momentum of the system has the same value about any axis parallel to the true rotational axis through C. Since $v_1 = \omega r'_1$ and $v_2 = \omega r'_2$, the magnitude of this total angular momentum is given by

$$L_c = (m_1 r'^2_1 + m_2 r'^2_2)\omega$$

Introducing the distance r between the particles, we have

$$r'_1 = \frac{m_2}{m_1 + m_2} r; \qquad r'_2 = \frac{m_1}{m_1 + m_2} r$$

Substituting these values in the expression for L_c gives

$$L_c = \frac{m_1 m_2}{m_1 + m_2} r^2 \omega = \mu r^2 \omega \tag{10.5}$$

where $\mu \ [= m_1 m_2/(m_1 + m_2)]$ is the reduced mass of the system. Equation (10.5) is an important result for the angular momentum of a so-called 'rigid rotator', and the quantity μr^2 is an example of what is called the *moment of inertia*, I, of a rigidly connected system. (We shall consider more complex cases later.) For such a system it is convenient to put

$$L_c = I\omega \tag{10.6}$$

where, in the present case, we have

$$I = \mu r^2 = m_1 r'^2_1 + m_2 r'^2_2$$

Let us now consider the more general case of a system of two particles moving with arbitrary velocities, as shown in Fig. 10.2(b). Again we shall refer the total angular momentum to an arbitrary origin O, and this time we make no assumption that the centre of mass C is at rest relative to O. Instead we assume that it may have some velocity v_c. We can always orient our diagram so that the origin O and the vector r from m_2 to m_1 lie in the plane of the paper, but the velocity vectors v_1, v_2 and v_c need not be confined to this plane. Let us now consider the total angular momentum defined by the vector sum of the contributions associated with m_1 and m_2 separately.

If the position vectors of m_1 and m_2 with respect to O are r_1 and r_2, as shown in the figure, we have

$$L_O = r_1 \times (m_1 v_1) + r_2 \times (m_2 v_2)$$

Let us now introduce the positions and velocities of the particles relative to the centre of mass:

$$r_1 = r'_1 + r_c; \qquad r_2 = r'_2 + r_c$$
$$v_1 = v'_1 + v_c; \qquad v_2 = v'_2 + v_c$$

Then we have

$$L_O = (r'_1 + r_c) \times m_1(v'_1 + v_c) + (r'_2 + r_c) \times m_2(v'_2 + v_c)$$

ANGULAR MOMENTUM

This can be rearranged into a sum of four terms as follows:

$$\begin{aligned}L_O = &(r_1' \times m_1v_1' + r_2' \times m_2v_2') \\ &+ (m_1r_1' + m_2r_2') \times v_c \\ &+ r_c \times (m_1v_1' + m_2v_2') \\ &+ r_c \times (m_1 + m_2)v_c\end{aligned}$$

Now it follows from the definition of the centre of mass that the second and third of the above terms vanish, for we have

$$m_1r_1' + m_2r_2' = 0$$
$$m_1v_1' + m_2v_2' = 0$$

The first term in the expression for L_O is the combined angular momentum, L_c, of the particles about C, and so we have

$$L_O = L_c + r_c \times Mv_c \tag{10.7}$$

where

$$L_c = r_1' \times (m_1v_1') + r_2' \times (m_2v_2')$$

Thus the total angular momentum about O is the net angular momentum about the centre of mass, plus the orbital angular momentum associated with the motion of the centre of mass itself. If the particles have a rigid connection and so rotate as a unit about C, we can use eqn (10.5) or (10.6) to give the explicit expression for the magnitude of L_c in eqn (10.7).

A study of the above analysis will make it clear that equivalent results hold good for a system of arbitrarily many particles; thus in eqn (10.7) we have a strong basis for the analysis of angular momentum in general. Notice in particular that if the centre of mass of an arbitrary system of moving particles is at rest, then the total angular momentum has the same value, equal to L_c, about any point.

For any two-particle system, rigidly connected or not, the value of L_c can be conveniently expressed in terms of the relative coordinate r of the two masses and their relative velocity v_{rel}. We take the expression for L_c in eqn (10.7):

$$L_c = r_1' \times (m_1v_1') + r_2' \times (m_2v_2')$$

and we first substitute for r_1' and r_2' in terms of r. From the defining equations

$$m_1r_1' + m_2r_2' = 0$$
$$r_2' - r_1' = r$$

we have

$$r_1' = -\frac{m_2}{m_1 + m_2}r \; ; \quad r_2' = \frac{m_1}{m_1 + m_2}r$$

It then follows that L_c can be written as follows:

$$L_c = \frac{m_1 m_2}{m_1 + m_2}(-r \times v_1' + r \times v_2')$$

i.e.

$$L_C = \mu r \times (v_2' - v_1') = r \times (\mu v_{\text{rel}}) \tag{10.8}$$

In the particular case of a rigid system rotating with angular velocity ω, we can further put

$$v_{\text{rel}} = \omega \times r$$

If ω is perpendicular to r this reduces eqn (10.5) to eqn (10.8); otherwise we have the slightly more complicated result

$$L_c = \mu r^2 \omega - \mu r(\omega \cdot r)$$

Angular momentum as a fundamental quantity

The preceding analysis has established that two connected masses, regarded as a single system with mass and size, have what can properly be described as an intrinsic angular momentum about the centre of mass. Regardless of the actual motion of the CM, one can identify this rotational property of the system. But angular momentum took on an even more basic aspect when it was discovered, in the development of quantum mechanics, that there was a natural unit of angular momentum, equal to Planck's constant h divided by 2π:

$$\text{Basic unit of angular momentum} = \frac{h}{2\pi}$$
$$= 1.054 \times 10^{-34} \text{ J s}$$

This is of course a very tiny unit, but it implies enormously high speeds of rotation in systems of atomic size. Let us consider one example. For many purposes a diatomic molecule, such as N_2, can be regarded as a rigid system such as we have discussed—two point masses a fixed distance apart. Nitrogen has two equal nuclei, each of mass about 2.3×10^{-26} kg, separated by about 1.1×10^{-10} m ($= 1.1$ Å). The moment of inertia (cf. eqn 10.5) is thus given by

$$I = 2m(s/2)^2$$
$$= 2(2.3 \times 10^{-26})(5.5 \times 10^{-11})^2 \text{ kg m}^2$$
$$\approx 1.4 \times 10^{-46} \text{ kg m}^2$$

If we put

$$I\omega = 1.054 \times 10^{-34} \text{ kg m}^2 \text{ s}^{-1}$$

then we find that

$$\omega \approx 7.5 \times 10^{11} \text{ s}^{-1}$$

The frequency (rps) corresponding to this would be $\omega/2\pi$, or about 10^{11} s^{-1}. Frequencies of this order are typical of molecules behaving as rigid rotators and can be studied through the techniques of microwave spectroscopy (the frequency just mentioned would correspond to a wavelength of 3 mm).

Even more fundamentally, it appears that all the elementary particles of the universe have an intrinsic angular momentum which is some integral multiple (including zero) of $h/4\pi$. In particular, our most familiar building blocks, nucleons and electrons, have just $h/4\pi$. At this level, however, the specification of what, if anything, is rotating becomes a moot question; one simply contents oneself with the fact of an intrinsic angular momentum that has the important property of being conserved in all the interactions and rearrangements of such particles. This conservation property of angular momentum in general is the subject of the next section.

Conservation of angular momentum

Many interesting experiments can be performed that illustrate the important property that the total internal angular momentum of a system of particles is conserved if external influences are absent. Some of the qualitative demonstrations are no doubt familiar to you—the speeding up of a whirling ice skater, for example, or an expert diver performing a somersault. A quite unskilled person can do similar tricks if he sits on a freely pivoted stool, gets himself turning slowly with a couple of dumbbells held at arm's length, and then pulls the dumbbells inward (Fig. 10.3a).

The conservation of internal angular momentum holds good, whatever internal rearrangements of the system take place. Some particularly nice consequences of this conservation can be shown if one has a good flywheel—e.g., a bicycle wheel with an extra loading of lead around the rim. For example, one person can sit on a pivoted stool (Fig. 10.3b) and another person can hand him the wheel after it has been set spinning with angular momentum L_w as shown (corresponding to clockwise rotation about a vertical axis pointing upward). The person on the stool is not himself rotating, but the system, person + stool, has the total internal angular momentum L_w. If the wheel is now inverted its rotational angular momentum about its own centre of mass is changed to $-L_w$. It follows that the system of two masses, M (the person) and m (the wheel), must acquire a clockwise rotation with a total rotational angular momentum of $+2L_w$ (Fig. 10.3c). If the wheel in this new orientation is handed to the assistant, who inverts it and hands it back, the total angular momentum is raised to $3L_w$. If the person on the stool again inverts

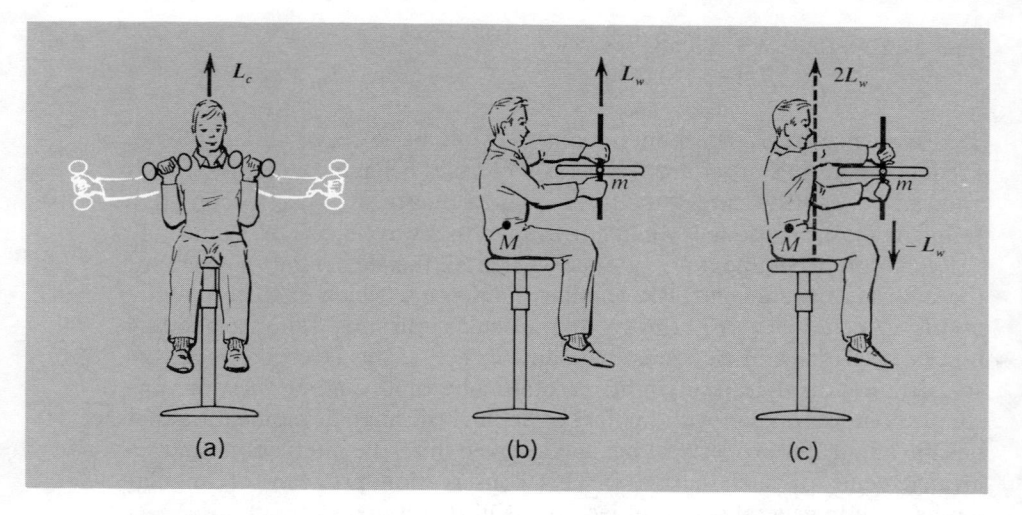

Fig. 10.3 Experiments on the conservation of angular momentum: (a) the person on the stool rotates faster if he pulls the dumbbells inward. (b) and (c) The man on the stool begins to rotate when he inverts the spinning wheel.

the wheel, the general rotation of $M + m$ is raised to $5L_w$. Thus angular momentum can be transferred back and forth in packets in such operations—although here we are going beyond the conservation of total angular momentum in a completely isolated system.

The formal proof of the conservation of angular momentum is not difficult. The total internal angular momentum (with respect to the centre of mass or to any other point in the CM frame) is given, according to eqn (10.7), by

$$L_c = \sum r_i' \times (m_i v_i') \qquad (10.9)$$

Let us consider the variation of L_c with time. Differentiating, we have

$$\frac{dL_c}{dt} = \sum v_i' \times (m_i v_i') + \sum r_i' \times (m_i a_i')$$

where a_i' is the acceleration of particle i relative to the CM. The first summation vanishes, because every product $v_i' \times v_i'$ is identically zero. In the second summation, we shall write a_i' as the vector difference between the true acceleration, a_i, of particle i, as measured in an inertial frame, and the acceleration $a_c (= dv_c/dt)$ of the centre of mass. (It is important to realize that a_c may exist even in a frame in which v_c is zero at some instant.) Thus we have

$$\frac{dL_c}{dt} = \sum r_i' \times m_i(a_i - a_c)$$

CONSERVATION OF ANGULAR MOMENTUM

$$= \sum r'_i \times (m_i a_i) - \left(\sum m_i r'_i\right) \times a_c$$

However, by the definition of the CM, the summation in the second term is zero. The first term is the total torque about C of all the forces acting on the particles, because $m_i a_i$ is the net force acting on any given particle. This force may be a combination of an external force F_i and a set of internal forces f_{ik}; thus we put

$$m_i a_i = F_i + \sum_k f_{ik}$$

Substituting this statement of $m_i a_i$ in the equation for dL_c/dt we therefore have

$$\frac{dL_c}{dt} = \sum r'_i \times F_i + \sum_i \sum_k r'_i \times f_{ik}$$

Now, as in our earlier discussion of the total linear momentum of a system of particles, we can arrange the double summation involving internal forces into pairs of terms, in this case of the type

$$r'_i \times f_{ik} + r'_k \times f_{ki}$$

If, however, we can assume that the forces of interaction between any two particles are equal, opposite forces along the line joining them (Fig. 10.4a), then each such pair of torques adds up to zero, because each force has the same lever arm CD with respect to C. Thus we arrive at a very simple equation:

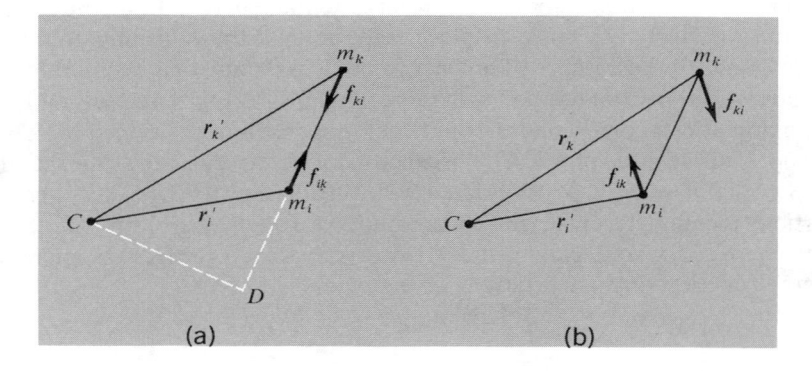

Fig. 10.4 (a) The equal and opposite internal forces between two particles can produce no torque about C if they are along the line joining the particles. (b) If the forces are not along the line joining the particles they produce a torque which is, however, nullified by other internal forces (see discussion in the text).

$$\frac{dL_c}{dt} = \sum r'_i \times F_i = \sum M'_i \qquad (10.10)$$

where M'_i is the torque exerted by F_i about the CM. Equation (10.10) is a very basic equation of rotational dynamics; we shall spell it out in words:

Regardless of any acceleration that the centre of mass of a system of particles may have as a result of a net external force exerted on the system, the rate of change of internal angular momentum about the CM is equal to the total torque of the external forces about the CM.

In particular, therefore, if the net torque about the centre of mass is zero, the internal angular momentum remains constant, whatever internal rearrangements may go on within the system.

[Our derivation of eqn (10.10) contains one weak link. This is the argument by which we conclude that the net torque of the internal interactions is zero. It is perfectly possible to imagine that the forces f_{ik} and f_{ki} are equal in magnitude and opposite in direction, thus conforming to Newton's third law, without having them act along the same line (see Fig. 10.4b). In this case they would constitute a couple with a resultant torque about C or any other point. The vanishing of the net torque of the internal forces, and the consequent conservation of total angular momentum if external torques are absent, is, however, a result that holds good in general; it does not depend on the limited assumption that the forces of interaction act along the lines joining pairs of particles. A powerful theoretical argument in support of this proposition —but requiring virtually no mathematics at all—can be made on the basis of the uniformity and isotropy of space. It runs as follows.

Consider first the conservation of the total *linear* momentum of a system of particles. This holds good if the total potential energy, V, of the system remains unaffected by linear displacement, because the vanishing of **grad** V corresponds to the absence of any net force. If we know that external forces are absent, the *invariance* of V with respect to linear displacements is more or less axiomatic; it corresponds to our belief that absolute position along a line has no significance in physics. In an exactly similar way, we can argue that the total potential energy, V, associated with the internal forces of a system of particles is completely insensitive to a rotation through an arbitrary angle θ of the system as a whole. Now just as we can evaluate a force from a potential-energy function through relations of the type

$$F_x = -\frac{\partial V}{\partial x}$$

so we can evaluate torques through relations of the type

$$M_z = -\frac{\partial V}{\partial \theta} \qquad (10.11)$$

where M_z is the torque about the axis z around which the rotation θ is imagined. Hence, if V is independent of θ, there can be no net torque, regardless of the detailed character of the internal interactions. Thus we can conclude that the conservation of total internal angular momentum of any isolated system must hold true in general.]

Moments of inertia of extended objects

For an arbitrary system of particles, the intrinsic rotational angular momentum is defined by the equation

$$L_0 = \sum r_i \times (m_i v_i) \tag{10.12}$$

(For simplicity we are here using r_i and v_i to denote position and velocity *with respect to the CM*.) If we apply eqn (10.12) to a system that is rotating as a unit with angular velocity ω and, in particular, to the case of an object that has a well defined geometrical symmetry and is rotating about an axis of symmetry through the CM, then the total rotational angular momentum is a vector in the direction of ω itself. We can put

$$L_0 = \left(\sum m_i R_i^2 \right) \omega \tag{10.13}$$

(R_i being the perpendicular distance of m_i from the axis of rotation)

where $\quad I = \sum m_i R_i^2 \tag{10.14}$

Equation (10.14) then defines the moment of inertia of the complete system about a given axis of symmetry.

If we consider a flywheel, which is an object with a well-defined symmetry, then its normal rotational axis is not its only axis of symmetry (see Fig. 10.5). If we label this normal rotational axis as z, then any other axis perpendicular to z and passing through the CM is also an axis of symmetry (though not of complete *rotational* symmetry). Thus we can define, in addition to I_z, two other moments of inertia, I_x and I_y, about a pair of axes perpendicular to z and to each other. These three quantities (I_x, I_y, I_z) are known as the *principal moments of inertia* of the object. In the case illustrated, we would of course have $I_x = I_y (\neq I_z)$.

It is important to recognize that it is only when the axis of rotation, as defined by the direction of ω, coincides with an axis corresponding to a principal moment of inertia that the rotational angular momentum vector L_c is parallel to ω itself and can be expressed simply as $I\omega$. We shall, however, be concerned mostly with situations in which this is the case. [It is a remarkable fact that for any object at all, even if it has no kind of symmetry, one can still find a set of three orthogonal rotation axes for which L and ω are parallel.

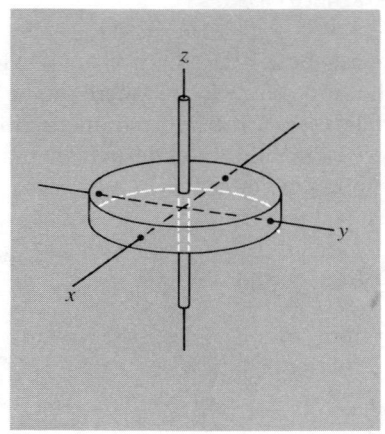

Fig. 10.5 Main spin axis and two other principal axes of inertia of an object with well-defined symmetry.

The existence of these axes allows us to identify three principal moments of inertia for an object of completely arbitrary shape.]

For an object composed of so many particles that it is effectively continuous, we can write the moment of inertia about a given symmetry axis as an integral instead of a summation:

$$I = \int r^2 \, dm \tag{10.15}$$

where dm is an element of mass situated at a perpendicular distance r from the axis through the centre of mass.

The moments of inertia of some geometrically simple bodies are shown in Fig. 10.6.

It may be noted that in each case the moment of inertia is the mass of the object, times the square of a characteristic linear dimension, times a numerical coefficient not very different from 1. It is hard to be wildly wrong in estimating the moment of inertia of a body, even without detailed calculation. It is quite common practice to write the moment of inertia simply as the total mass times the square of a length k that is called the *radius of gyration* about the axis in question:

$$I = Mk^2 \tag{10.16}$$

Two theorems concerning moments of inertia

The calculation of moments of inertia can often be simplified with the help of two theorems that we shall now present. The first of them applies to any kind

(a) $I = MR^2$ (ring)

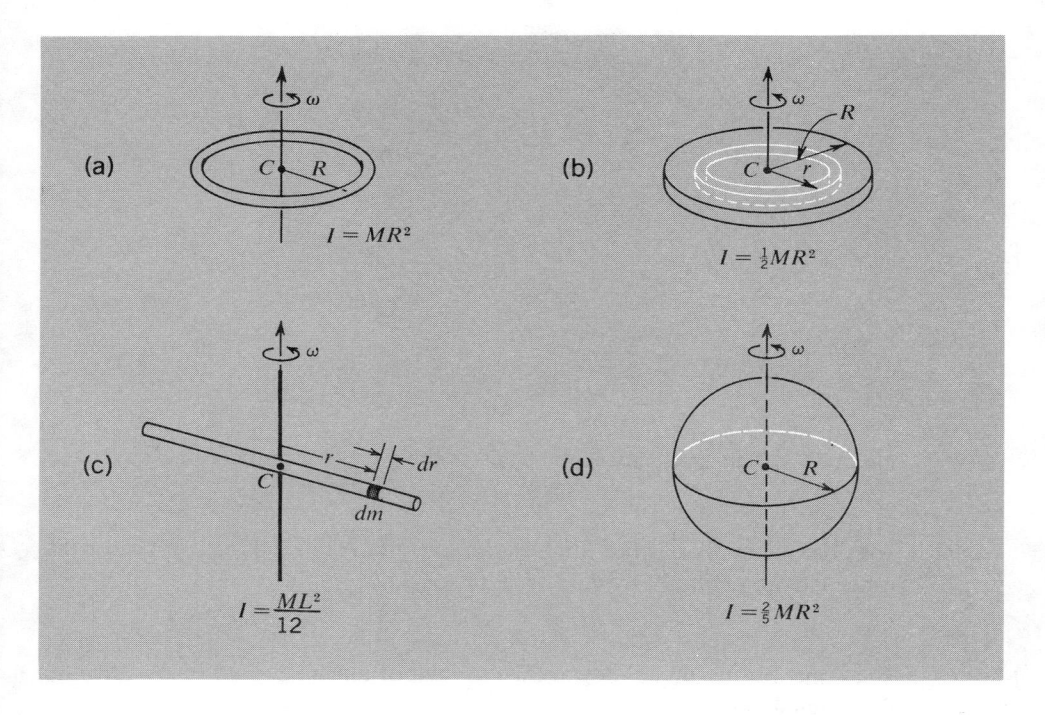

(b) $I = \tfrac{1}{2}MR^2$ (disc)

(c) $I = \dfrac{ML^2}{12}$ (bar or rod)

(d) $I = \tfrac{2}{5}MR^2$ (solid sphere)

Fig. 10.6 Geometrically simple objects with exactly calculable moments of inertia: (a) ring, (b) disc, (c) bar or rod, and (d) solid sphere.

of object; the second is applicable only to objects that can be treated, to some approximation, as flat objects of negligible thickness.

The theorem of parallel axes

When we speak of '*the* moment of inertia' of an object we normally mean the moment of inertia about an axis of rotational symmetry drawn through the centre of mass. There are, however, many situations in which the actual physical axis of rotation does not pass through the CM. In such cases one can make use of a theorem that directly relates the moment of inertia I_0 about the given axis to the moment of inertia I_c about a parallel axis through the centre of mass.

Figure 10.7 illustrates the situation. Suppose that the given axis is perpendicular to the plane of the paper and passes through the point O. Imagine a parallel axis through C, and consider the object as being built up of a set of thin slices parallel to the plane of the paper. The vector distance from the axis through O to the axis through C is a constant, h. Within any one slice we can

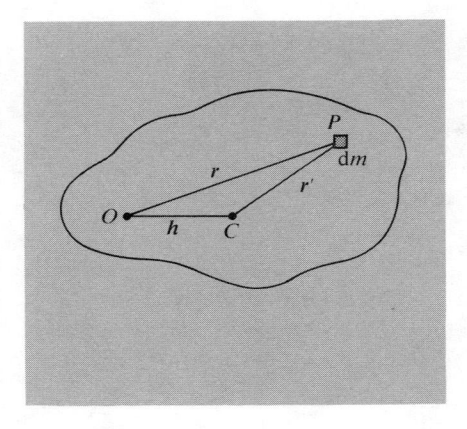

Fig. 10.7 Diagram to show the basis of the parallel-axis theorem.

consider elements of mass such as dm at point P in the figure, a vector distance r from O and r' from C. We have

$$r = h + r'$$

The contribution dI_O of dm to the moment of inertia about the axis through O is then given by

$$dI_O = r^2\, dm = (h + r') \cdot (h + r')\, dm$$
$$= h^2\, dm + 2h \cdot r'\, dm + r'^2\, dm$$

The last term on the right is, however, just the contribution of dm to the moment of inertia about the parallel axis through C. Thus we have

$$dI_O = h^2\, dm + 2(h \cdot r')dm + dI_c$$

We can now carry out the summation or integration of all such contributions, first within the slice and then over all the slices that build up to the complete object. This gives us

$$I_O = h^2 \int dm + 2h \cdot \int r'\, dm + I_c$$

By the definition of the centre of mass, the middle term on the right is zero. The first term is simply the total mass, M, of the object times h^2. Thus we finally have the following result:

(parallel-axis theorem) $I_O = I_c + Mh^2$ (10.17a)

where h is the distance between the axes. If we choose to express I_c according to eqn (10.16) in terms of the mass M and the radius of gyration k, we can write eqn (10.17a) in the alternative form:

(parallel-axis theorem) $\quad I_O = M(k^2 + h^2)$ \hfill (10.17b)

It may be seen from eqns (10.17a) and (10.17b) that the moment of inertia of an object about an arbitrary axis is always greater than its moment of inertia about a parallel axis through the CM.

Theorem of perpendicular axes

Suppose now that we have a flat object of arbitrary shape, cut out of thin sheet material (Fig. 10.8a). Let us take an arbitrary origin O in the plane of the object, and a z axis perpendicular to it. Consider now the moment of inertia, I_z, of the object about the z axis. An element of mass dm, a distance r from the axis, makes the contribution $r^2\,dm$, and we have

$$I_z = \int r^2\,dm$$

However, since r lies in the xy plane, we have

$$r^2 = x^2 + y^2$$

Thus we can put

$$I_z = \int x^2\,dm + \int y^2\,dm$$

Since the object is flat, however, the first term on the right simply defines the moment of inertia I_y of the object about the y axis, and th e second term correspondingly is I_x. Thus we have

(flat objects) $\quad I_z = I_x + I_y$ \hfill (10.18)

This is known as the perpendicular-axis theorem. Its usefulness can be illustrated by two different kinds of examples:

Example 1. An object is in the form of a uniform rectangular plate (Fig. 10.8b). What is its moment of inertia about an axis through its centre and perpendicular to its plane?

Using the result shown in Fig. 10.6(c),

$$I_x = \frac{Mb^2}{12}; \quad I_y = \frac{Ma^2}{12}$$

It follows at once that

$$I_z = \frac{M(a^2 + b^2)}{12}$$

Example 2. A uniform disc has mass M and radius R. What is its moment

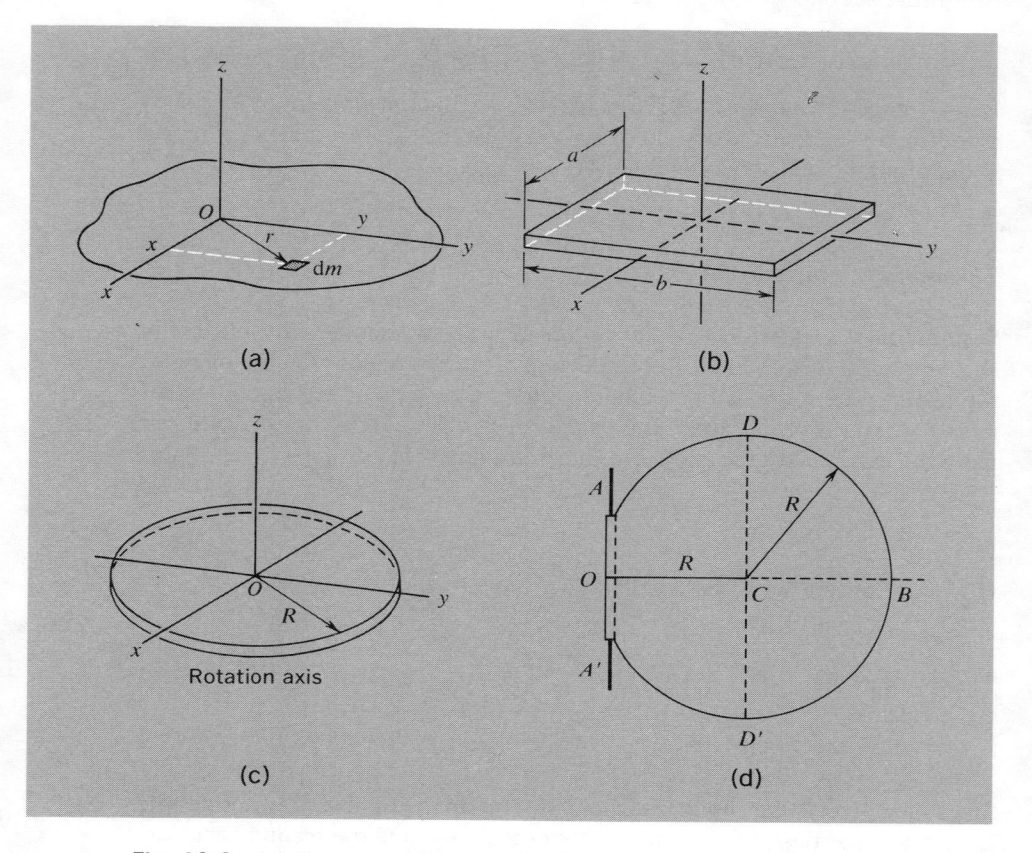

Fig. 10.8 (a) Diagram to show the basis of the perpendicular-axis theorem. (b) and (c) Flat objects to which the perpendicular-axis theorem can be usefully applied. (d) Calculation of the moment of inertia of a circular lid about the axis AA' exploits both the parallel-axis and the perpendicular-axis theorems.

of inertia about an axis along one of its own diameters, say the x axis in Fig. 10.8(c).

This illustrates a more elegant use of the perpendicular-axis theorem. The direct calculation of the moment of inertia of a disc about a diameter would be quite awkward. We know, however, that the moment of inertia has the same value about any diameter. We also know, from Fig. 10.6(b), the moment of inertia I_z about the axis perpendicular to the disc through its centre. Thus we can at once put

$$I_x + I_y = 2I_x = I_z = \frac{MR^2}{2}$$

Therefore,

$$I_x = \frac{MR^2}{4}$$

Finally, in Fig. 10.8(d) we show a situation in which we can exploit both of the above theorems. A circular disc (e.g. the lid of a cylindrical tank) is pivoted about an axis AA' that lies in the plane of the disc and is tangent to its periphery. What is the moment of inertia of the disc about AA'?

Beginning with the moment of inertia $MR^2/2$ about the axis through C perpendicular to the disc (see Fig. 10.6b), we first use the perpendicular-axis theorem to deduce that the moment of inertia about the axis DD' is $MR^2/4$. We can then use the parallel-axis theorem, eqn (10.17), to deduce that the moment of inertia about AA' is given by

$$I_{AA'} = \frac{MR^2}{4} + MR^2 = \frac{5MR^2}{4}$$

Such a result is not worth memorizing for its own sake; the important thing is to realize that in these two theorems we have a powerful way of extending a few basic results to handle a whole variety of more complicated situations.

Kinetic energy of rotating objects

A rotating system, of course, has kinetic energy by virtue of its rotation about its centre of mass. Thanks to the general validity of eqn (10.3) this can be calculated separately and added to any kinetic energy associated with motion of the centre of mass itself. When a 'rigid' object rotates with angular velocity ω about an axis, a particle within it of mass m, distance r from the axis, has a speed ωr and hence a kinetic energy $\frac{1}{2}m\omega^2 r^2$. The total kinetic energy of rotation is thus given by

$$K_{\text{rot}} = \tfrac{1}{2}\omega^2 \sum mr^2 = \tfrac{1}{2}I\omega^2 \tag{10.19}$$

Hence, if the CM has a speed v_c with respect to the laboratory, the total kinetic energy as measured in the laboratory is given by

$$K = \tfrac{1}{2}I\omega^2 + \tfrac{1}{2}Mv_c^2 \tag{10.20}$$

For an object that rolls along the ground, there will be a purely geometrical connection between v_c and ω (e.g. for a wheel of radius R, $v_c = \omega R$). In such a case the kinetic energy can be expressed in terms of ω (or v_c) alone.

An important aspect of rotational kinetic energy is that a large rotating object is an energy reservoir of possibly very large capacity. The use of flywheels as energy-storage devices in this sense is an important feature of all sorts of machines, giving to such systems a much improved stability with respect to sudden changes of load. One of the most impressive examples of the use of flywheels for energy storage is in the National Magnet Laboratory at MIT. There are two flywheels, each being an assembly of circular plates of steel, 190 in. in diameter, with a mass of 85 tons. They are part of a generating system for producing extremely strong magnetic fields. The normal speed of rotation of each flywheel is 390 rpm. From these figures we have (for each flywheel)

$$M = 7.7 \times 10^4 \text{ kg}; \quad R = 2.4 \text{ m}$$
$$I = \tfrac{1}{2}MR^2 \approx 2 \times 10^5 \text{ kg m}^2$$
$$\omega = 2\pi \times 390/60 \approx 40 \text{ rad s}^{-1}$$

Therefore,

$$K = \tfrac{1}{2}I\omega^2 \approx 1.6 \times 10^8 \text{ J}$$

When one of these flywheels is used as a power source, its speed of rotation can be lowered from 390 to 300 rpm in 5 s. This means that about 40% of the stored energy can be drawn upon, at the rate of about 8% per second. The power output corresponding to this is close to 15 MW—enough, while it lasts, to equal the total rate of electrical energy consumption of a town of about 20 000 inhabitants.

Angular momentum conservation and kinetic energy

The conservation of rotational angular momentum has some interesting implications for the total kinetic energy of a system that changes its shape or size. To take simplest possible example, consider a system of two masses rotating about their centre of mass (Fig. 10.9a). We shall assume that the

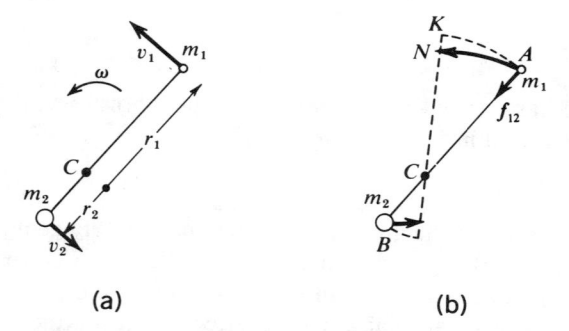

Fig. 10.9 (a) Pair of objects rotating at fixed distances about their centre of mass. (b) Rotational displacements accompanied by a change of radial distance between the objects.

centre of mass is stationary in our frame of reference, but in the absence of external forces it does not have to be defined by a fixed pivot or anything of that kind.

Suppose that the system is rotating with an angular velocity ω about an axis perpendicular to the plane of the diagram. Then if the distances of the masses from C are r_1 and r_2, respectively (with $m_1 r_1 = m_2 r_2$), we have

$$v_1 = \omega r_1; \quad v_2 = \omega r_2$$

We can then proceed at once to write down expressions for the total angular momentum and the total kinetic energy:

$$L_c = (m_1 r_1^2 + m_2 r_2^2)\omega$$
$$K = \tfrac{1}{2}(m_1 r_1^2 + m_2 r_2^2)\omega^2$$

It is helpful to introduce the reduced mass μ $[= m_1 m_2/(m_1 + m_2)]$, the relative separation r $(= r_1 + r_2)$, and the moment of inertia I $(= \mu r^2)$. We then have

$$L_c = \mu r^2 \omega = I\omega \quad (= \text{const.})$$
$$K = \tfrac{1}{2}\mu r^2 \omega^2 = \tfrac{1}{2} I \omega^2$$

Combining these two equations, we arrive at the following result:

$$K = \frac{L_c^2}{2\mu r^2} = \frac{L_c^2}{2I} \tag{10.21}$$

Consider now what happens if the particles draw closer together under some mutual interaction, e.g. the pull of a spring or elastic cord that connects them. The value of L_c remains constant, but r and the moment of inertia I decrease; hence the values of ω and K must increase. Where does the extra energy come from? Clearly, it has to be supplied through the work done by the internal forces that pull the masses together. One can feel this very directly if one does the experiment of sitting on a rotating stool with two weights at arm's length and then pulling the weights inward towards the axis (cf. Fig. 10.3a). One very simple way of interpreting this from the standpoint of the rotating frame itself is to consider the small change of K associated with a small change of r. From eqn (10.21) we have

$$\Delta K = -\frac{L_c^2}{\mu r^3} \Delta r$$

Substituting $L_c = \mu r^2 \omega$, this becomes

$$\Delta K = -\mu \omega^2 r \Delta r$$

But $\mu \omega^2 r$ is the magnitude of the centrifugal force that is trying (from the standpoint of the *rotating* frame) to make the masses fly apart. A force equal and opposite to this must be supplied to hold the masses at a constant separation r, and an amount of work equal to this force times the magnitude of the displacement Δr is needed to pull the masses towards one another. (Note that, in the case we have assumed, Δr is negative and so ΔK is positive.)

From the standpoint of a *stationary* observer, of course, the change of kinetic energy can be understood in terms of the fact that if r is changing (Fig. 10.9b) the radial forces of interaction, f_{12} and f_{21}, have components along the paths of the masses (e.g. the curve AN for m_1) so that work is done, which would not be the case if the masses remained on circular arcs (e.g. AK in Fig. 10.9b) with the radial forces always perpendicular to the displacements and velocities.

The fact that the increase in rotational kinetic energy of a contracting system must come from the internal interactions places restrictions on the possibility that such contraction can occur at all in a particular case. If the increase that would be called for in the kinetic energy is greater than the work that could supplied by the internal forces, then the contraction cannot take place. Especially interesting situations of this type may arise in the gravitational contraction and condensation of a slowly rotating galaxy or star. The mutual gravitational energy, V, of any system of particles is always negative and becomes more so as the linear dimensions of the system shrink and the particles of the system come closer together. (Remember that, for a two-particle system, $V = -Gm_1m_2/r$.) Thus, qualitatively at least, we recognize a source of the extra kinetic energy needed. There is more to it than this, however, because the magnitudes of the corresponding changes ΔK and ΔV, for a given change ΔR in the radius of the system, are not automatically equal. The relationship between them will depend on the magnitude of L_c and on the precise distribution of matter in the system. If ΔK would be larger than ΔV the total energy would be required to increase, and this simply could not happen. If ΔK were less than ΔV, however, there would be a surplus of released gravitational energy that could be disposed of by developing random particle motions (heating) and by radiation of heat and light into space.

With the help of certain extreme simplifying assumptions the discussion can be made quantitative. Suppose, in particular, that the contraction occurs in such a way that the system merely undergoes a change of linear scale without altering the *relative* distributions of density or velocity. This means that if the linear dimensions shrink by a factor n, the density at distance r/n from the centre is n^3 times the original density at r. The moment of inertia, I, is then simply proportional to the square of any characteristic dimension of the system, e.g. its outer radius R if it can be considered as being a spherical object with an identifiable boundary. Thus the kinetic energy K ($= L^2/2I$) varies as $1/R^2$. The assumed uniform contraction also increases the potential energy ($\sim 1/r$) between every pair of particles by the factor n. It follows that the total (negative) potential energy V varies as $1/R$. Under these assumptions, therefore, the total energy E of the system is given by an equation of the form

$$E(R) = \frac{A}{R^2} - \frac{B}{R} \qquad (10.22)$$

The constant A is given dimensionally, and perhaps even in order of magnitude, by the combination $L^2/2M$, where M is the total mass. Similarly, B is proportional to GM^2, where G is the universal gravitation constant. (The gravitational *self*-energy of a sphere of matter is of the order of $-GM^2/R$).

The two contributions to E, and their sum, are shown graphically as functions of R in Fig. 10.10. The whole situation is very reminiscent of our discussion of energy diagrams in Chapters 6 and 7, and it is clear that we can calculate a radius R_m that corresponds to an equilibrium configuration of minimum energy:

$$\frac{dV}{dR} = -\frac{2A}{R^3} + \frac{B}{R^2} = 0$$

$$R_m = \frac{2A}{B}$$

If we put $A \approx L^2/2M$ and $B \approx GM^2$, we then have

$$R_m \approx \frac{L^2}{GM^3} \qquad (10.23)$$

which would indicate the way in which the linear dimensions of similar galaxies might depend on the mass and the total angular momentum.

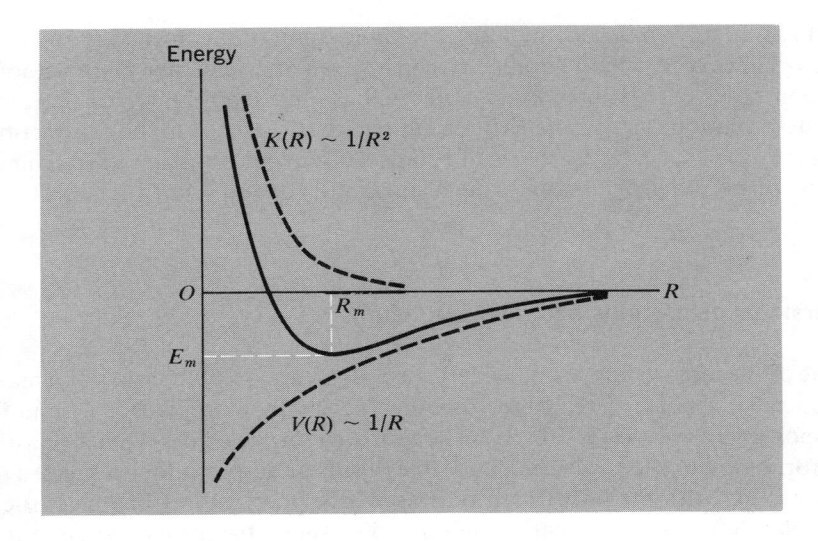

Fig. 10.10 Dependence of kinetic, potential and total energy on radius for a rotating system held together by gravitational attraction.

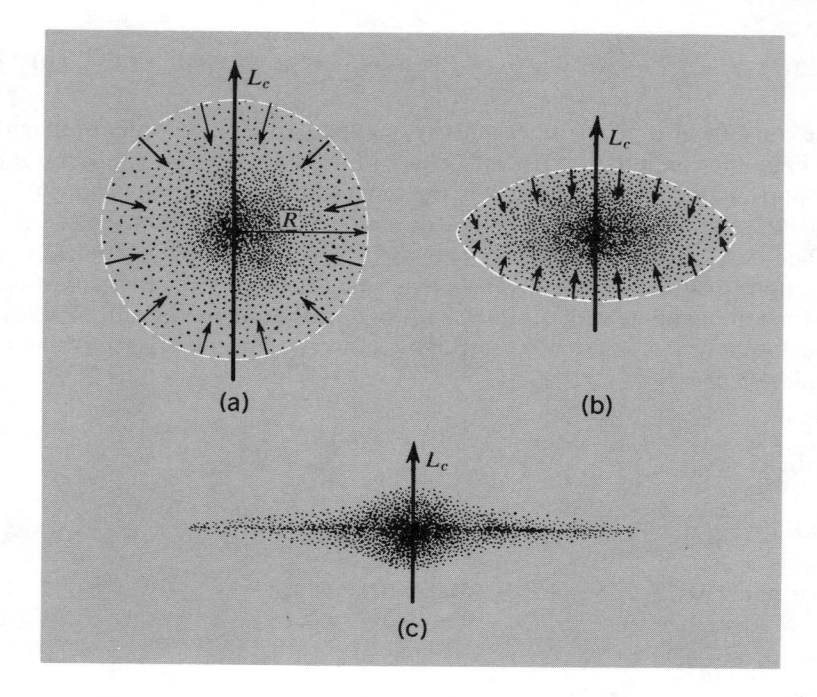

Fig. 10.11 Successive stages of contraction of a rotating gas cloud to form a disc-shaped galaxy.

Figure 10.11 indicates the more probable trend of a contracting rotating mass. Since contraction parallel to the direction of L_c can take place without any increase of rotational kinetic energy, it is quite reasonable that this type of deformation can continue after a limit has been reached to the contraction radially inward toward the axis of rotation. One can certainly understand in these terms the progression through stages (a), (b) and (c) of Fig. 10.11.

Torsional oscillations and rigid pendulums

One of the most valuable and widely used physical systems consists of a mass suspended from a wire, fibre, or other device that provides a torque in response to a twist (Fig. 10.12a). The restoring torque comes from an elastic deformation of the suspension and, like the linear deformations discussed in connection with the linear harmonic oscillator problem in Chapter 3, such angular deformations usually result in a restoring effect proportional to the deformation.

These torsion devices are often used in static measurements—for example

TORSIONAL OSCILLATIONS AND RIGID PENDULUMS

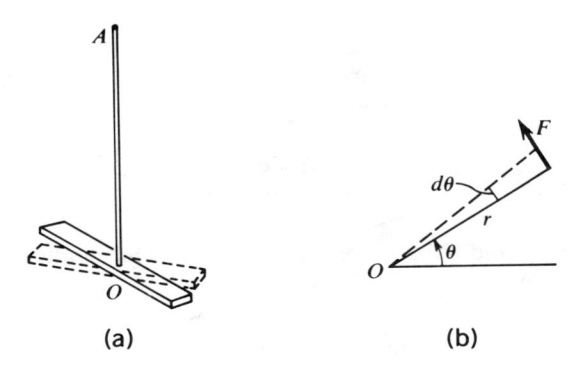

Fig. 10.12 (a) Simple torsional pendulum. (b) Diagram to indicate the work done in twisting a torsion fibre.

in ammeters, where a steady current passing through the instrument may be used (with the help of a permanent magnet) to produce a steady torque, leading to a steady deflection that is the meter reading of the current. But the free torsional oscillation of such a system is also of interest and importance. The analysis of this oscillatory motion is very conveniently made in terms of the constant total energy of the rotating system.

We shall suppose that the suspended mass has a moment of inertia I about the axis defined by the torsion fibre. Let the angle of deflection around this axis (we shall call it the z axis) be θ. Then the kinetic energy K is given by

$$K = \frac{1}{2} I \left(\frac{d\theta}{dt} \right)^2 \tag{10.24}$$

The potential energy V is the work done in twisting the ends of the suspension through an angle θ relative to one another. The restoring torque M_z, assumed proportional to θ, is given by

$$M_z = -c\theta \tag{10.25}$$

where c is the *torsion constant* (measured in m N rad^{-1} or dimensionally equivalent units).

It is not hard to guess that, by analogy with a stretched spring, a system that obeys eqn (10.25) will lead to a storage of potential energy proportional to θ^2. To make this quantitative we need only consider one simple idea. Suppose we are looking down the axis of a torsion wire (Fig. 10.12b) to which is attached a lever arm of length r. A force F is applied at right angles to the end of the lever, just sufficient to overcome the torque M_z. This means that

$$rF = -M_z$$

Suppose that the angle of twist is increased by $d\theta$. Then the end of the lever moves through a distance $r\,d\theta$, and the work done by F is given by

$$dW = Fr\, d\theta = -M_z\, d\theta$$

Hence the total potential energy stored in the system, in going from its normal configuration to a twist θ, is given by

$$V(\theta) = \int_0^\theta c\theta\, d\theta = \tfrac{1}{2}c\theta^2 \tag{10.26}$$

Combining eqns (10.24) and (10.26), we see that the total mechanical energy of the system is given by

$$\tfrac{1}{2}I\left(\frac{d\theta}{dt}\right)^2 + \tfrac{1}{2}c\theta^2 = E \quad (=\text{const.}) \tag{10.27}$$

which is the familiar form of a harmonic-oscillator equation. The period will be given by

$$T = 2\pi\left(\frac{I}{c}\right)^{1/2} \tag{10.28}$$

It is worth noting, by the way, that the relation between M_z, as given by eqn (10.25), and $V(\theta)$, as given by eqn (10.26), satisfies the general relation between potential energy and torque that we cited earlier eqn (10.11) in our discussion of angular-momentum conservation:

$$M_z = -\frac{\partial V}{\partial \theta}$$

Situations very similar to those of the torsional oscillator are represented by a so-called *rigid pendulum*—an arbitrary object free to swing about a horizontal axis, as shown in Fig. 10.13(a). Let us suppose that this axis is a distance h from the centre of mass. Normally, therefore, the centre of mass C is a vertical distance h below the axis through O, but if the system is displaced through an angle θ, the CM moves along a circular arc of radius h. This causes the CM to rise through the distance $h(1 - \cos\theta)$; if θ is small the consequent increase of gravitational potential energy is given by

$$V(\theta) \approx \tfrac{1}{2}Mgh\theta^2 \tag{10.29}$$

The *kinetic* energy of the system is equal to the kinetic energy associated with the linear velocity of the CM, plus the energy of rotation about the CM:

$$K = \tfrac{1}{2}Mv_c^2 + \tfrac{1}{2}I_c\left(\frac{d\theta}{dt}\right)^2$$

[The term representing the rotational energy about C in this equation embodies an important feature. If the object has the angular velocity $d\theta/dt$ about its true axis of rotation through O, every point in it also has the angular velocity $d\theta/dt$ about a parallel axis through C, or through any other point for

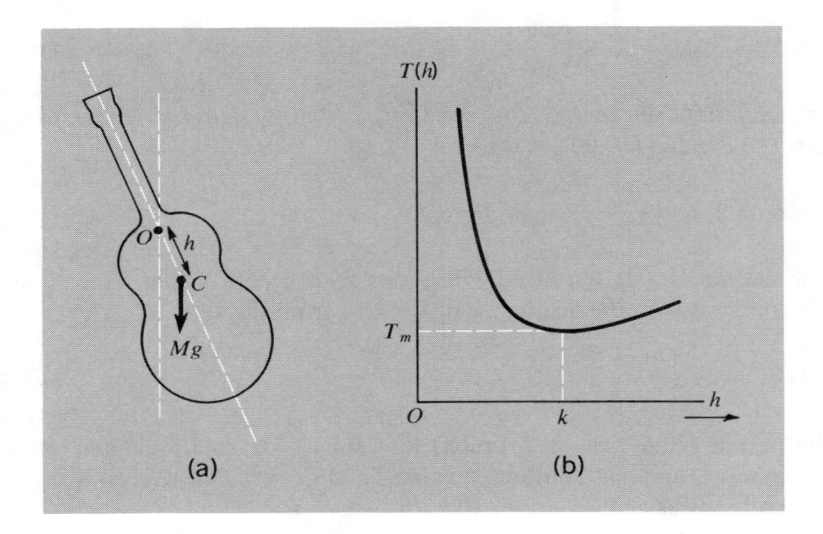

Fig. 10.13 (a) Example of a rigid pendulum. (b) Period of oscillation of a rigid pendulum as a function of the distance h between the CM and the point of suspension.

that matter. One can properly speak of *the* angular velocity of a rotating object without reference to a specific axis of rotation. Any line drawn on a rotating disc, for example, has the same rate of angular displacement as one of the radii.]

Returning now to the expression for K, we can put

$$v_c = h \frac{d\theta}{dt}$$

Thus we can write

$$K(\theta) = \tfrac{1}{2}(Mh^2 + I_c)\left(\frac{d\theta}{dt}\right)^2$$

or

$$K(\theta) = \tfrac{1}{2}I_O \left(\frac{d\theta}{dt}\right)^2$$

where I_O is the moment of inertia of the object about the axis through O, and the whole motion of every point in the object is expressed, as we know it can be, in terms of pure rotation about this axis.

For our present purposes, it is most illuminating to write I_c in the form Mk^2, where k is the radius of gyration. If we do this, we have

$$K(\theta) = \tfrac{1}{2}M(h^2 + k^2)\left(\frac{d\theta}{dt}\right)^2 \tag{10.30}$$

The equation of energy conservation, given by adding the results of eqns (10.29) and (10.30), is thus

$$\tfrac{1}{2}M(h^2 + k^2)\left(\frac{d\theta}{dt}\right)^2 + \tfrac{1}{2}Mgh\theta^2 = E \quad (=\text{const.}) \tag{10.31}$$

This defines simple harmonic vibrations with a period that depends in a systematic way on the distance h of the CM from the axis:

$$T(h) = 2\pi\left(\frac{h^2 + k^2}{gh}\right)^{1/2} \tag{10.32}$$

This period would become infinitely long for $h = 0$ (rotational axis passing through CM) and has a minimum value T_m for $h = k$. The over-all variation of T with h is as shown in Fig. 10.13(b).

For any given value of h, the period of oscillation corresponds to that of an 'equivalent simple pendulum' of length l such that

$$l = \frac{h^2 + k^2}{h}$$

Linear and rotational motions combined

Near the beginning of this chapter we developed the two results which, between them, provide the basis for analysing the motion of extended objects under any circumstances. These results are as follows:

1. The rate of change of linear momentum is equal to the resultant external force. Expressed in terms of the motion of the centre of mass, this statement becomes

$$\mathbf{F}_{\text{net}} = M\frac{d\mathbf{v}_c}{dt} = M\mathbf{a}_c \tag{10.33}$$

where M is the total mass.

2. The rate of change of angular momentum about the centre of mass is equal to the resultant torque of the external forces about the CM:

$$\mathbf{M}_c = \frac{d}{dt}(\mathbf{L}_c) \tag{10.34}$$

In the present section we shall limit ourselves to cases in which both the torque and the angular motion are about an axis parallel to the axis of symmetry of the object. This means that we can put $L_c = I_c\omega$, and eqn (10.34) takes on the following special form:

(special case, $L_c \parallel \omega$) $\quad M_c = I_c \dfrac{d\omega}{dt} = I_c \alpha$ \hfill (10.35)

where α is a vector representing the angular acceleration.

Let us at once consider a specific situation in which eqns (10.33) and (10.34) are applicable. An aircraft is just touching down. When one of the landing wheels first makes contact with the ground, it has a large horizontal velocity v_0 but no angular motion; therefore, it is bound to skid at first. Anyone who has watched a plane landing will have seen the initial puff of smoke from burning rubber resulting from this violent skidding. After touchdown there are forces on the wheel applied at its two contacts with the external world—its axle and the place where the wheel touches the ground (see Fig. 10.14a). The forces at the axle pass through the CM and so can exert no torque. The normal component, N, of the force of contact with the ground also passes through the CM; furthermore, since N is vertical, it does not affect the horizontal motion of the wheel. Thus we have two forces to consider in analysing the forward motion of the wheel and its rotational motion. These are a force F, pushing the wheel forward at the axle, and a frictional force $\mathcal{F}$ acting backward, as shown in Fig. 10.14(a). By eqn (10.33) we then have

$$F - \mathcal{F} = Ma_c$$

We can guess that a_c is negative, because the wheel remains attached to the plane (we hope) and the reaction force $-F$ applied by the wheel to the plane represents an unbalanced force that is acting to decelerate the plane as a whole. If we assume that a_c is constant, we then have

$v_c = v_0 + a_c t \quad (a_c < 0)$ \hfill (10.36)

When we look at the rotational component of the motion, we see that the only force that produces a torque about C is the frictional force $\mathcal{F}$. Further-

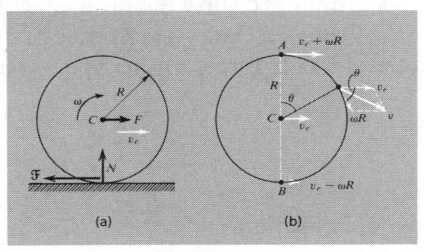

Fig. 10.14 (a) Forces and motions for a landing wheel of an aircraft. (b) The velocity of any point on a wheel is the superposition of the linear motion of the centre and the rotation about the centre, as shown for an arbitrary point P and for the special cases represented by the top and bottom points, A and B.

more, although $\mathcal{J}$ acts to slow down the linear motion, its torque is in such a direction as to speed up the angular motion. We have, in fact, by eqn (10.35),

$$M_c = R\mathcal{J} = Mk^2 \frac{d\omega}{dt} = Mk^2\alpha$$

where k is the radius of gyration of the wheel and R its actual radius. If we assume further that the angular acceleration α is constant, we have

$$\omega = \alpha t \tag{10.37}$$

As long as the skidding goes on, eqns (10.36) and (10.37) operate separately to define the linear and angular velocities of the wheel. At any instant, the resultant velocity of any point on the rim of the wheel is the vector sum of the horizontal velocity v_c of C and a velocity of magnitude ωR along the tangent, as shown in Fig. 10.14(b). This then allows us to identify the condition for the skidding to stop. Skidding means the existence of relative motion between the ground and the point on the wheel that is instantaneously in contact with it. Since the ground defines our rest frame for this problem, the cessation of skidding requires that the velocity of the lowest point, B, on the wheel becomes zero. This velocity, v_B, is, however, the resultant of v_c forward and ωR backward. Thus we have

$$v_B = v_c - \omega R$$

Skidding therefore stops, and rolling begins, when we have

$$v_c = \omega R$$

Using eqns (10.36) and (10.37), we see that this occurs at a time t such that

$$v_0 + a_c t = \alpha R t$$

By using the dynamical equations that define the actual values of a_c and α, we can find the time t and hence the amount by which v_c has been reduced from its initial value v_0 at this instant. To solve the problem completely, we would also have to consider the linear deceleration imposed on the total mass of the plane by air resistance and by the forces of the type $-\mathcal{J}$ due to all the wheels together.

Background to gyroscopic motion

Everybody is intrigued by gyroscopic devices, and probably everybody feels that their behaviour somehow flies in the face of the usual rules of mechanics, even though intellectually one knows that this cannot be the case. It cannot be

BACKGROUND TO GYROSCOPIC MOTION

denied, however, that gyroscopic motions often seem surprising and bizarre, and this of course is the main source of their fascination.

The prime requirement for the appreciation and understanding of gyroscopic behaviour is a full awareness of angular momentum as a vector. And with angular momentum as the central quantity, one must also learn to think primarily in terms of torques rather than forces, and in terms of angular rather than linear velocities and accelerations. Once one has achieved this, the phenomena fall far more readily into a rational pattern.

It is basic to the analysis of gyroscopic problems that we recognize the possibility that a single object may have simultaneous contributions, along different axes, to its total rotational angular momentum. An essential preliminary to this, which we have not needed to consider previously, is the full implication of angular velocity itself as a vector. If it is to be possible to speak of *the* angular velocity vector of a rotating object, then the instantaneous linear velocity, $\omega \times r$, of any point in the object must also be describable as the vector sum of the linear velocities due to component angular velocities ω_x, ω_y and ω_z along orthogonal axes. The validity of this description also requires that, if only one component of ω—say ω_x—were present, the motion of the object would be pure rotation about the axis (x) in question.

To see how these ideas do work correctly in a simple specific case, consider the situation shown in Fig. 10.15. A uniform rectangular board is made to rotate with angular velocity ω about an axis (in the plane of the board) making an angle δ with an axis of x drawn parallel to one pair of edges. Consider the motion of a point P with coordinates described equivalently by (x, y) or (r, θ). The instantaneous velocity of P, as given by $\omega \times r$, is vertically upward from the plane of the diagram and its magnitude is given directly by the product of ω with the perpendicular distance PN from P to the rotation axis. Thus we have

$$v = \omega r \sin(\theta - \delta)$$
$$= \omega r \sin \theta \cos \delta - \omega r \cos \theta \sin \delta$$

However, by resolving ω and r along the x and y axes, we have

$$\omega_x = \omega \cos \delta; \quad \omega_y = \omega \sin \delta$$
$$x = r \cos \theta; \quad y = r \sin \theta$$

This means that the expression for v can be rewritten as follows:

$$v = \omega_x y - \omega_y x$$

We see that, physically, this is exactly what we would get from the superposition of two separate rotational motions with angular velocities ω_x and ω_y about the x and y axes, respectively.

Let us proceed now to the angular momentum of the whole board about the axis of ω. The linear momentum of an element of mass dm at P is $v\,dm$,

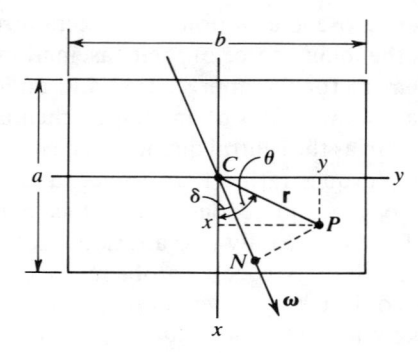

Fig. 10.15 Angular velocity as a vector. The rotational dynamics of the rotating board can be completely analysed in terms of the components of ω along the x and y axes.

and its contribution to the angular momentum about ω is $v\,dm$ multiplied by the distance PN. Since v itself is equal to ω times PN, we see that the magnitude of the rotational angular momentum L_ω about the axis of ω is given by the following integral:

$$L_\omega = \int dm\,\omega r^2 \sin^2(\theta - \delta)$$

Expanding this we have

$$L_\omega = \omega \cos^2\delta \int y^2\,dm + \omega \sin^2\delta \int x^2\,dm$$
$$- 2\omega \sin\delta \cos\delta \int xy\,dm$$

Now the first two integrals are the definitions of the moments of inertia, I_x and I_y, respectively, about the principal axes x and y. And the third integral (an example of what is called a *product of inertia*) vanishes, as one can see from the fact that for each element of mass dm at (x, y) there is an equal element at $(x, -y)$. Thus we have

$$L_\omega = I_x \omega \cos^2\delta + I_y \omega \sin^2\delta$$

Introducing the components ω_x and ω_y of ω, this can be written

$$L_\omega = I_x \omega_x \cos\delta + I_y \omega_y \sin\delta \tag{10.38}$$

This shows that the angular momentum about the axis of ω is precisely what we would get from the projections along ω of separate angular momenta $I_x\omega_x$ and $I_y\omega_y$ about the x and y axes. [In general, the vector combination of the angular momenta $I_x\omega_x$ and $I_y\omega_y$ will also have a component perpendicular to

BACKGROUND TO GYROSCOPIC MOTION

ω, and the *total* vector angular momentum of the board is the combination of L_ω with this other component. We shall not interrupt the present discussion to consider this other component of L.]

Finally, let us look at the total rotational kinetic energy. This is defined by

$$K = \int \tfrac{1}{2} v^2 \, dm$$

Putting $v = \omega r \sin(\theta - \delta)$, this becomes

$$K = \tfrac{1}{2}\omega^2 \int r^2 \sin^2(\theta - \delta) \, dm$$

We see that the expression for K involves precisely the same integral that appeared in the calculation of L_ω, and we therefore have

$$K = \tfrac{1}{2}(\omega^2 \cos^2 \delta) I_x + \tfrac{1}{2}(\omega^2 \sin^2 \delta) I_y$$

Substituting $\omega \cos \delta = \omega_x$, $\omega \sin \delta = \omega_y$, we thus arrive at the result

$$K = \tfrac{1}{2} I_x \omega_x^2 + \tfrac{1}{2} I_y \omega_y^2 \tag{10.39}$$

We see, then (at least in the special case discussed above), how the dynamics of a rotating object is analysable into separate contributions associated with component rotations about the principal axes.

Although we shall not take the matter any further here, it is not difficult to show that results of the form that we have developed are true in general. The essential starting point is again the vector property of angular velocity, and we shall close this discussion with a few additional remarks about that. For a rigid object, pivoted at the CM or at some other fixed point, the motion of any given point in the object is confined to the surface of a sphere. Any given change in position can be produced by rotations about three chosen axes. It is, however, impossible to represent these finite angular *displacements* as rotation vectors with x, y and z as axes, because the resultant of two successive displacements depends on the order in which they are made (Fig. 10.16a shows a rather extreme example). Thus there does not exist a unique angular displacement, characterized by an axis direction and a magnitude, that represents the combination of two or more individual rotations. However, the definition of angular *velocity*, like that of linear velocity, is based on infinitesimal displacements during a time δt that becomes zero in the limit. When one considers the combination of displacements of this type (Fig. 10.16b) one finds that the vector sum *is* unique, regardless of the order of addition, and can always be represented by a single infinitesimal rotation through the angle $\omega \, \delta t$, where ω is the vector sum of $\mathbf{i}\omega_x, \mathbf{j}\omega_y$ and $\mathbf{k}\omega_z$. (In our diagram we have assumed, for simplicity, that $\omega_z = 0$.) If we choose to spell out the calculation of the linear velocity of any point P in terms of components, we thus have

EXTENDED SYSTEMS AND ROTATIONAL DYNAMICS

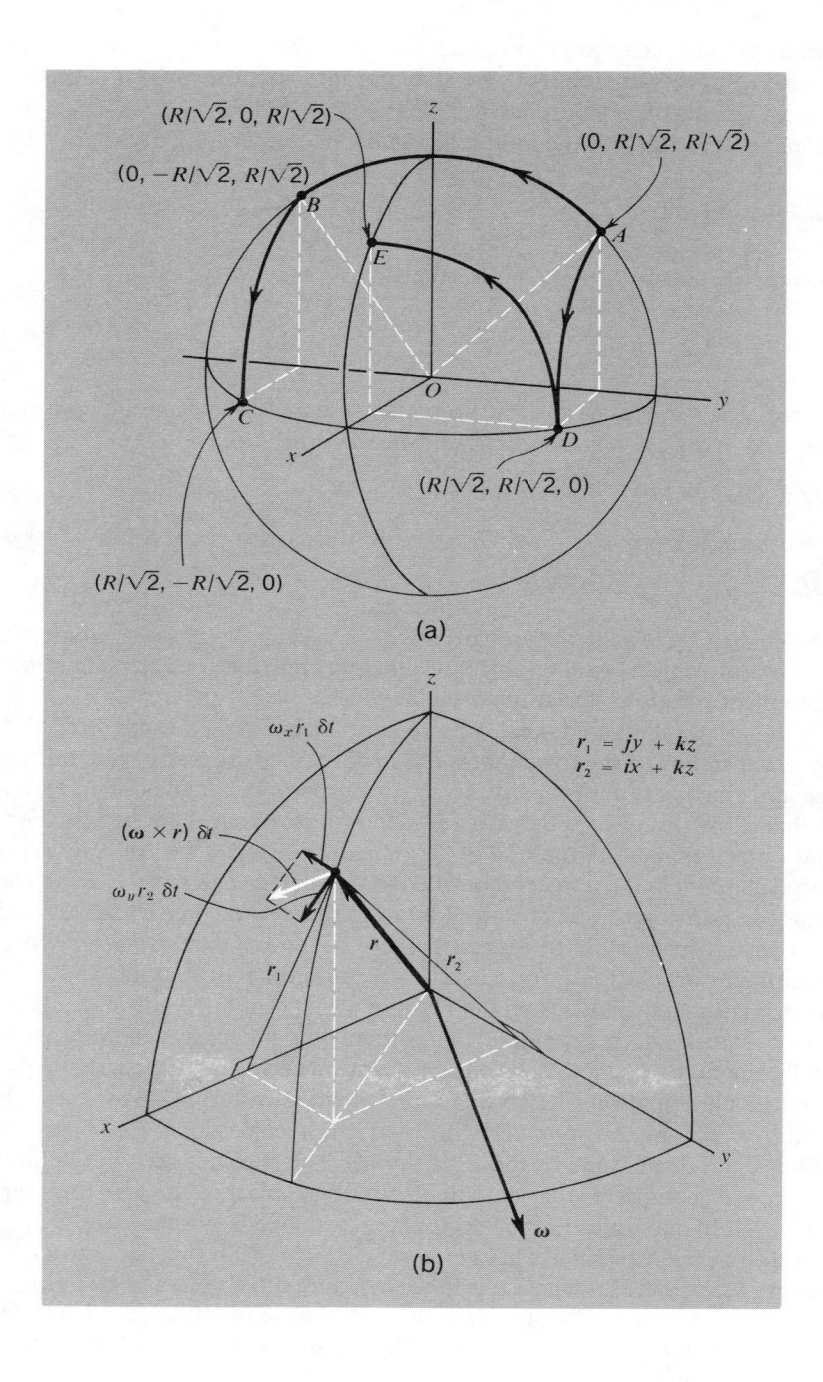

(a)

- $(R/\sqrt{2}, 0, R/\sqrt{2})$
- $(0, -R/\sqrt{2}, R/\sqrt{2})$
- $(0, R/\sqrt{2}, R/\sqrt{2})$
- $(R/\sqrt{2}, R/\sqrt{2}, 0)$
- $(R/\sqrt{2}, -R/\sqrt{2}, 0)$

(b)

$r_1 = jy + kz$
$r_2 = ix + kz$

$\omega_x r_1 \, \delta t$
$(\omega \times r) \, \delta t$
$\omega_y r_2 \, \delta t$

$$v = (i\omega_x + j\omega_y + k\omega_z) \times (ix + jy + kz)$$

Using the relations $i \times i = 0$, $i \times j = -j \times i = k$, and so on, it is easy to establish that this equation gives us

$$v = i(\omega_y z - \omega_z y) + j(\omega_z x - \omega_x z) + k(\omega_x y - \omega_y x)$$

It may be noted that the last term corresponds exactly to the value of v that we obtained in our special example of the rotating board (for which we took $\omega_z = 0$ and $z = 0$ everywhere). It is often convenient, as with any vector cross product expressed in orthogonal components, to write the general expression for v as a determinant:

$$v = \begin{vmatrix} i & j & k \\ \omega_x & \omega_y & \omega_z \\ x & y & z \end{vmatrix}$$

Gyroscope in steady precession

A gyroscope is basically just a flywheel that is mounted so that it has three different possible axes of rotation, which can if desired be made orthogonal to one another. The first axis is the normal spin axis of the flywheel itself; the other two allow this spin axis to tilt in any direction. Figure 10.17 shows how this can be achieved by mounting the flywheel inside a pair of freely pivoted gimbal rings. Our concern will not be with the details of the arrangement but with the dynamics of the response of the flywheel to torques of various kinds, granted that the complete latitude in angular position exists.

One of the simplest and most striking tricks that one can do with a gyroscope is to start it spinning about a horizontal axis and then set one end of its axis down on a pivot, as shown schematically in Fig. 10.18(a). The gyroscope then begins to *precess*, i.e. its axis OA, instead of slumping downwards, proceeds to move around so that the extreme end A settles down (after some initial irregularities) into a horizontal circular path at constant speed. We can explain this as a direct consequence of the torque acting on the flywheel.

Fig. 10.16 (a) Finite angular displacements do not commute. Starting from the point A, two successive 90° clockwise rotations about the x and y axes lead to the very different points C and E, depending on the order in which the rotations are made. (b) The infinitesimal displacements on which the definition of angular velocity is based do have the commutative property and justify the treatment of ω as a vector. The small displacement $(\omega \times r) \delta t$ can be obtained as the resultant of small rotations of the position vector r about the x and y axes. The vectors r_1 and r_2 are drawn perpendicular to the x and y axes, respectively, from the tip of the r vector.

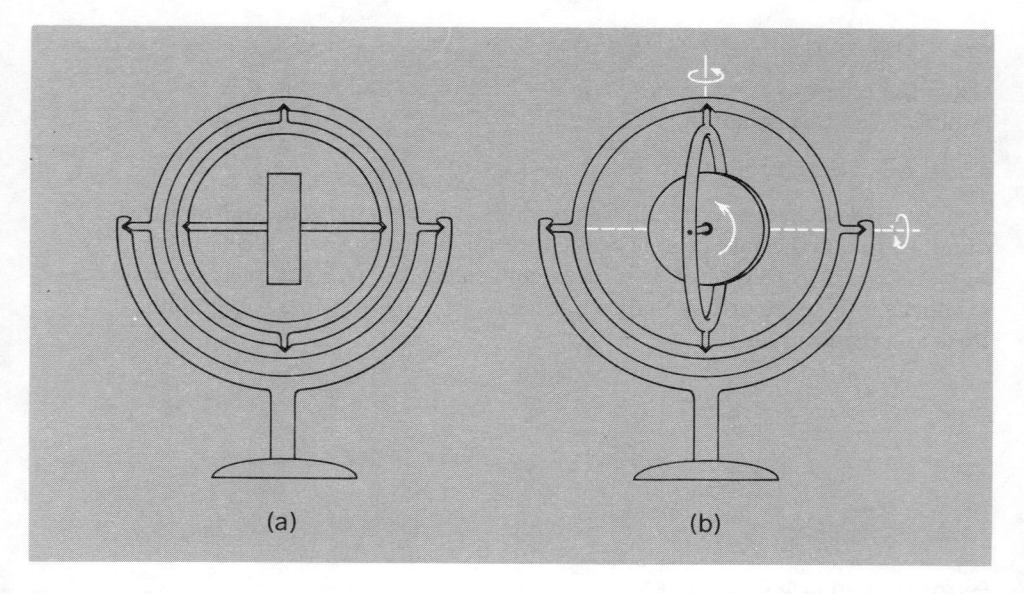

Fig. 10.17 (a) Gyroscope with its gimbal rings lying in the same plane. (b) Gyroscope with its gimbal rings perpendicular, showing the availability of three mutually perpendicular axes of rotation.

The flywheel, supported at one end of its axis, is subjected to a vertical downward force F_g at its CM and an equal, opposite force at the pivot O (Fig. 10.18b). These two forces constitute a couple whose axis is horizontal and at right angles to the spin axis. In a time δt the torque of this couple adds an amount of angular momentum $T\,\delta t$ at right angles to L. The result of this is to change the direction of L without changing its magnitude. If we denote by φ the angle between the axis of the flywheel and some standard horizontal reference axis (x) we have (see Fig. 10.18c)

$$T\,\delta t = L\,\delta\varphi$$

If the distance from the pivot O to the centre of mass is l, we have

$$T = F_g l = Mgl$$

where M is the mass of the flywheel. Putting $L = I\omega$, we then find that

$$\Omega = \frac{d\varphi}{dt} = \frac{Mgl}{I\omega} \tag{10.40}$$

where Ω is the angular velocity of precession about the vertical (z) axis, as shown in Fig. 10.18(c). Equation (10.40) gives quantitative expression to the property of gyroscopic stability. The necessary condition is to make $I\omega$ large, so that the precessional angular velocity Ω is small under an applied torque

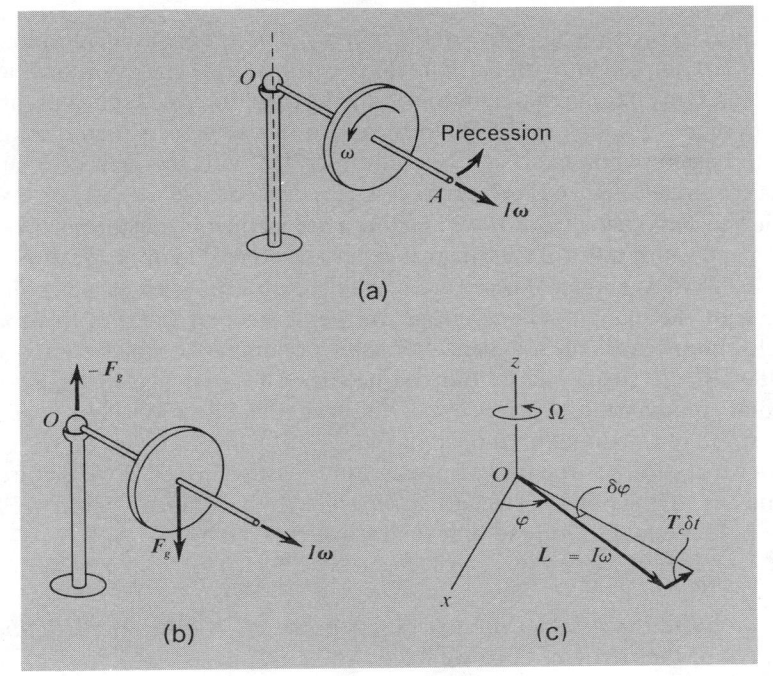

Fig. 10.18 (a) Gyroscope in steady precession. (b) Forces providing a gravitational torque. (c) Addition of angular momentum to a flywheel at right angles to the initial spin axis, resulting in precession.

and can, within certain limits, be made negligible in a practical gyroscopic system.

Notice the relationship between the directions of the various vectors involved here. The spin angular momentum is horizontal. The torque vector is also horizontal and is always perpendicular to L. The precession is described by an angular velocity vector Ω directed along the z axis, perpendicular to both L and T. Vectorially, the result can be written

$$T = \frac{dL}{dt} = \Omega \times L \tag{10.41}$$

Atoms and nuclei as gyroscopes

We mentioned earlier the intrinsic 'spin' angular momentum that is possessed by elementary particles. A large fraction of all atomic nuclei and neutral atoms also have such angular momentum, in amounts corresponding to simple multiples of the natural unit $h/4\pi$. This gives to these particles a natural gyroscopic stability. The spin angular momentum is, however, always accompanied by intrinsic magnetism; it is as if the atom or nucleus contains a tiny bar magnet pointing along the direction of its spin axis, and in the presence of a magnetic field this leads to steady precessional motion.

The basic phenomenon can be well described with the help of a simple classical model. Figure 10.19 portrays a spherical particle with its spin axis at angle θ to a magnetic field. If we imagine a bar magnet inside the particle, the north and south poles of this magnet experience forces in opposite directions parallel to B, as shown. These forces produce a torque with its axis pointing up out of the plane of the diagram. In the absence of the spin this torque would simply cause the magnet (after some preliminary oscillations) to align itself with the field, like a compass needle. The existence of an intrinsic angular momentum L, however, changes the situation and leads to precession in the direction indicated.

An expression for the rate of precession is easily derived if we picture the magnet as having poles of strength m separated by a distance l. The force F on each pole is equal to mB, so that the torque is given by

$$T = mBl \sin \theta$$

We can write the product ml as a single quantity, μ—the *magnetic dipole moment* of the particle. Thus we have

$$T = \mu B \sin \theta \tag{10.42a}$$

If we write the magnetic field and the magnetic moment as vectors, we can describe the torque completely by the equation

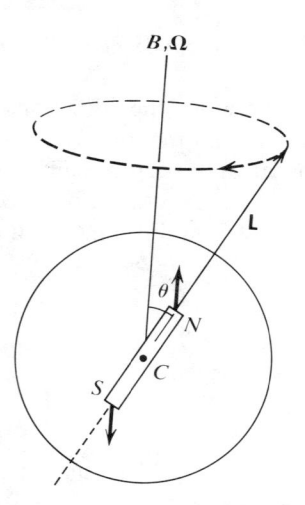

Fig. 10.19 Atoms and nuclei are like spinning magnets and precess in a magnetic field.

$$T = \mu \times B \tag{10.42b}$$

Setting the torque equal to the rate of change of angular momentum due to a precessional angular velocity Ω, we then have

$$\mu B \sin \theta = \Omega L \sin \theta$$

or

$$\Omega = \frac{\mu B}{L} \tag{10.43}$$

It is interesting to note that the angular velocity of precession is independent of the orientation θ.

The existence and the rate of this precession can be detected by picking up the tiny electrical signals that the rotating atomic or nuclear magnet can cause, by electromagnetic induction, in a coil placed nearby. Of course the signal from an individual particle is almost inconceivably small, but by using a sample containing vast numbers of identical particles, all precessing in the same way, the effect becomes measurable. The detection of nuclear spin magnetism in this way is known as *nuclear magnetic induction* and was first studied by F. Bloch and E.M. Purcell (independently, and by different methods) in 1946. They shared the Nobel prize in 1952 for this work.

The precession of the equinoxes

It is fitting that we should end this book with another of the great astronomical problems for which Newton first supplied the explanation. This is the slow precession of the earth, which behaves as a gyroscope under the torques due to the gravitational pulls of the sun and the moon.

The story begins with the ancient astronomers, who through their amazingly careful observations had discovered that the celestial sphere of 'fixed' stars seems to be very gradually turning from west to east with respect to a reference line defined by the intersection of the celestial equator with the ecliptic (Fig. 10.20(a)). This reference line also defines the positions of the equinoxes, when the sun lies in the equatorial plane of the celestial sphere (and of the earth) so that day and night are of equal length over the whole earth. Since only relative motions are involved, the phenomenon could be described as a slow *westward* drift or precession of the equinoctial points themselves. The amount of this precession is about 50" of arc per year, corresponding to a complete precessional period of about 26 000 years.

Figure 10.21 presents the basis for describing the phenomenon using the dynamics of gyroscopic precession. The earth's spin axis makes an angle θ to the normal to the plane of the ecliptic, so that the earth's equatorial bulge

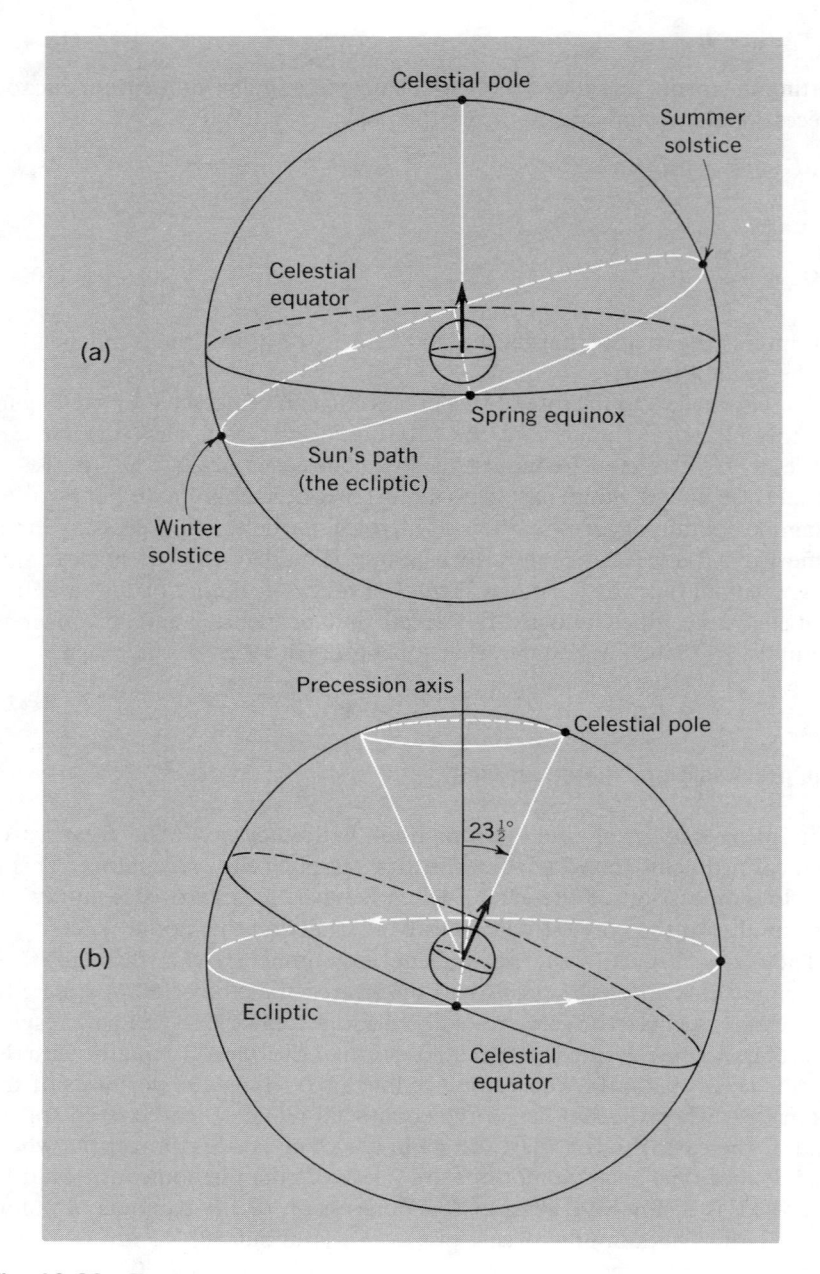

Fig. 10.20 Precession of the equinoxes as described (a) in terms of a movement of the ecliptic around the celestial sphere, and (b) in terms of the conical path traced out by the earth's spin axis with respect to a fixed axis perpendicular to the ecliptic.

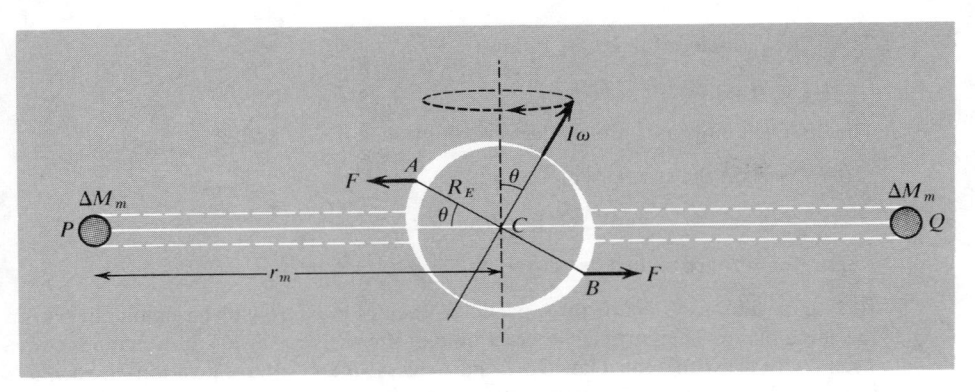

Fig. 10.21 Origin of precessional torque due to gravitational attraction between the earth's equatorial bulge and a ring representing the effective distribution of the moon's mass around its orbit.

($\Delta R/R_E \approx 1/300$) is oriented unsymmetrically, as shown. Now the period of the precession is, as we have seen, immensely long; thus, from the standpoint of the earth, both the sun and the moon go through very many orbits within a time (e.g. 100 years) in which the direction of the spin axis hardly changes. This means, in effect, that the mass represented by the sun or the moon is smeared out uniformly around its orbit as seen from the earth. In other words, the earth's gravitational environment is just like two rings of material; the one representing the moon is indicated in Fig. 10.21.

The origin of the precession is now clear. The earth's bulge in the vicinity of A experiences a net force towards the left of the diagram, and the bulge near B experiences an equal force to the right. Together these give a torque whose axis points upwards from the plane of the diagram. Since the earth's steady rotation from west to east means a spin angular momentum I_ω directed as shown, the result is a precession in which the tip of the spin angular momentum vector traces out a circular path from east to west. It is one more tribute to the genius of Newton that he not only gave this qualitative explanation, but also showed quantitatively that the precessional period so calculated agreed with observation. The analysis involves the same relative effects of moon and sun as one has in the theory of the tides (pp. 214–217).

Problems

10.1 (a) Devise a criterion for whether there is external force acting on a system of two particles. Use this criterion on the following one-dimensional system. A particle of mass m is observed to follow the path

$$x(t) = A \sin(\omega t) + L + vt$$

The other particle, of mass M, follows

$$X(t) = B \sin(\omega t) + Vt$$

The different constants are arbitrary except that $mA = -MB$.

(b) Try it on the system with

$$x(t) = A \sin(\omega t) \quad \text{and} \quad X(t) = B \sin(\omega t + \varphi)$$

where A and B are related as before, and $\varphi \neq 0$.

10.2 Consider a system of three particles, each of mass m, which remain always in the same plane. The particles interact among themselves, always in a manner consistent with Newton's third law. If the particles A, B and C have positions at various times as given in the table, determine whether any external forces are acting on the system.

Time	A	B	C
0	(1, 1)	(2, 2)	(3, 3)
1	(1, 0)	(0, 1)	(3, 3)
2	(0, 1)	(1, 2)	(2, 0)

10.3 Two skaters, each of mass 70 kg, skate at speeds of 4 m/s in opposite directions along parallel lines 1.5 m apart. As they are about to pass one another they join hands and go into circular paths about their common centre of mass.
 (a) What is their total angular momentum?
 (b) A third skater is skating at 2 m/s along a line parallel to the initial directions of the other two and 6 m off to the side of the track of the nearer one. From his standpoint, what is the total angular momentum of the other two skaters as they rotate?

10.4 A molecule of carbon monoxide (CO) is moving along in a straight line with a kinetic energy equal to the value of kT at room temperature (k = Boltzmann's constant = 1.38×10^{-23} J/K). The molecule is also rotating about its centre of mass with a total angular momentum equal to $h/2\pi$ ($= 1.05 \times 10^{-34}$ J s). The internuclear distance in the CO molecule is 1.1 Å. Compare the kinetic energy of its rotational motion with its kinetic energy of translation. What does this result suggest about the ease or difficulty of exciting such rotational motion in a gas of CO molecules at room temperature?

10.5 A uniform disc of mass M and radius R is rotating freely about a vertical axis with initial angular velocity ω_0. Then sand is poured onto the disc in a thin stream so that it piles up on the disc at the radius r ($< R$). The sand is added at the constant rate μ (mass per unit time).
 (a) At what rate are the angular velocity and the rotational kinetic energy varying with time at a given instant?
 (b) After what length of time is the rotational kinetic energy reduced to half of its initial value? What has happened to this energy?

10.6 Estimate the kinetic energy in a hurricane. Take the density of air as 1 kg/m³.

PROBLEMS

10.7 A useful way of calculating the approximate value of the moment of inertia of a continuous object is to consider the object as if it were built up of concentrated masses, and to calculate the value of Σmr^2. As an example, take the case of a long uniform bar of mass M and length L (with its transverse dimensions much less than L). We know that its moment of inertia about one end is $ML^2/3$.

(a) The most primitive approximation is to consider the total mass M to be concentrated at the midpoint, distant $L/2$ from the end. You will not be surprised to find that this is a poor approximation.

(b) Next, treat the bar as being made up of two masses, each equal to $M/2$, at distances $L/4$ and $3L/4$ from one end.

(c) Examine the improvements obtained from finer subdivisions, e.g. 3 parts, 5 parts, 10 parts.

10.8 (a) Calculate the moment of inertia of a thin square plate about an axis through its centre perpendicular to its plane. (Use the perpendicular-axis theorem.)

(b) Making appropriate use of the theorems of parallel and perpendicular axes, calculate the moment of inertia of a hollow cubical box about an axis passing through the centres of two opposite faces.

(c) Using the result of (a), deduce the moment of inertia of a uniform, solid cube about an axis passing through the midpoints of two opposite faces.

(d) For a cube of mass M and edge a, you should have obtained the result $Ma^2/6$. It is noteworthy that the moment of inertia has this same value about *any* axis passing through the centre of the cube. See how far you can go towards verifying this result, perhaps by considering other special axes—e.g., an axis through diagonally opposite corners of the cube or an axis through the midpoints of opposite edges.

10.9 (a) A hoop of mass M and radius R *rolls* down a slope that makes an angle θ with the horizontal. This means that when the linear velocity of its centre is v its angular velocity is v/R. Show that the kinetic energy of the rolling hoop is Mv^2.

(b) There is a traditional story about the camper-physicist who has a can of bouillon and a can of beans, but the labels have come off, so he lets them roll down a board to discover which is which. What would you expect to happen? Does the method work? (Try it!)

10.10 A uniform rod of length $3b$ swings as a pendulum about a pivot a distance x from one end. For what value(s) of x does this pendulum have the same period as a simple pendulum of length $2b$?

10.11 (a) A piece of putty of mass m is stuck very near the rim of a uniform disc of mass $2m$ and radius R. The disc is set on edge on a table on which it can roll without slipping. The equilibrium position is obviously that in which the piece of putty is closest to the table. Find the period of small-amplitude oscillations about this position and the length of the equivalent simple pendulum.

(b) A circular hoop hangs over a nail on a wall. Find the period of its small-amplitude oscillations and the length of the equivalent simple pendulum.

(In these and similar problems, use the equation of conservation of energy as a starting point. The more complicated the system, the greater is the advantage that this method has over a direct application of Newton's law.)

10.12 A uniform cylinder of mass M and radius R can rotate about a shaft but is restrained by a spiral spring. When the cylinder is turned through an angle θ from its equilibrium position, the spring exerts a restoring torque M equal to $-c\theta$. Set up an equation for the angular oscillations of this system and find the period, T.

10.13 Assuming that you let your legs swing more or less like rigid pendulums, estimate the approximate time of one stride. Hence estimate your comfortable walking speed in miles per hour. How does it compare with your actual pace?

10.14 A wheel of uniform thickness, of mass 10 kg and radius 10 cm, is driven by a motor through a belt (see the figure). The drive wheel on the motor is 2 cm in radius, and the motor is capable of delivering a torque of 5 m N.

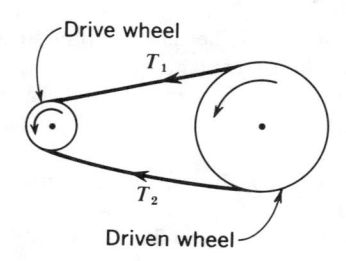

(a) Assuming that the belt does not slip on the wheel, how long does it take to accelerate the large wheel from rest up to 100 rpm?

(b) If the coefficient of friction between belt and wheel is 0.3, what are the tensions in the belt on the two sides of the wheel? (Assume that the belt touches the wheel over half its circumference and is on the point of slipping.)

10.15 A possible scheme for stopping the rotation of a spacecraft of radius R is to let two small masses, m, swing out at the ends of strings of length l, which are attached to the spacecraft at the points P and P' (see the figure). Initially, the masses are held at the positions shown and are rotating with the body of the spacecraft. When the masses have swung out to their maximum distance, with the strings extending radially straight out, the ends P and P' of the strings are released from the spacecraft. For given values of m, R, and I (the moment of inertia of the spacecraft), what values of l will leave the spacecraft in a non-rotating state as a result of this operation? Apply the result to a spacecraft that can be regarded as a uniform disc of mass M and radius R. (Put in some numbers, too, maybe.) [*Hint*: Just use conservation of angular momentum and energy.]

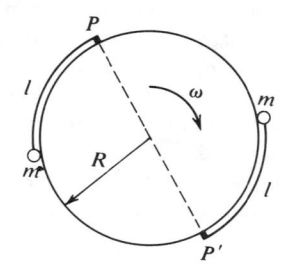

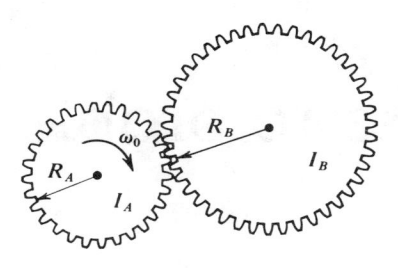

10.16 Two gear wheels, A and B, of radii R_A and R_B, and of moments of inertia I_A and I_B, respectively, are mounted on parallel shafts so that they are not quite in contact (see the figure). Both wheels can rotate completely freely on their shafts. Initially, A is rotating with angular velocity ω_0, and B is stationary. At a certain instant, one shaft is moved slightly so that the gear wheels engage. Find the resulting angular velocity of each in terms of the given quantities. (*Warning:* Do not be tempted into a glib use of angular momentum conservation. Consider the forces and torques resulting from the contact.)

10.17 A section of steel pipe of large diameter and relatively thin wall is mounted as shown on a flat-bed truck. The driver of the truck, not realizing that the pipe has not been lashed in place, starts up the truck with a constant acceleration of 0.5 g. As a result, the pipe rolls backward (relative to the truck bed) without slipping, and falls to the ground. The length of the truck bed is 5 m.

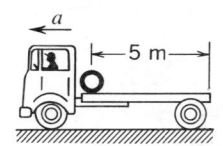

(a) With what horizontal velocity does the pipe strike the ground?
(b) What is its angular velocity at this instant?
(c) How far does it skid before beginning to roll without slipping, if the coefficient of friction between pipe and ground is 0.3?
(d) What is its linear velocity when its motion changes to rolling without slipping?

10.18 A man kicks sharply at the bottom end of a vertical uniform post which is stuck in the ground so that 6 ft of it are above ground. Unfortunately for him the post has rotted where it enters the ground and breaks off at this point. To appreciate why 'unfortunately' is the appropriate word, consider the subsequent motion of the top end of the post.

10.19 In most cars the engine has its axis of rotation pointing fore and aft along the car. The gyroscopic properties of the engine when rotating at high speed are not negligible. Consider the tendency of this gyroscopic property to make the front end of the car rise or fall as the car follows a curve in the road. What about the corresponding effects for a car with its engine mounted transversely? Try to make some quantitative estimates of the importance of such effects. Consider whether a left-hand curve or a right-hand curve might involve the greater risk of losing control over the steering of the car.

Answers to problems

Chapter 1
1.2 (b) $(x, y, z) = R(\cos \alpha \cos \beta, \sin \beta, \sin \alpha \cos \beta)$.
1.4 $R_E \approx (25{,}000/2\pi) \approx 4000$ miles.
1.7 (a) $\mathbf{r}(t) = Bt\mathbf{i} + AB^2 t^2 \mathbf{j}$; (b) $v = B[1 + (2ABt)^2]^{1/2}$.
1.8 $t_{AB}(x) = [(y_A^2 + x^2)^{1/2}/v_1] + \{[y_B^2 + (l-x)^2]^{1/2}/v_2\}$.
1.9 (a) $|\mathbf{v}| = 20\sqrt{3}$ km/hr, directed 120° west of north;
  (b) Minimum separation of 2.6 km occurs at about 13:17 hours.
1.10 Putting $T_0 = 2l/V$, the results are: (a) $T_0/(1 - v^2/V^2)$;
  (b) $T_0/(1 - v^2/V^2)^{1/2}$;
  (c) $T_0[1 - (v \sin \theta/V)^2]^{1/2}/(1 - v^2/V^2)$.
1.11 (b) The cutter reaches the ship a distance $Dv/(V^2 - v^2)^{1/2}$ down the coast from the port and a time $DV/v(V^2 - v^2)^{1/2}$ after leaving port.
1.12 $\theta = 87°$ implies f (= sun's distance/moon's distance) = 19. Correct value of θ is 89°51', corresponding to $f = 385$. For $\Delta\theta = \pm 0.1°$, $150 \leq f \leq 1300$.
1.13 (b) $t_0 = (2l/g)^{1/2}$, $t = t_0$ (i.e. second object is not dropped until separation is l).
1.14 Overtaking car travels 775 ft approx; minimum distance ≈ 1523 ft ≈ 0.3 mile.
1.15 (a) $y_{\text{heaven}} \approx 1.9 \times 10^6$ km, $v_{\text{entry}} \approx 620$ km/sec (assuming 17.5 hr of daylight, which is the length of the longest day of the year in England, where Milton wrote).
  (b) $y \approx 9 \times 10^4$ km, $v \approx 10$ km/sec.
1.16 (a) In first interval, $v_x = 4.3$ m/sec, $v_y = -2.7$ m/sec;
  In second interval, $v_x = 4.9$ m/sec, $v_y = -2.1$ m/sec;
  (b) $a_x = a_y = +30$ m/sec^2.
1.17 (b) $v \approx 380$ m/sec.
1.19 $2(1+\sqrt{35})$ m ≈ 13.8 m behind thrower.
1.21 (a) $\mathbf{r}(t=8) - \mathbf{r}(t=6) = -0.91\mathbf{i} + 2.80\mathbf{j}$,
  where $\mathbf{r}(t=0) = 2.5\mathbf{j}$; $\mathbf{r}(t=2.5) = 2.5\mathbf{i}$.
  (b) $\mathbf{v}(t=4) = (\pi/2)[-\mathbf{i}\cos(\pi/5) - \mathbf{j}\sin(\pi/5)]$;
  $\mathbf{a}(t=4) = (\pi^2/10)[-\mathbf{i}\sin(\pi/5) + \mathbf{j}\cos(\pi/5)]$.
1.22 (a) 3×10^{-3} g; (b) 6×10^{-4} g; (c) 8×10^{21} g;
  (d) 40 g (all approx.).
1.23 (a) $a_x = (\ddot{r} - r\dot{\theta}^2)\cos\theta - (2\dot{r}\dot{\theta} + r\ddot{\theta})\sin\theta$.
  $a_y = (\ddot{r} - r\dot{\theta}^2)\sin\theta + (2\dot{r}\dot{\theta} + r\ddot{\theta})\cos\theta$;
  (b) $a_r = \ddot{r} - r\dot{\theta}^2$, $a_\theta = 2\dot{r}\dot{\theta} + r\ddot{\theta}$.
  Each dot above a variable denotes a differentiation with respect to time; e.g. $\dot{r} = dr/dt$, $\ddot{\theta} = d^2\theta/dt^2$.
1.24 (a) $v = (\sqrt{3}/8)$ m/sec, $a = (5/8)$ m/sec^2;
  (c) $t = \frac{2}{5}[\pi - \cos^{-1}(4/5)]$ sec ≈ 1 sec.

Chapter 2
2.3 (a) $(\bar{x}, \bar{y}) = (1/W)[a(w_2 + w_3), b(w_3 + w_4)]$.
2.6 (a) 100 lb.
2.7 $2(8\sqrt{3} - 1)/3 \approx 8.6$ ft.
2.8 (a) $\theta = \cos^{-1}(r/R)$.
2.9 (a) $W/20$; (b) $(W-w)/20$.
2.11 Yes — just barely.
2.12 (a) $F_{\text{avg}} \approx 1.96 \times 10^4$ N; (b) $h = 0.5$ m.
2.13 (a) $F_x = 0, F_y = -k$; (b) $F_x = -kx, F_y = -ky$;

300

ANSWERS TO PROBLEMS   301

      (c) $F_x = -k, F_y = 0$.
2.14 (a) $F(t=0) = 5$ N; (b) $F_{max} = \pm 10$ N.
2.15 (b) $F(t) = mA\alpha^2 (i \sin \alpha t + j \cos \alpha t)$; a mass attached to the rim of a rolling wheel.
2.16 (a) $v_{max} = (Tl/m)^{1/2}[1 - (mg/T)^2]^{1/2}$;
      (b) $v_{max} = [(T-mg)l/m]^{1/2}$.

Chapter 3
3.1 (a) $a = 0.6$ m/sec^2; (b) $F_C = 3.2$ N.
3.2 (a) $T_A = 2250$ N; $T_B = T_C = 1125$ N; (c) $F_C = 375$ N.
3.4 9 m/sec.
3.5 (a) $T(l) = F(1 - l/L)$, where l is the distance from the pulled end. Doing it vertically does not change the value of $T(l)$.
     (b) $T(l) = [M + m(1 - l/L)]a$;
         $T(l) = [M + m(1 - l/L)](a + g)$.
3.6 7 ft.
3.7 2800 volts.
3.8 $v_{max} = 3(g/l)^{1/2}/2\pi$.
3.9 (a) 12°; (b) 44 m/sec ≈ 97 mph.
3.10 (a) $t_1 = (mlT/F^2)^{1/2}$; (b) $a(t_1/2) = (T^2 + 16F^2)^{1/2}/4m$;
      (c) $s = 3lT/2F$.
3.11 (a) $v_{min} = (gr/\mu)^{1/2}$; (b) $\varphi = \tan^{-1} \mu$;
      (c) 9 m/sec = 20 mph; 31°.
3.12 (a) $r \le 0.03$ mm; (b) $r \ge 2$ mm; (c) 9.5 m/sec; 4.6 m.
3.13 About 1 millisec.
3.14 50 cm/sec, $\frac{1}{10}$ sec.
3.15 (a) $2(T/mL)^{1/2}$; (b) $[TL/mD(L-D)]^{1/2}$.
3.16 (a) $gT^2/4\pi^2$; (b) $\mu gT^2/4\pi^2$.

Chapter 4
4.1 $T_s \approx 2.88 \times 10^{-4} T_E \approx 2\frac{1}{2}$ hr.
4.2 $5\sqrt{10}/4 \approx 4.0 < T_A/T_E <, 3\sqrt{3} \approx 5.2$.
4.3 (a) $R - R_E \approx 1,700$ km; (b) $t = \frac{5}{11}$ hr ≈ 27 min + 20 sec.
4.4 $R - R_J \approx 9 \times 10^7$ m ≈ $1.3 R_J$.
4.5 $\Delta r/r \approx 5.1 \times 10^{-6}$; or $\Delta r \approx 210$ m.
4.6 No, $\Delta g \approx 3.5 \times 10^{-2}$ mgal.
4.7 $T = (3\pi/\rho G)^{1/2}$; $T_{H_2O} \approx 1.2 \times 10^4$ sec.
4.8 $\rho_{sun} \approx 1.3 \times 10^3$ kg/m^3.
4.9 (a) $M \approx 10^5$ kg, $R = 1.9$ m; (b) No; (c) About 5 m/sec.
4.11 (a) $T = (7.4 \times 10^5)(D^3/M)^{1/2}$; (b) $\frac{10}{11}$.
4.12 $M_{total} \approx 10^{11} M_{sun}$.
4.13 $\Delta T \approx 2.5 \times 10^{-3}$ sec.

Chapter 5
5.2 $v = (v_0/4)(5 - 2\sqrt{2})^{1/2} \approx 0.37 v_0$,
    $\theta = \tan^{-1}[1/(2\sqrt{2} - 1)] \approx 29°$.
5.3 $F_{av} = 36$ N.
5.4 $v = 73$ ft/sec = 22 m/sec, F = 350 N ≈ 80 lb.
5.5 (a) $v_0 = [Mg/\pi R^2 \rho]^{1/2}$. (b) $v_0 = 15.5$ m/sec.
5.6 (a) $a = (\mu v_0/M_0) - g$; (b) $\mu = 7 \times 10^3$ kg/sec.
5.7 (b) $v_2 = v_0 \ln \left[\dfrac{n}{rn + (1-r)} \right]$; (c) $n = \sqrt{N}$;
     (d) $v_{max} = 2v_0 \ln \left[\dfrac{N^{1/2}}{rN^{1/2} + (1-r)} \right]$;
     (e) $v = v_0 \ln \left[\dfrac{N}{rN + (1-r)} \right]$.
5.9 (a) $n = 2$; (b) $F = 2A\rho v^2$; (c) $F = \pi r^2 \rho v^2$.
5.10 (a) $m_2/m_1 = 3$; (b) $v_{cm} = u_1/4$; (c) $\frac{3}{4}(\frac{1}{2}m_1 u_1^2)$;
     (d) $\frac{9}{16}(\frac{1}{2} m_1 u_1^2)$.
5.12 (c) $m_n = 0.12 m_0$, $v_n = 2.8 v_0$, $KE_n = 0.95 KE_0$.

5.13 (a) implies $v_1 = 150$ mph; (b) $\frac{5}{8}$.
5.14 (a) $v_1 = \frac{3}{4}u$, $v_2 = \frac{1}{4}u$; (b) $E_0/8$ is lost.
5.15 (a) $M_n = (1.2 \pm 0.4)$ amu;
 (b) $v_{initial} = (3.1 \pm 0.4) \times 10^7$ m/sec.
5.16 (a) $v_N^* = 12.7 \times 10^6$ m/sec, $v_{C-12}^* = 1.06 \times 10^6$ m/sec;
 (b) $v_N^{lab} = 10^7$ m/sec, $\theta = 132°$; (c) $n = 50$.
5.17 (b) $\theta = \tan^{-1}[(M-m)/(M+m)]^{1/2} \frac{1}{2}\cos^{-1}$ m/M.

Chapter 6
6.1 (a) 5×10^4 N; (b) 5×10^5 J; (c) 2.5×10^5 J.
6.2 (a) $mu^2/2$; $m(u^2 + 2uv)/2$, where m is the mass of the ball.
6.3 (a) $F\omega R$.
6.4 (a) 1.95×10^6 ft lbs $(= 1.33 \times 10^6$ J);
 (b) 118 min $\approx$ 2 hr; (c) 7.80×10^6 ft lb.
6.7 (a) 2.4×10^{12} J; (b) 3.8×10^6 man-days $\approx 10^4$ man-years.
6.8 (a) $Um/M(M+m)g = H_0$;
 (b) $(2gh)^{1/2}$, $(2gh)^{1/2} + 2Um/M(M+m)]_{1/2}$;
 (c) $H_0 + 2(H_0 h)^{1/2}$, which is higher than in (a) by $2(H_0 h)^{1/2}$.
6.9 10^{-5} eV.
6.10 $(k_1/k_2)^{1/2}[(1 + 2mv^2 k_2/k_1^2)^{1/2} - 1]^{1/2}$.
6.12 (a) $(m_1 + m_2)g$.
6.13 (b) 4 J; (c) $x < 0$ m, $x > 2$ m.
6.14 About 39 sec.
6.15 (b) $r - (Mg/k)$;
 (c) $\Delta T = (\pi/\omega_0) + (2/\omega_0) \sin^{-1}\{1/[1 + (\omega_0 T_0/2)^2]^{1/2}\}$,
 where T_0 = period of bouncing for perfectly rigid floor; $\omega_0 = \sqrt{k/M}$. Note that when $\omega_0 T_0 \gg 1$ (that is, when $H \gg Mg/k$), $\Delta T = \pi/\omega_0$; when $\omega_0 T_0 \ll 1$, $\Delta T = 2\pi/\omega_0$.
6.16 (a) $V(x) = +(k_1 x^2/2) - (k_2 x^3/3)$; (b) $k_2 = (k_1/2b)$;
 (d) $v = (k_1 b^2/3m)^{1/2}$.
6.17 $(3mv^2/2\alpha)^{1/3}$.
6.18 $(2 + \sqrt{3})(mb^2/3V_0)^{1/2}$.
6.19 (b) $A = (k/8l_0^2)$.
6.22 (a) $(3mv^2/2) + (kL^2/8)$; (b) v; (c) $(3kL^2/8m)^{1/2}$;
 (d) $2\pi(2m/3k)^{1/2}$.
6.23 (a) 5.4 eV; (b) 1.9×10^{13} Hz.
6.24 (b) $r_0 = (nB/6A)^{1/(n-6)}$; (c) $D = (n - 6)A/nr_0^6$.

Chapter 7
7.2 (a) $(gl)^{1/2}$; (b) $2mg$; (c) $2(gl)^{1/2}/(n+1)$;
 (d) $\cos\theta = 1 - (n-1)^2/2(n+1)^2$.
7.3 (a) $(l/\mu)[m/(M+m)]^2$; (b) $\mu_{critical} = 3m/2M$.
7.4 (a) $mg(5 + 2h/r)$ directed vertically upward;
 (b) $g(3 + 2h/r)$.
7.5 $(3 + \sqrt{3})$m.
7.6 (a) $\cos^{-1}(2/3)$; (b) $\cos^{-1}[(2 + v_0^2/gR)/3]$; (c) 5.8 m.
7.7 (a) 9 mm; (b) 5.4 sec.
7.8 It would lose nearly 2 sec per week.
7.10 (a) $M = 2\pi R^2 h\rho_s/[(\Delta T/T) + (h/R)]$; (b) 10 sec/day.
7.11 (a) For $0 < r < R$, $F(r) = 0$; For $R < r < 2R$,
 $F(r) = GMm/r^2$; For $2R < r$, $F(r) = 3GMm/r^2$.
 (b) $-2GMm/R$; (c) $2(GM/R)^{1/2}$.
7.12 (b) $T_{tunnel} = (3\pi/\rho G)^{1/2} = 1$ hr. 49 min;
 (c) $T_{satellite} = T_{tunnel}$.
7.13 According to this criterion, the earth could retain all four of these gases; the moon, only N_2 and CO_2; Mars, all except H_2. It should be noted that an era of higher temperature could have resulted in the loss of heavier gases.
7.15 (a) $v_{escape} = (5GM/2R)^{1/2}$. (b) The launch point has a speed $0.12 v_{escape}$ relative to the center of mass of the system, so the needed launch speed can be reduced.
7.16 (a) $F(z) = -2GMmz/(r^2 + z^2)^{3/2}$;

(b) $V(z) = -2GMm/(r^2+z^2)^{1/2}$;
(c) For $z \gg r$, $F(z) \approx -2GMm/z^2$, $V(z) \approx -2GM/z$;
For $z \ll r$, $F(z) \approx -2GMmz/r^3$,
$V(z) = -2GM(1-z^2/2r^2)/r$;
(d) $T_P = T_0/2\sqrt{2}$.

7.17 (a) $F_N(r) = [-\lambda e^{-r/r_0}/(r_0 r)][1+(r_0/r)]$;
(b) $F_N(r=1.4F) \approx -4.2 \times 10^3$ N,
$F_C(r=1.4F) \approx 1.2 \times 10^2$ N;
(c) $F_N(r) = 10^{-2} F_N(r_0)$ for r just over 5F; the coulomb force is about 10 N there.

7.19 (a) $\theta = 45°$; (b) $\theta = \frac{1}{2}\tan^{-1}(-1/\mu)$; (c) $\theta \approx 50.8°$.

7.20 (a) $g = 2\pi G\sigma\{1 - [h/(R^2+h^2)^{1/2}]\}$;
(b) For $R = 2h$, $g = 0.55(2\pi G\sigma)$; For $R = 5h$, $g = 0.80(2\pi G\sigma)$; For $R = 25h$, $g = 0.96(2\pi G\sigma)$;
(c) 0.08 sec/yr.

Chapter 8
8.1 (a) 840 N; (b) 700 N; (c) 560 N; (d) 560 N, 700 N, 840 N;
(e) The lift has an acceleration of 20/7 m/sec² directed upward. The direction of the lift's motion is not determined.

8.2 (a) 0.5 m; (b) 0.75 m.
8.3 (b) $a_{slip} = 3\mu g/8$.
8.4 Speed is 79 m/sec ($\approx$ 175 mph), 1180 m of runway is used ($\approx \frac{3}{4}$ mile).
8.5 $a_{max} = 6.7 \times 10^4$ m/sec².
8.6 (a) $\tan^{-1}(a/g)$; the apple dropped in a straight line at an angle $\tan^{-1}(a/g)$ forward of "straight down". Thus, if $(h/d) < (a/g)$, it hit the floor; otherwise it hit the wall. (b) The balloon tilted *forward* at an angle $\tan^{-1}(a'/g)$ to the (upward) vertical.
8.7 (a) $a_{3m} = g/3$ downward; (b) $F_P = 2mg$.
8.8 (a) It is in equilibrium at $\tan^{-1}(a/g)$ back from vertically downward;
$T = 2\pi[l/(a^2+g^2)^{1/2}]^{1/2}$.
(b) It is in equilibrium when string is normal to track (i.e., θ back from vertical);
$T = 2\pi[l/(g\cos\theta)]^{1/2}$.
(c) It is in equilibrium when string is vertically down, provided that $a < g$;
$T = 2\pi[l/(g-a)]^{1/2}$.
8.9 (a) $\omega \geq [(a+g)/R]^{1/2}$; (b) $\omega \geq [(a^2+g^2)^{1/2}/R]^{1/2}$.
8.10 (b) $\omega_{max} = (2kS_{max}/\rho)^{1/2}$;
(c) $\omega_{max} \approx 5.2 \times 10^3$ sec^{-1}, or about 50,000 rpm.
8.11 (a) $g_{eff} \approx 160,000 \times$ normal 'g'; (b) $F = 9.5 \times 10^{-14}$ N;
(c) $v = 0.1$ mm/sec.
8.12 $15/\pi$ rpm.
8.13 $F_{net} = 3mg_R(\Delta R/R)$. (Note that F_{net} acts in the direction of the radial displacement ΔR.)

Chapter 9
9.1 (a) $v = (ke^2/mr)^{1/2}$; (b) $l = (ke^2 mr)^{1/2}$;
(c) $r_n = n^2 h^2/4\pi^2 ke^2 m$;
(d) $V(r_n) = -4\pi^2 k^2 e^4 m/n^2 h^2$;
thus $E(r_n) = -2\pi^2 k^2 e^4 m/n^2 h^2$;
(e) $r_1 \approx 0.5 \times 10^{-10}$ m ≈ 0.5 Å; $-E_1 \approx 14$ eV.
9.2 $KE_{final}/KE_{initial} = h^2/l^2$.
9.3 $m_B = v_0^2(D^2-d^2)/2Gd$.
9.5 (a) $r_{max} = 16R/7$; (b) $r_{max} = 9R/7$.
9.9 (b) $v_P = 3v_0$; (c) $\alpha = \cos^{-1}(3/5)$.
9.10 $[1 \pm (\sqrt{3}/2)]r_1$.
9.11 (a) $E = -3GMm/16r$; $l = m(GMr)^{1/2}/2$.
9.12 $F \sim -1/r^5$.
9.13 (a) $(r_{ap}/r_E) = [1+(\Delta v_1/v_0)]^2/\{2-[1+(\Delta v_1/v_0)]^2\}$, where r_{ap} is the aphelion distance in AU, v_0 is the earth's orbital speed, and Δv_1 is the increment.
(b) $(v_{ap}/v_0) = [1+(\Delta v_1/v_0)](r_E/r_{ap}) = \{2-[1+(\Delta v_1/v_0)]^2\}/[1+(\Delta v_1/v_0)]$;
Thus, $[(\Delta v)_{total\ needed}/v_0] = (\Delta v_1/v_0) + (v_{ap}/v_0) = [1-(\Delta v_1/v_0)]/[1+(\Delta v_1/v_0)]$

9.14 The year has lengthened about 20 millisec in 5000 years.

Chapter 10

10.3 (a) ± 420 J s; the sign depends on the sense of the rotation
 (b) ± 2310 J s if the third skater travels in same direction as the nearer one was initially skating; ± 1470 J s if the third skater travels in the opposite direction to the nearer one.

10.4 $KE_{rot} \approx 0.01\, KE_{trans}$.

10.5 (a) $d\omega/dt = -[\mu r^2 I_0/(I_0 + \mu r^2 t)^2]\omega_0$;
 $dE_{rot}/dt = -[\mu r^2 I_0/(I_0 + \mu r^2 t)^2][I_0 \omega_0^2/2)$, where $I_0 = MR^2/2$.
 (b) $t = I_0/\mu r^2 = (M/\mu)(R^2/2r^2)$.

10.8 (a) $I = Ma^2/6$, plate of edge a, mass M;
 (b) $I = 5Ma^2/18$, box of edge a, mass M;
 (c) $I = Ma^2/6$, cube of edge a, mass M.

10.10 $x = b$.

10.11 (a) $T = 2\pi(3R/g)^{1/2}$; the equivalent simple pendulum has length $3R$.
 (b) $T = 2\pi(2R/g)^{1/2}$; the equivalent simple pendulum has length $2R$.

10.12 $T = 2\pi R(M/2c)^{1/2}$.

10.14 (a) $(\pi/30)$ sec ≈ 0.105 sec; (b) $T_1 \approx 82$ N; $T_2 \approx 32$ N.

10.15 $l = R[\sqrt{1 + (I/2mR^2)} - 1]$;
 for $I = MR^2/2$; $l = R[\sqrt{1 + (M/4m)} - 1]$.

10.16 $\omega_{Af} = \omega_0/[1 + (I_B R_A^2/I_A R_B^2)]$; $\omega_{Bf} = -(R_A/R_B)\omega_{Af}$.

10.17 (a) 5 m/sec in the direction of the truck's motion;
 (b) $\omega = -(5 \text{ m/sec})/r$, where r is the radius of the pipe;
 (c) $5/4\mu = 4.16$ m; (d) 0.

Index

Acceleration 13
  centrifugal 202
  centripetal 16, 46, 50, 73
  in polar coordinates 16, 47, 224
  invariance of 36
  radial 17, 47, 224, 239
  related to force 44
  transverse 18, 47, 224
Action and reaction
  in collisions 96, 101
  in contact of objects 26
  in jets and rockets 101
Air resistance 54, 184
Alpha particle scattering 247
Amplitude 64, 136
Angular frequency 51, 139
Angular momentum 225, 258
  and centrifugal potential
      energy 229, 234
  and kinetic energy 274
  as a fundamental quantity 262
  conservation 225, 263, 266
  internal (intrinsic) 259
  many-particle system 258
  orbital 225, 235, 259
  quantized 262
  total 258
  vector additivity 285
Angular velocity 17
  vector properties 203
Apogee (Aphelion) 242
Apple
  the moon and 73
Atoms, gyroscopic behaviour 291

Banking of curves 46
Basin vortex 210
Binary stars 88
Bode's law 86
Bounded orbits 232
Bursting speed (rotating object) 52

Cartesian coordinates 3
Central force 161, 223, 228
  conservative property 163
Central-force motion
  effective PE curves 229, 234
  energy conservation in 228
  law of equal areas 71, 225, 238, 239
  radial part 229
Centre of gravity 131
Centre of mass 107
  and collision processes 109
  kinetic energy of 109, 257
  motion of 109, 117, 257
Centre of mass frame (CM frame) 107
  angular momentum in 264
  kinetic energy in 108, 258
Centrifugal force 200, 203, 207, 229, 275
Centrifugal potential energy 229, 234
Centrifuge 221 prob
Centripetal acceleration 16, 46, 50, 73
Charged particles
  motion in magnetic fields 49
Circular motion 15
  acceleration in 16
  charged particle in magnetic
      field 48
  dynamics 46
  energy conservation in 155
Circular motion (non-uniform) 18, 154
Collisions
  and frames of reference 109, 112
  and KE 108

and momentum conservation 95
  elastic and inelastic 107
  explosive 107
  invariance of KE change 110
  nuclear 112
  perfectly elastic 110, 112
  two-dimensional 107
  with external forces 116
  zero-momentum frame 105
Conservation
  of angular momentum 118, 225
  of energy 118, 124, 126, 180, 228
  of linear momentum 95, 118
Conservative field 176
  motion in 179
Conservative forces 128, 152, 161, 173
Constraints 154
Coordinates
  Cartesian 3
  polar 5
  rectangular 3
  spherical polar 4
Coordinate systems 3
Coriolis forces 203, 207
  and cyclones 209
  and deviation of falling object 208
Coulomb's law 247
Cross product 29, 49, 204, 226
Cyclotron 51
Cyclotron frequency 51

Displacements 3, 5
  relative 6, 11
Dissipative forces 176, 183
Dot product 8, 153

Earth
  as gyroscope 293
  mass 79
  mean density 80
  radial variation of density 170
Earth as rotating frame 207
  deviation of falling body 208
  effect on g 207
  formation of cyclones 209
Earth-moon system
  precession of equinoxes 293
  tide production 221

Earth satellites 78
  orbit decay 184
Einstein, Albert 1
  equivalence principle 81, 198
  general theory 81, 90, 198
  special theory 2, 33
  theory of gravitation 90
Electric field 180
Electric force
  and motion of charged particles 59, 180
Electrons
  in combined B and E fields 181
  in magnetic field 48, 180
Ellipse
  geometry of 239
Elliptic orbits 239
Energy 124, 125
  conservation 124, 126
  conservation in central force motion 228
  gravitational self-energy 277
  kinetic (see Kinetic energy)
  potential (see Potential energy)
  units of 126
Energy method
  one-dimensional problems 129, 135, 137
Equal areas, law of 71, 225, 238, 239
Equilibrium
  rotational 25, 27
  stable 140
  static 25
  translational 25
Equinoxes, precession of 293
Equipotentials 177
Equivalence principle 81, 198
Equivalent simple pendulum 282
Escape speeds 171
Euclidean geometry 1
Explosive collisions 107

Fields 176, 198
  electric 180
  gravitational 177
  magnetic 49, 180
'Fixed' stars 217
Flywheel 273, 289

Force 25
    and inertial mass 32
    as rate of change of momentum 97
    central 161, 223, 228
    centrifugal 200, 203, 207, 229, 275
    centripetal 46, 201
    components 26
    conservative 128, 152, 161, 175, 228
    conservative, criteria 173
    Coriolis 203, 207
    electric 59, 180
    fluid jet 101
    frictional 54
    inertial 193, 196, 197
    magnetic 49, 181
    vector nature 25
Forces
    equilibrium 25
    linear superposition 33
    polygon of 25
    resistive 54
    resolution of 46
Foucault pendulum 193
Frame of reference 2, 11, 193
    'fixed' stars 217
    inertial 31, 193, 217
    linearly accelerated 194, 196
    and Newton's laws 196
    rotating 200
    rotating, earth as 207
        free fall in 207
        inertial forces in 203
    zero-momentum 105
Free fall
    and air resistance 57
    and weightlessness 83
    on rotating earth 207
Friction 54
    coefficient 68 prob
    dry 54
    fluid 54
    fluid (Newtonian theory of) 120

g 81
    altitude dependence 80
    as gravitational field 177
    latitude dependence 82
    local variations 80

G 79
Galaxies
    contracting, dynamics of 276
    masses of 90
    rotation of 90, 276
Galilean transformations 35
Geodesic 91
Gravitation
    Einstein's theory 90
    law of 73
    Newton's theory 73
    outside the solar system 87
    universal 70
Gravitational attraction
    sphere 75, 168
    spherical shell 165
Gravitational forces
    conservative property 161
    gradient of, and tides 216
Gravitational mass 81, 198
Gravity
    acceleration due to, g 81, 82, 177
    equivalence to accelerated
        frame 198
    force of 81
Gyroscope 284
    steady precession 289

Harmonic motion 62, 137
Harmonic oscillators 64, 134, 137, 280
Heaven
    free fall from 21 prob
Hooke's law 62, 134
Hyperbolic orbits 246, 247

Impact parameter 234, 247
Impulse 97
Inelastic collisions 107
Inertia
    principle of 31
Inertial forces 193, 196, 197
Inertial frame(s) 31, 193
    dynamical equivalence of 193
    fundamental 217
Inertial mass 32, 33, 81, 100, 198
    speed dependence 33

Invariance
    of acceleration   36
    of Newton's laws   34
    rotational, and torque   266
    translational, and force   266
Inverse square law (gravitation)   73, 164
    deduced from elliptic orbit   239
    from Kepler's third law   72
    types of orbit   236, 243
Isolation diagram   45, 52

'Jerk' (as unit)   34
Jet propulsion   101
Joule (unit)   125

Kepler's laws   71, 88
Kepler's second law   71, 239
    implies central force   225
Kepler's third law   70, 86, 238, 240
    explained by Newton   72
Kinematic equations   14
Kinetic energy (KE)
    in CM frame   109, 258
    in collisions   108
    of many-particle system   255
    of rolling objects   273
    of rotating objects   273, 276
    of two-body system   108

Laboratory frame   108
Law of areas   71, 225, 238, 239
Laws of motion   31, 32, 97
Length (units of)   10
Lever, law of   27
Light, speed of   10
Linear oscillator   64, 141
    as two-body problem   141
Lines of force   49, 177

Mach number   11
Magnetic field   49
Magnetic force   49
    and motion of charged particles   50
Many-particle system   255
    angular momentum   258
    centre of mass frame   255
    kinetic energy   257

Mass
    and weight   81
    gravitational   81, 198
    inertial   32, 33, 81, 100, 198
    speed dependence   33
Mercury, orbit precession   91
Metre (def)   10
Milky Way   90
Millikan experiment   58
Moment   27, 30
Moments of inertia   260, 281, 286
    principal   267
    special theorems   268
Momentum, angular (see Angular momentum)
Momentum, linear   32
    conservation   95, 97, 99, 266
    vector character   95
Moon
    and apple   73
    and precession of equinoxes   295
    and tide production   210
    as falling object   73
Motion   1
    accelerated   13
    against resistive force   54
    governed by viscosity   58
    in conservative field   179
    laws of   31, 32, 97
    linear and rotational combined   282
    of charged particles   180
    relative   11
    rotational   267, 273
    simple harmonic   62, 135, 137
    straight line   13
    under central force   223
    uniform circular   15, 17
    uniformly accelerated   14

Neptune, discovery   84
Neutrino   118
Neutrons
    elastic collisions   112
Newton
    and universal gravitation   73
    concepts of space and time   1
    on moon's motion   73

INDEX

on proportionality of weight to
    mass   81
  precession of equinoxes   293
  theory of fluid resistance   120 prob
  theory of tides   212
Newton's first law   31
Newton's second law   32
  deductive and inductive uses   44
  invariance of   34
  logical status   37
  simple applications   44
Newton's third law   97, 101
  limitations to   98
Nuclear reactions
  dynamics of 113
Nuclei
  gyroscopic behaviour   291

Oil drop experiment   58
Orbits
  bounded   232
  circular   236
  elliptical   239
    energy in   242, 246
  families of   241
  hyperbolic   246, 247
  parabolic   246
  precession of   233
  unbounded   233
  under inverse square attraction   243
Oscillations
  harmonic   62, 137, 280
  torsional   278

Parabolic orbits   246
Parabolic trajectories   195, 257
Parallel axis theorem   269
Parallel forces   28
Pendulum, rigid   280
Pendulum, simple   157
  as harmonic oscillator   159
  energy conservation   155
  period vs amplitude   160
Perigee   245
Perihelion
  precession   91
Periodic motion   132
Perpendicular-axis theorem   271
Perturbations   84

Phase   64
Planets
  motions of   71
  orbital data (table)   71
Pluto   87
Polar angle   4
Polar coordinates
  plane   5
  spherical   4
  velocity and acceleration in   16, 47, 224
Polygon of forces   25
Potential   177
Potential energy (PE)   127
  centrifugal   229, 234
  effective, in central force
    motion   229, 234
  gravitational   154
  scalar character   165
  spring   134
Potential energy gradient   128, 179
Power   125
Precession
  of atoms and nuclei   291
  of equinoxes   293
  of gyroscope   289
  of orbits   233
  of perihelion of Mercury   91
Principle of equivalence   81, 198
Products of inertia   286
Proton-proton collisions   110

Radius of gyration   268, 281, 283
Reduced mass   109, 143, 260, 275
Reference frames   2, 11, 193
Relative motion   11
Relativity
  Einstein's general theory   81, 90, 198
  Einstein's special theory   2, 33
  Newtonian   34
Resistance, viscous   58
Resisted motion   54
  growth and decay   60
Rest mass   33
Rigid pendulum   280
Rockets   102
Rolling objects   273

Rotating frame of reference   200, 203
Rotating objects
   fracture of   52
   kinetic energy of   273, 276
Rotational equilibrium   27
Rotational motion   267, 273
   combined with linear   273
Rutherford scattering   247

Satellites   78, 184
Scalar product   7, 153
Scattering of alpha particles   247
Second (def)   10
Simple harmonic motion (SHM)   62
   by energy method   137
   of torsional oscillator   278
Simple pendulum   157, 160
   as harmonic oscillator   159
   isochronism   160
Small oscillations   140
Space   1
   curvature   91
   geometry   1, 91
Special relativity   2, 33
Sphere
   gravitational attraction   75, 168
   moment of inertia   269
Spherical coordinates   3
Spherical shell
   gravitational attraction   76, 165
Spring constant   134
Stars
   binary   88
   'fixed'   217
   orbits of   88
Static equilibrium   25
Straight-line motion   13
Sun
   influence on tides   214
   radial density variation   170
Superposition of forces   33

Terminal speed   55, 57, 60, 61
Tides   210
   equilibrium theory   212
   height of   214
Time   1
   unit of   10
Time constant   61

Torque   27
   and change of angular
      momentum   265, 285, 290
   and potential energy   280
   vector character of   30
Torsion constant   279
Torsional oscillations   278
Trajectories
   helical   51
   parabolic   22, 257

Ultracentrifuge   221 prob
Unbounded orbits   233
Unit vectors   3
   time dependence   17, 224
Units   9
   of length and time   10
Universal gravitation   70
   constant of   79
   law of   73
Uranus
   discovery of   84
   perturbations of   84

Vector product (cross product)   29, 49,
      204, 226
Vectors
   addition and subtraction   5
   components   8
   cross product   29, 49, 204, 226
   resolution of   8
   scalar product   7, 153
Velocity   10
Velocity, angular   17
   as a vector   203
Velocity, linear
   as derived quantity   10
   in polar coordinates   16, 224
   instantaneous   11
   relative   11
Viscous resistance   58

Watt (unit)   125
Weight   81
   proportionality to mass   81
Weightlessness   83
Work   125, 126
   in closed path   174

Zero-momentum frame   105, 259